Exercise Therapy for Recovery from Hemiplegia

Kazumi Kawahira • Megumi Shimodozono
Tomokazu Noma
Editors

Exercise Therapy for Recovery from Hemiplegia

Theory and Practice of Repetitive Facilitative Exercise

Springer

Editors
Kazumi Kawahira
Laboratory of Repetitive Facilitative Exercise
Kirameki Terrace Health Care Hospital
Kagoshima city, Kagoshima, Japan

Megumi Shimodozono
Rehabilitation and Phy. Med.
Kagoshima University
Kirishima, Japan

Tomokazu Noma
Department of Rehabilitation
Nihon Fukushi University
Handa, Aichi, Japan

Translators
Hiroko Itoh
Geoffrey Tozer
Edward Fisher
Walk100Phyisio Chiryoshitsu
Tokyo, Japan

ISBN 978-981-19-0788-3 ISBN 978-981-19-0789-0 (eBook)
https://doi.org/10.1007/978-981-19-0789-0

This Springer imprint is published by the registered company Springer Nature Singapore Pte Ltd.
The registered company address is: 152 Beach Road, #21-01/04 Gateway East, Singapore 189721, Singapore

Preface for the Third Edition in Japanese

RFE (Repetitive Facilitative Exercise), which the authors advocate is the neural net contracture therapy, achieves and repeats the patients' intended movement in order to reconstruct/strengthen the nerve pathway needed to accomplish the voluntary movements. RFE explored a new theory focusing on standing on the intact lower limb for therapy during gait training. RFE has its principle in the selective reconstruction/strengthening (facilitation of learning) of damaged nerve paths, not only for cases with hemiplegia in the upper limbs, in particular finger movements that were not responsive to conventional treatment, but also for the hemiplegic in the lower limbs as well as those with walking impairment for which the effect of conventional treatment was limited. Thus, it may be an effective rehabilitation technique for paralysis and sensory disorder due to a cerebral stroke, spinal injury, or peripheral nerve damage.

In recent years, RFE has proven effective not only in the recovery phase but also in acute and chronic phases. The nerve paths can be strengthened by RFE. Moreover, the effect of combining therapy using RFE with continuous electric stimulation, transcranial magnetic stimulation, botulinum therapy, robot, etc., which enhances the effect of RFE, has each been reported. In rehabilitation after neurological regenerative medicine, RFE, which can facilitate the targeted neural pathways, and combination therapy (neural modulation), which makes RFE more effective, will be essential.

RFE can reconstruct and strengthen the target nerve pathways selectively. RFE development that strengthens the target nerve pathways effectively was supported by the idea of localized brain functions, information processing, and the plasticity after the brain was injured. They were clarified with basic studies on monkeys and clinical studies on humans using fMRI, PET, and MEG. These facts provided a new perspective and logical development for the therapies to facilitate the recovery from hemiplegia, bringing about a drastic change in the exercise as well as occupational therapies for paralysis and apraxia due to cerebral stroke and other central nervous disorders. It replaced the conventional method where the disorder is interpreted based on reflex theory and set up a treatment plan. We began to consider scientific treatment emphasizing the construction of nerve pathways by forming the synapse and improving transmission efficiency, based on localized brain functions, a central program, and a neural net. No new treatment can be developed without this method, as it is indispensable for developing new treatment.

This book has been written to assist the therapist's understanding of RFE. First, it briefly explains the localized brain functions, voluntary movements, plasticity of the brain function, sprouting of nerves and other structures, and strengthening mechanisms for nerve pathways. It then presents a proven effect from RFE and describes the basic manual exercise techniques and combination therapies.

In this edition, we placed particular emphasis on (1) extra content written by Dr. Megumi Shimodozono and Dr. Tomokazu Noma, (2) color photos, figures, and tables to elaborate the RFE treatment technique and update the video as Electronic Supplementary Material (ESM), (3) reasonable treatment strategy by adding scientific evidence on treatment effects, and (4) explanations over treatment for walking disorders.

However, we decided to select only those procedures that are important in daily clinical practice as ESM. We could not present all the procedures as there are a tremendous number of them. Moreover, there are many new papers being published. We appreciate your understanding with our limitations.

Many people supported the publication of this book and we express our sincere gratitude toward Ms. Ikumi Nagamori, an occupational therapist who appears as a model in the photos and movies; Dr. Seiji Etoh, Dr. Atsuko Ogata, and Dr. Shuji Matsumoto, who provided their research data; Dr. Tazu Aoki, who provided proofreading on chapter one; OG Wellness Technologies Co., Ltd.; and ITO Co., Ltd., who provided the photo shooting devices.

Shibuya, Tokyo, Japan Kazumi Kawahira
April 2017

Preface for the First Edition in Japanese

A recent development in the study of brain science has clarified that neural stem cells also exist in the central nervous system of adults and that the plasticity of the brain is larger and faster than assumed previously. The clinical treatment of functional reconstruction, and regenerative medicine using neural stem cells, is becoming a reality. On the other hand, the response to rehabilitation medicine/treatment has not been adequate yet. Rehabilitation medicine/treatment should be most enthusiastic in developing a technique to reconstruct and strengthen nerve paths.

In spite of efficient reconstruction, nerve path strengthening is required for recovery from hemiplegia. CI therapy (constraint-induced movement therapy) is regarded as a cutting-edge therapy, which may be a result of insufficient effort by the persons involved in rehabilitation medicine. CI therapy is inefficient and demands repeated trial and error from the patient.

A "facilitation technique" has been used for the rehabilitation of central paralysis, aiming at achieving the patients' intended movements more easily. However, its scientific evidence is insufficient due to a lack of treatment details and assessment methodology. *The Japanese Guideline for the Management of Strokes, 2004* classified the facilitation technique as Grade C1: Meaning it could be considered for practice but there is inadequate scientific evidence, even if its period of literature retrieval is considered, which is limited to the years 1960–2002.

For this version, I wanted to publish this book in order to change the situation. I hope I can contribute to establishing a "real facilitation technique" which medical doctors striving in the advanced treatment and research of selective thrombolytic therapy, neural stem cell transplants, and other fields can trust as the rehabilitation technique that reconstructs and strengthens nerve paths in a short period of time.

I received much support during the publication process. In particular, I wish to express my sincere gratitude toward Emeritus Professor Nobuyuki Tanaka for his guidance on the theory, Mr. Tomokazu Noma, Occupational Therapist for his photo shootings, Ms. Ayako Hitouji and Ms. Saori Baba for their modeling, and Dr. Seiji Etoh, Dr. Shuji Matsumoto, and Dr. Megumi Shimodozono for providing research data.

I also truly appreciate Mr. Satoshi Ohno, Mr. Mitsuo Ando, and IGAKU SHOIN for their patience over my repeated corrections.

Kirishima, Japan
February 2006

Kazumi Kawahira

Contents

Part I

Basics

Theory of Repetitive Facilitative Exercise

1

Kazumi Kawahira, Megumi Shimodozono, and Tomokazu Noma

1.1 Different Functions in Different Parts of the Brain

The cerebral cortex has an area of 2600 cm^2 and contains approximately 28×10^9 neurons and another 28×10^9 glial cells, which form the neural pathways. Each individual neuron can form thousands of links with other neurons, creating a network of over 10^{12} synapses. The cortex has several parts, and each of these parts has its own function. The minimal unit is a cortical column—a cylinder measuring about 0.5 mm in diameter, where cells related to a specific function are grouped (Fig. 1.1). Information enters different parts of the cortex and is processed in parallel. The two hemispheres of the brain exchange information across the cerebral cortex via commissural fibers that run through the corpus callosum.

Supplementary Information The online version contains supplementary material available at [https://doi.org/10.1007/978-981-19-0789-0_1].

K. Kawahira (✉)
Laboratory of Repetitive Facilitative Exercise, Kirameki Terrace Health Care Hospital, Kagoshima city, Kagoshima, Japan

M. Shimodozono
Department of Rehabilitation and Physical Medicine, Kagoshima University, Kagoshima, Japan

T. Noma
Department of Rehabilitation, Nihon Fukushi University, Handa, Aichi, Japan

K. Kawahira et al. (eds.), *Exercise Therapy for Recovery from Hemiplegia*, https://doi.org/10.1007/978-981-19-0789-0_1

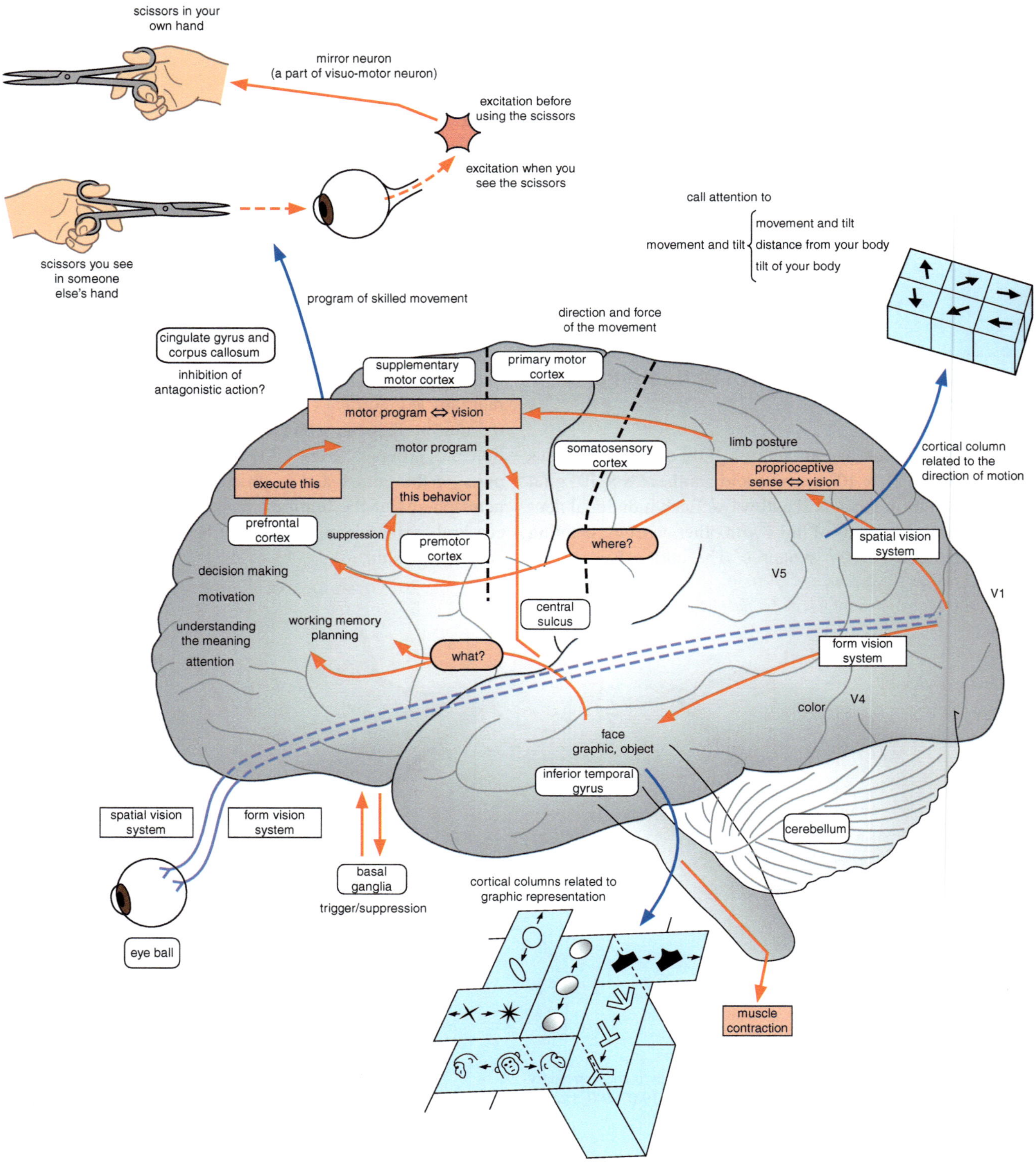

Fig. 1.1 Processing visual information and voluntary movement. The figure above shows the mechanism of processing the visual information required to make an appropriate movement. The basal ganglia and limbic cortex cannot be seen from the surface of the cerebrum, but both structures are involved in motivation and emotional aspect. Some neurons in the prefrontal cortex that memorize a specific motor program are excited only by visual information related to that program. These neurons are called "visuo-motor neurons." Some neurons that are programmed for the finger movements involved in using scissors become excited simply by the sight of someone else using scissors. These neurons are called "mirror neurons." Due to the existence of visuo-motor neurons, we can make the appropriate movements to manipulate an object without consciously thinking about each step of the movement

1.2 Voluntary Movement

The mechanisms of voluntary movement have been explained by the central program theory and reflex theory. Here we will review the central program theory to promote your scientific understanding of Repetitive Facilitative Exercise.

1.2.1 Mechanism of Voluntary Movement

Consider the process that occurs when you use a pair of scissors (Fig. 1.1). When you look at the scissors, information about the shape and color is sent to the primary visual area (V1) and then to the temporal lobe via the ventral pathway (a system that recognizes form). Information about position and motion is sent to the primary visual area and then to the parietal lobe through the dorsal pathway (a system that recognizes the spatial relationship of the object to the viewer). All of this information is processed in these respective areas [1]. The temporal association area contains cells that respond to faces, hands, etc. The parietal association cortex contains cells that respond to position and motion and cells that convert visual information into somatosensory [2]. All of this information is also sent to the frontal lobe, basal ganglia, and cerebellum.

How do the areas in the frontal lobe function when you use scissors? First, the prefrontal cortex interprets the information from the temporal lobe and parietal association cortex and makes a decision to "use the scissors." Then the prefrontal cortex gives a command to the "program for movement using the scissors" [3, 4], which is then processed automatically in the premotor region [5]. The primary motor area makes adjustments to the direction of movement, muscle power, etc. to execute the movement according to the selected "program for movement using scissors" (Fig. 1.2). The cerebellum makes corrections based on differences between the planned movement and actual movement, and amends the "program" so that the movement can be performed as planned. By repeating this process, the cerebellum plays an important role in forming an internal model to enable movement without requiring further feedback each time [6]. The "program" is processed mainly in the premotor area when based on external triggers or by the supplementary motor area when based on consecutive movements you are already skilled in [7]. Excitation generated by the "movement program" is transmitted from the motor area to the spinal anterior horn cells to eventually realize the movement.

1.2.2 Central Programming of Voluntary Movements and Reflex Theory

In patients with hemiplegia following a stroke, the pyramidal tract has become damaged, causing a separation of the lower central nervous system (program) from the upper central nervous system. As a result, positive signs such as spasm, synergistic movement or postural reflex are observed. Suppression of these positive signs, emphasized in the conventional neuromuscular facilitation method, generates neither new connections to the lower central nervous system nor new programs within the undamaged cerebral cortex or intact hemisphere.

If using a method that only focuses on suppression of spasms (normalization of muscle tonus, control of the suppression level of the nerve path in the spastic muscle), then there is no repetition of a movement pattern that can be obtained by the patient's own effort. As a result, the nerve path that connects the cerebral cortex and spinal anterior horn cells is not reconstructed or strengthened. Neither the program of the cerebral cortex nor the internal model in the cerebellum is facilitated. Therefore, this method has not shown positive treatment results and does not contribute to recovery from paralysis.

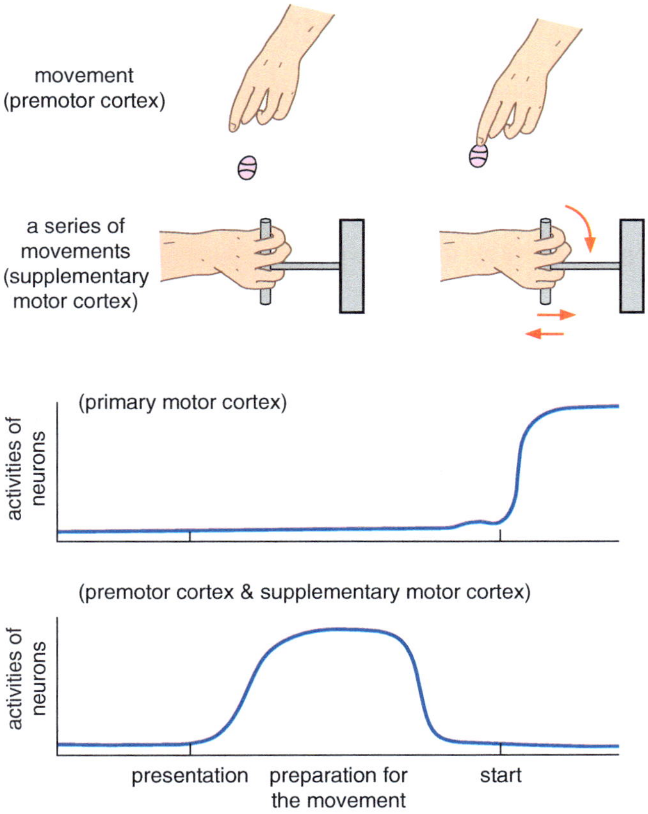

Fig. 1.2 Development of movement and neuronal activities in the primary motor cortex, premotor cortex, and supplementary motor cortex. The primary motor cortex gets activated right before the onset of movement, whereas excitation of the premotor cortex and supplementary motor cortex occurs before activation of the primary motor cortex. For example, cells in the premotor cortex become engaged when you move small seeds but not when you handle a large ball. In lever handling movements, the cells of the supplementary motor cortex involved in a sequence of "rotate, push, and pull" are not activated in a sequence of "push, pull, and rotate"

1.3 The Process of Learning Movements

During memory formation in the brain, formation and strengthening of certain neural circuits occur based on two processes:

1. In the neural circuit that repeatedly transmits excitation (or activity), the efficiency of synapse transmission is increased by improving the binding of neurotransmitters.
2. Synchronously excited neural pathways connect (Fig. 1.3). When learning a new movement, the premotor cortex, primary motor cortex, and cerebellum play important roles.

As hemiplegia is caused by damage to the neural path connecting the primary motor cortex to the spinal anterior horn cells, it is necessary for recovery from hemiplegia to reconstruct the regions related to movement, especially the pathway connecting the cortex to the spinal anterior horn cells. For effective learning, therapists must focus on promoting and strengthening the connection of this neural pathway.

1.3.1 "Learning Without Error"—Repetitive Excitatory Transmission to the Target Neural Pathway

The method called "learning without error" effectively allows an amnesiac patient to remember something new [8]. Suppose we prepare a photo card of person A, a photo card of person B, a name card for person A, and a name card for person B to help the patient memorize the names of A and B. The change to the neural pathway is shown in Fig. 1.4.

If the therapist shows the photo of A and B and repeatedly asks the patient "Is this A or B?" in an attempt to help the patient memorize the names of these two people, the excitatory transmission in the brain is repeated through the "face cell A → name cell A" pathway for the correct answer, but also is repeated through the "face cell A → name cell B" pathway for the wrong answer. Since the training also includes the wrong answer, it is less likely that the synapses are strengthened only in the "face cell A → name cell A" pathway for the correct answer, and thus it becomes more difficult for the patient to memorize the name of person A.

On the other hand, in "learning without error," the therapist repeatedly shows the photo of A along with the name of A, and asks the patient, "This is A, right?" to reconfirm. In the patient's brain, excitatory transmission is repeated only along the pathway for the "face cell A → name cell A." Therefore, only the synapses in the "face cell A → name cell A" pathway are strengthened, and the patient can memorize the name of A.

In recovery from hemiplegia as well, reconstruction and strengthening of the neural pathway, needed to realize the intended movement, requires "learning without error." Excitation should be repeatedly transmitted only along the neural pathway that must be reconstructed or strengthened. For recovery from paralysis after brain injury, reconstruction and strengthening of the neural pathway in the cerebrum, where there are many more neural pathways than in the spinal cord, plays an important role. Here we show how the selective strengthening of the target neural pathway occurs to understand how movement is recovered after paralysis. As shown in Fig. 1.5, in order to enable the patient who can only perform synergistic movement to flex his/her elbow separately, the therapist must instruct the patient to flex his/her

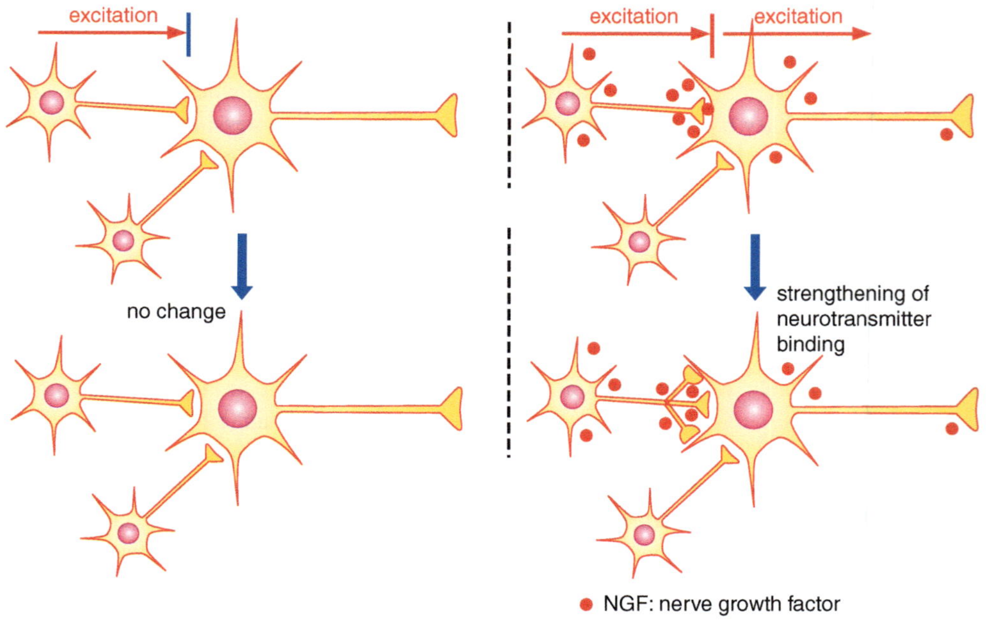

Fig. 1.3 Formation and reconstruction of a neural pathway: improvement of excitatory transmission across the synapse/strengthening of NGF release and binding

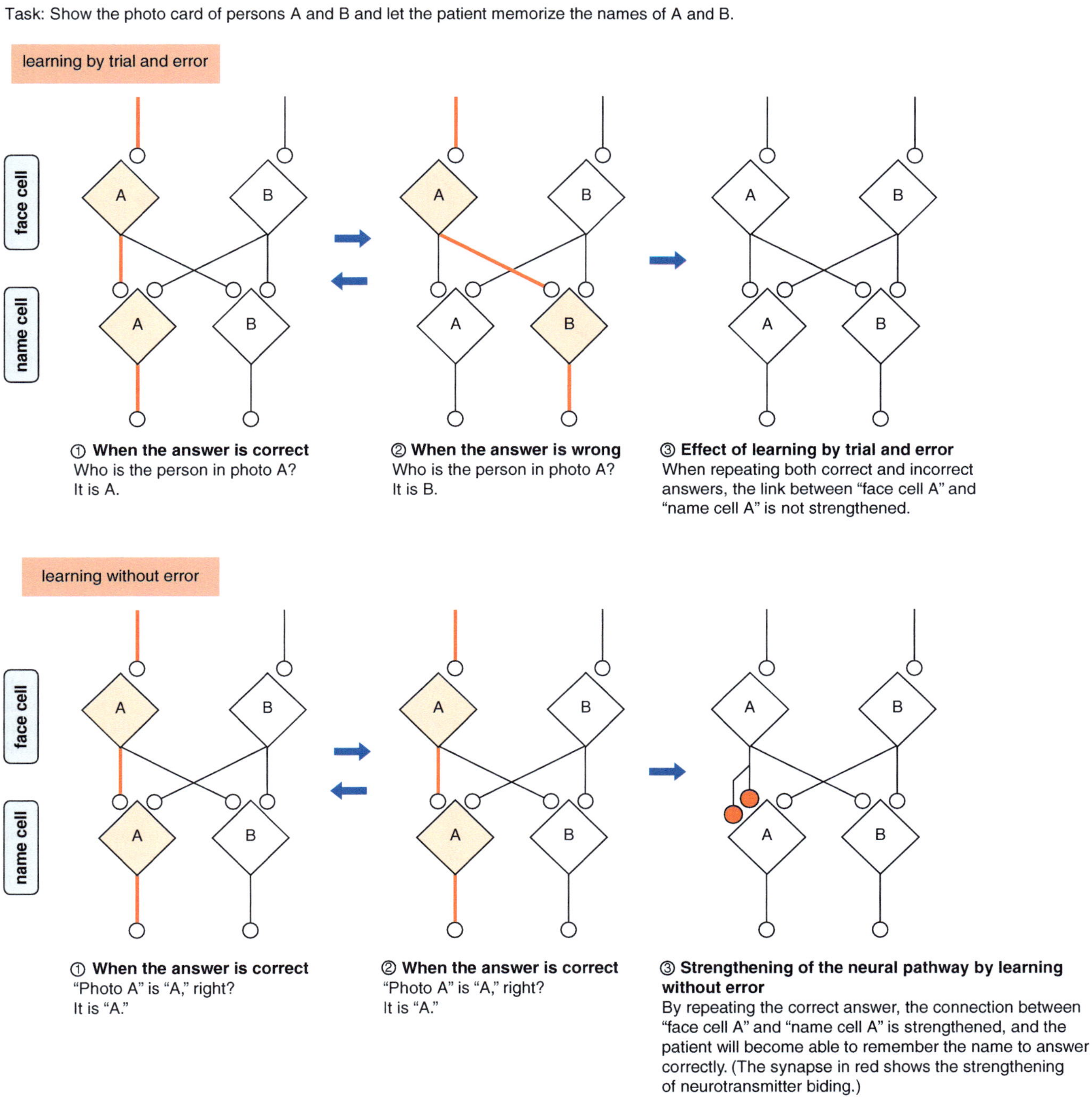

Fig. 1.4 Learning without error and changes in the neural pathway

elbow while applying facilitative intervention to raise the level of excitation transmitted along the neural pathway, realize elbow flexion, and then repeat the movement. Without facilitative intervention to induce movement, excitation will only be transmitted along the most excitable neural pathway and only synergistic movement will be activated. Non-synergistic movement will not be evoked.

To regain the programmed movement lost due to a stroke—that is, using the correct combination of necessary muscles, and to contract the muscles in the right order—the patient needs to try to realize the movement pattern while having Repetitive Facilitative Exercise help the neural pathway and neurons memorize the movement pattern. This is the only effective way [9]. If a patient has difficulty in walk-

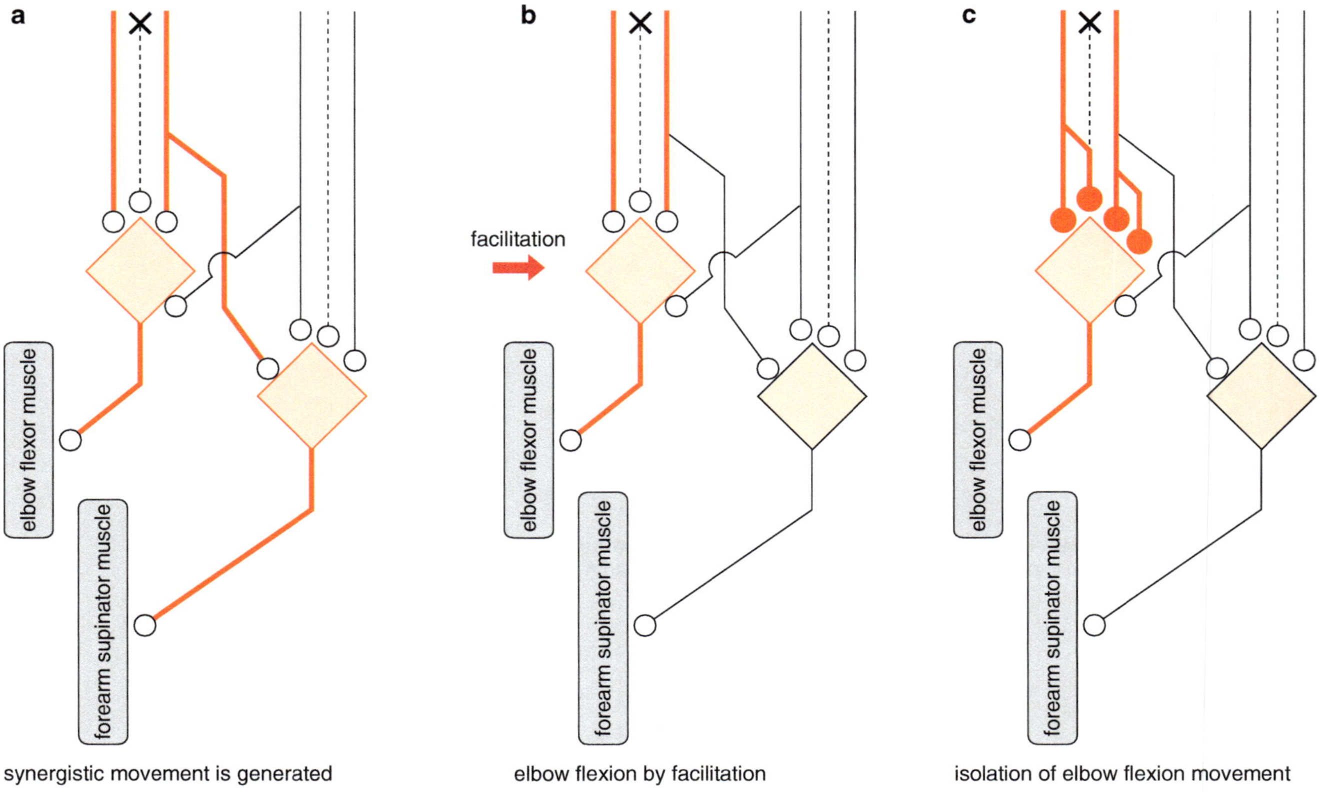

Fig. 1.5 Reconstructing and strengthening the neural pathway by repetitive facilitation. The mechanism that generates synergistic movement depends on strong connections including interneurons between particular groups of muscles at the level of the spinal cord (*filled black circle*). We selected elbow flexion and forearm supination as an example to explain synergistic flexion movement in the upper limb. Bold lines show the excitatory transmission. (**a**) This diagram is highly simplified. The descending pyramidal tract (*cross symbol*) has been damaged by a stroke. When the patient tries to flex the elbow, excitation in the intact pyramidal tract expands to the neural pathway in the spinal cord controlling synergistic muscle movement. This causes not only elbow flexion but also forearm supination resulting in the combined generation of synergistic movement. (**b**) If facilitation therapy is applied so that excitation is transmitted only to the neural pathway leading elbow flexion repeatedly, then only the neural pathway for elbow flexion will be strengthened. (**c**) Once the neural pathway for elbow flexion (*red filled circle*) has been strengthened, the patient will be able to flex his/her elbow without the assistance of facilitative intervention, and elbow flexion will be isolated from synergistic movement

ing due to synergistic movement and spasticity following a stroke, we should first place the patient in the supine position. In that position, movements will be least affected by synergy and spasticity and the movement patterns for walking will be more easily reproduced. Following this, the patient should be asked to stand up and repeat the walking pattern both alone and with the therapist's assistance. During the walking exercise, it is important to use parallel bars, orthosis, a cane, or weight-bearing equipment (to lift and hold the body). The best walking pattern should be performed repeatedly (Fig. 1.6).

The central nervous system memorizes only what the patient has done. In the treatment of hemiplegia, the plan of therapy should distinguish between these two approaches to treatments: exercises that isolate the target movement from synergistic movement, and exercises requiring coordinated combinations of movement such as walking.

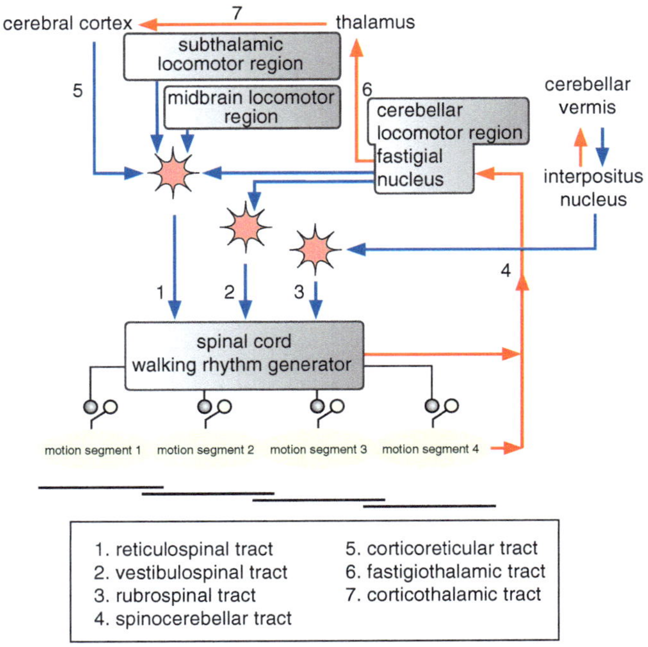

Fig. 1.6 Central pattern generator. For patients with severe damage to the spinal cord, stimulating the central pattern generator in the spinal cord is one of the methods of restoring the walking pattern. The damage of neural pathway 5 is a common cause for walking impairment in patients with hemiplegia following stroke or with spinal cord damage. However, for patients with hemiplegia due to stroke, more emphasis should be placed on Repetitive Facilitative Exercise, functional electric stimulation, and spasticity suppression with vibration stimulation aiming to isolate synergistic movements in the affected lower limb (improved voluntariness) and acquisition of the movement pattern necessary for walking. This approach is preferable since the central pattern generator cannot generate a walking pattern that exceeds the ability to voluntarily move the lower limbs while lying, sitting, or standing

1.3.2 Improvement of Excitatory Transmission Efficiency Across Synapses and Strengthening Links Between Neurons by Excitatory Transmission

Each time the synapse transmits excitation, transmission across the synapse becomes more efficient. With repeated excitatory transmission, excitation becomes easily transmitted and the functional connections in the neural circuit are strengthened. To maintain the strengthened connection, structural change should occur. However, such structural change does not occur in the target synapse if the excitation is alternately transmitted to different circuits.

Here we show an example of disturbance of movement learning caused by interference of the links of the neural pathway for the movement (Figs. 1.7 and 1.8) [10]. When performing the exercise for the lever movement, the patient tries to follow a target trajectory under the condition of receiving an external force to the arm toward a particular direction.

If trying to perform the same movement learning again after an interval, the movements will have become closer to the target movement compared to those in the first trial, showing that learning has progressed.

On the other hand, if we add lever movement exercise under the application of an irregular force that interferes with learning just after the first trial, the effects of the first trial on the movement learning will not remain when tested several hours later.

If the interval between the first trial and the task involving learning interference exceeds 4–6 h, the interference with learning decreases and the structural connection to the neural pathway related to the target movement is strengthened. If the interval time is more than 5 min, some of the learning effects will remain. However, in actual clinical practice, the time for treatment is limited and the movement patterns used by Repetitive Facilitative Exercise do not interfere with the learning process. Repetitive Facilitative Exercise requires only 30 s to 1 min of rest, which will also contribute to maintaining the patient's concentration.

Structural connections are markedly reinforced while sleeping. To promote the steady accumulation of the effects of exercise, good quality sleep is important. Movement training using tasks that include learning interference tasks such as mirror image or point symmetry also show greater progression after sleep [11] (Fig. 1.9).

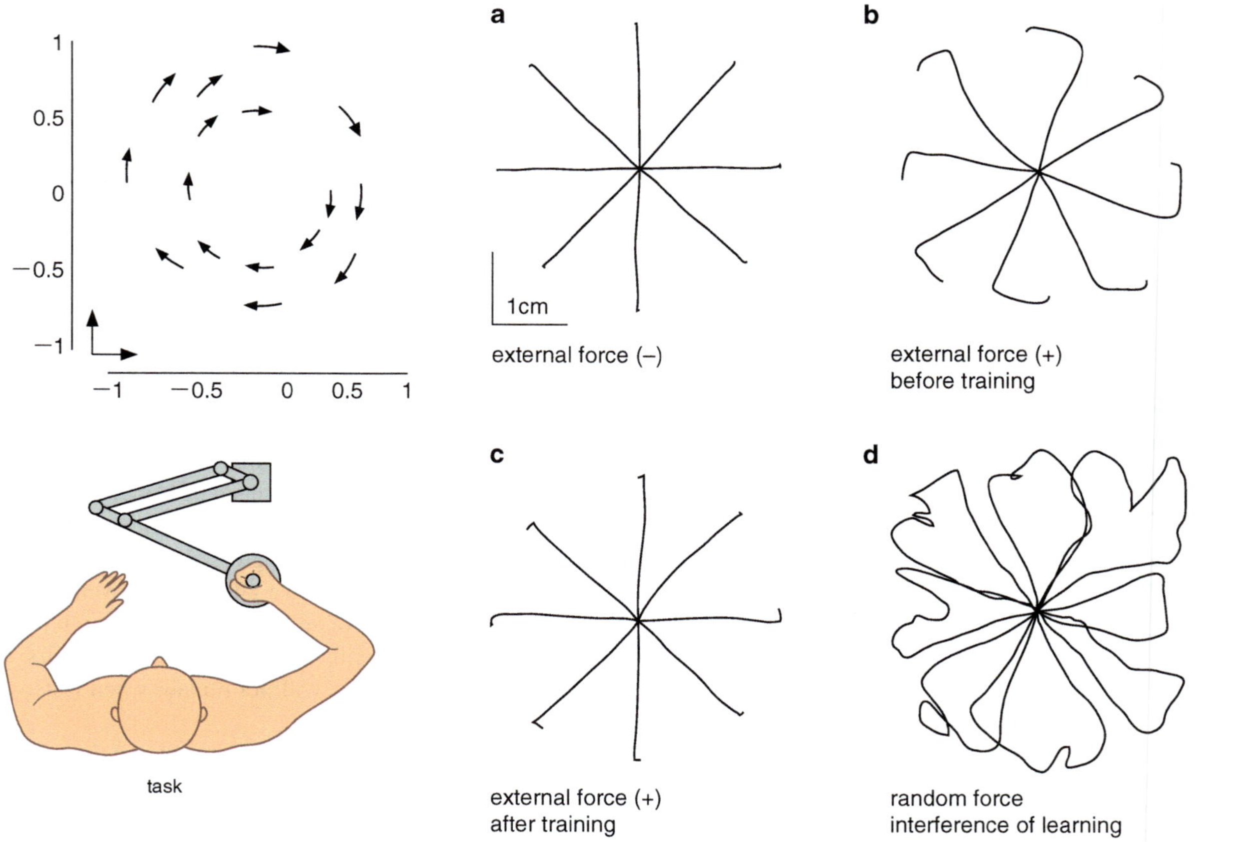

Fig. 1.7 Movement task with interference of learning [10]. Moving the lever along the target trajectory while irregular force (*arrows*) is applied. (**a**) The patient can perform the movement as intended when there is no force applied. (**b**) The movement deviates from target trajectory when resistance is applied to the lever. (**c**) After repeating the movement while force is applied, the patient becomes able to control her power to offset the resistance and perform the movement as intended. (**d**) If the strength and direction of the force are unexpectedly changed, the patient cannot successfully perform the target movement despite repeated training

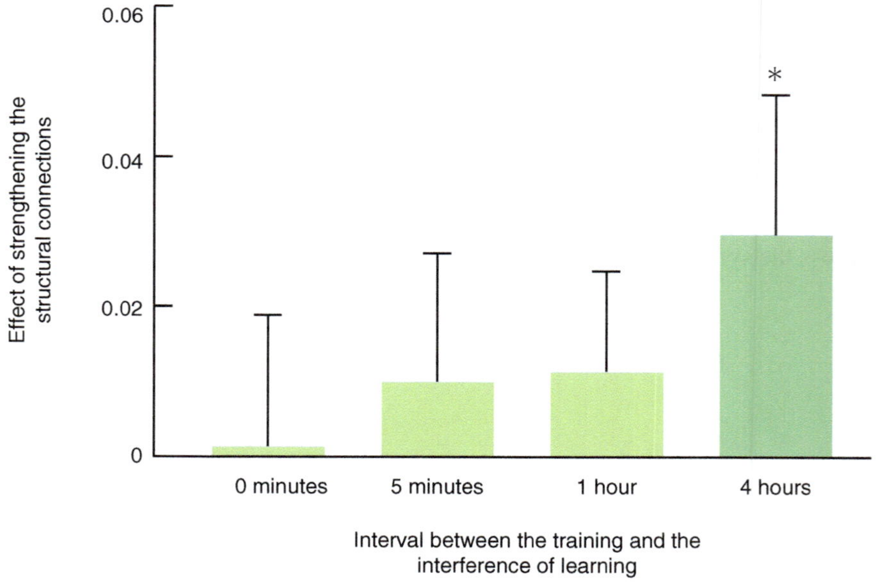

Fig. 1.8 Learning the movement and strengthening the structural connections. The patient performed the training while force was applied to the lever and became able to move the lever as intended (Fig. 1.7c). If non-uniform resistance was applied and learning interfered with (Fig. 1.7d) immediately afterward, the effect of the training did not persist. When the patient took a five-minute break after the initial training session (**c**) and then performed additional tasks involving leaning interference (**d**), the effect persisted. When the interval was 4 h, the learning interference decreased. This suggests that the structural connection continues to strengthen over time

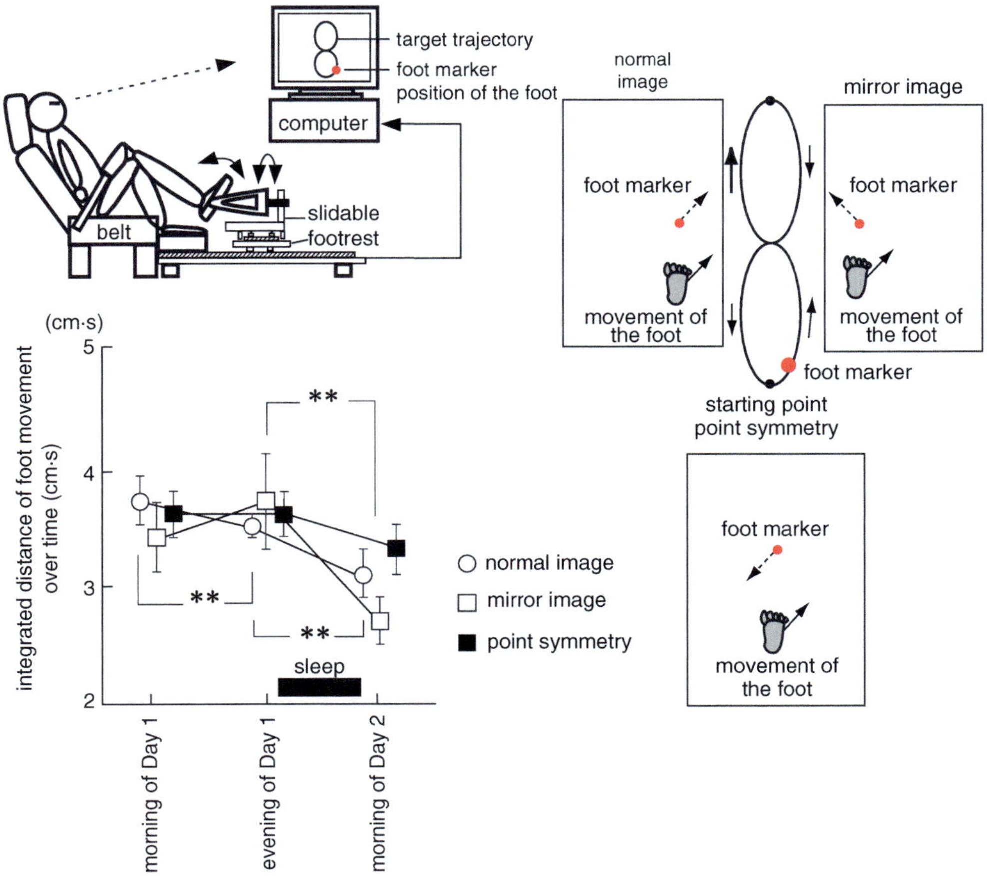

Fig. 1.9 Movement learning facilitated by sleep. Not only the patients who performed the training using the normal image but also those who underwent learning interference using a mirror image or point symmetry immediately after training showed improvement of task performance after they slept

1.4 Proof of Plasticity

Plasticity occurs basically in a use-dependent manner [12]. To be more precise, plasticity depends on the neural pathway reconstructed and strengthened by the repeated excitatory transmission. When considerable stimulation is given to the fingertips, the receptive field of the somatosensory area in the brain enlarges remarkably [13]. When sensory inputs into the foreleg of a monkey are blocked [14] or when a human upper limb is amputated [15], the receptive field for these body parts will be transformed into a receptive field for a neighboring body part. Furthermore, when the receptive field for the finger is ablated, a neighboring area in the cerebral cortex will be transformed into a receptive field for the finger (Fig. 1.10) [16]. One example for this fact is that the occipital lobe processes visual information in a healthy person but is involved in reading Braille with the fingertips in a person blind at birth [17].

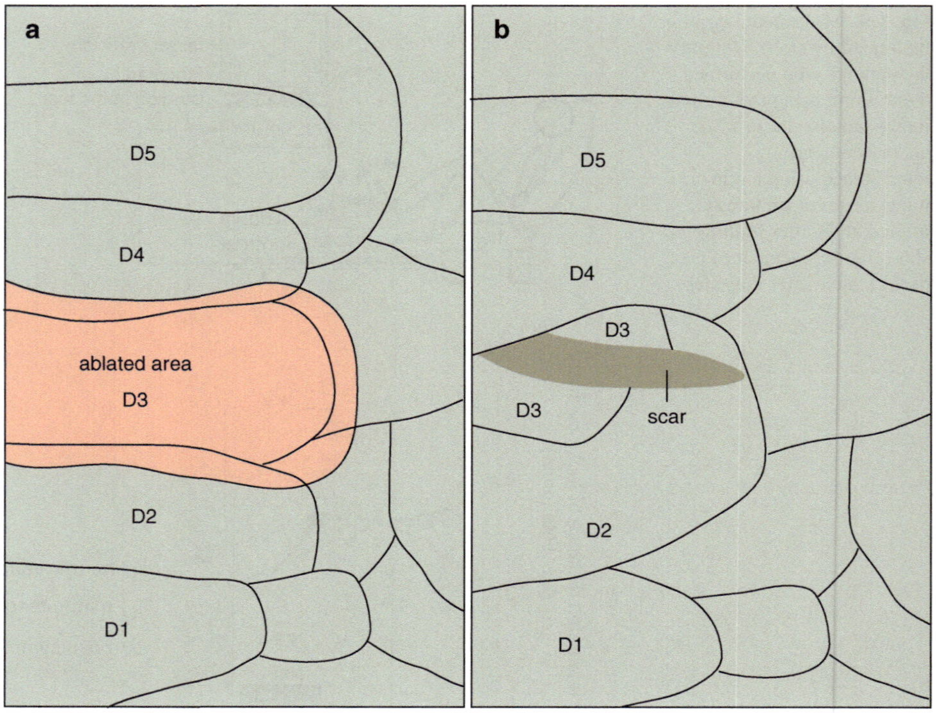

Fig. 1.10 Plasticity was detected in a monkey after the ablation of a part of the sensory area [16]. (**a**) The receptive field for the third finger was ablated in a monkey. During surgery, the monkey's fingers were stimulated while cell activities in the sensory area were recorded, confirming the receptive fields for each finger, then the receptive field for the third finger was removed. (**b**) After a few months, a new receptive field for the third finger was reconstructed in the adjacent cortex to the ablated area (*scar*)

1.5 The Mechanism of Functional Recovery and Plasticity

Achieving functional recovery requires the involvement of the intact hemisphere of the cerebrum and the intact area adjacent to the affected area. Plasticity of the cerebral cortex in particular, plays an important role in this process.

1.5.1 Plasticity of the Brain

Plasticity is defined as the ability of a neuron which is already specified to a particular function to newly acquire a different function. Although it was previously considered that the adult mature brain did have less plasticity, it has recently been demonstrated that even a mature brain can demonstrate great plasticity. Moreover, it was shown that even in the mature brain, neuroblasts migrate from the lateral cerebral ventricle to the association cortex and that neural stem cells are still present. The conventional understanding that "no new neurons are produced after birth" has now been disproven. Although it may be possible that replacement by newly generated neurons may be involved in improvement of the central nervous system or functional recovery after injury, the plasticity of intact mature neurons plays a major role in recovery from hemiplegia following a stroke. Surprisingly, when the entire cortex of both cerebral hemispheres of a monkey was incrementally removed over a two-year period, the monkey could maintain its walking ability if sufficient training for both range of joint motion and walking patterns was provided after each successive surgery [18]. This data shows the significance of central nervous system plasticity and the importance of starting rehabilitation from the early stages of the post-injury period.

1.5.2 Reconstructing the Functions in the Intact Hemisphere

Although there are individual variations, about 85% of the pyramidal fibers from the cerebral hemisphere cross over to the contralateral side of the body, while 15% remain on the ipsilateral side. Fibers crossing over to the contralateral side control the upper limbs, lower limbs, and trunk; in particular, the distal parts of the upper limbs and lower limbs are strongly controlled by neurons originating on the contralateral side of the hemisphere [19] (Figs. 1.11 and 1.12). If one side of the cerebral hemisphere is removed during the neonatal period, although its function will be almost fully recovered leaving only slight paralysis or sensory dysfunction, any fine motor control of the fingers will be poor, showing recovery of functions controlled by fibers on the ipsilateral side is limited.

However, while the percentage of crossover is calculated on histological studies, functional neural pathways arising from a large amount of synaptic connections within the brain or spinal cord are not considered. Therefore, the actual limits on functional recovery remain unclear. Paralysis caused by damage to the corticospinal tract in the spinal cord can be improved if the propriospinal neurons can form a neural pathway (Figs. 1.13 [20] and 1.14 [21]).

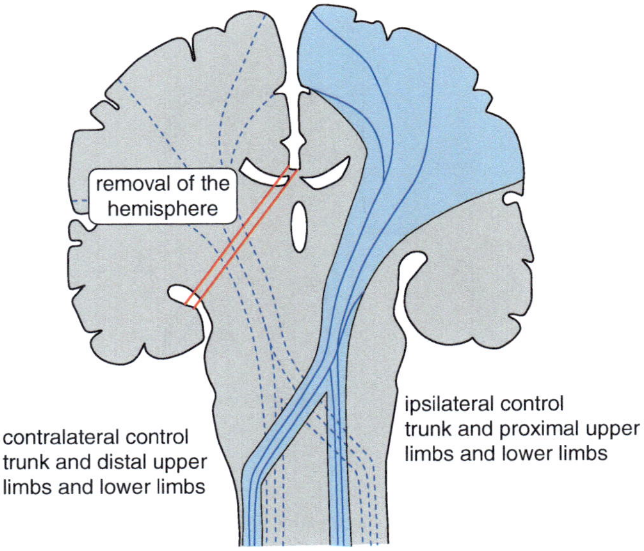

Fig. 1.11 Functional recovery and corticospinal tract (Video 1.5). Most of the corticospinal tract from one side of the cerebral hemisphere controls the contralateral side, but the remaining part controls the ipsilateral side. Even if one side of the cerebral hemisphere is removed or the corticospinal tract (indicated by the dashed line) is completely damaged, the tract in the intact hemisphere (indicated by the solid line) will establish control of the ipsilateral side and improve the paralysis

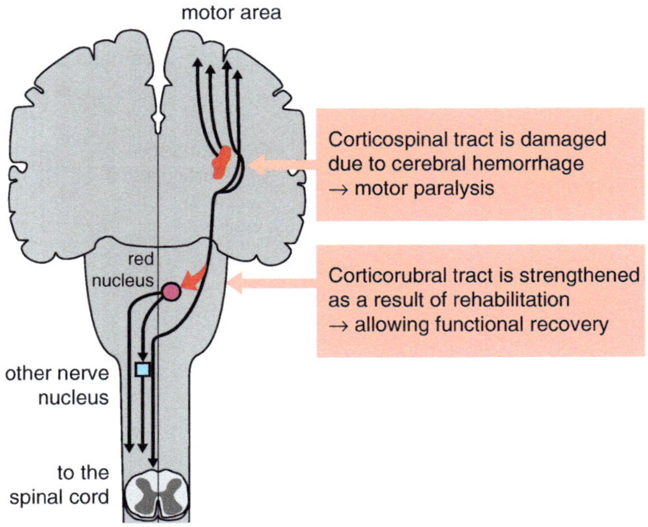

Fig. 1.12 Experiment showing reconstruction of the corticospinal tract [19]. For the recovery of hemiplegia following a stroke, the motor path from the cerebral cortex to spinal anterior horn cells must be reconstructed. When intensive rehabilitation was given to a rat following a stroke, an increase in nerve fibers extending from the motor cortex to the red nucleus in the brain stem was observed, suggesting strengthening of the neural pathway of corticorubral tract is involved in the recovery of motor functions

Fig. 1.13 Functional recovery with the assistance of propriospinal neurons in the spinal cord. Some interneurons in the spinal cord create neural circuits with the motor neurons of other spinal cord segments and are involved in functional recovery. (Note: The inhibited circuit [*red dotted line*] becomes functional [*red solid line*])

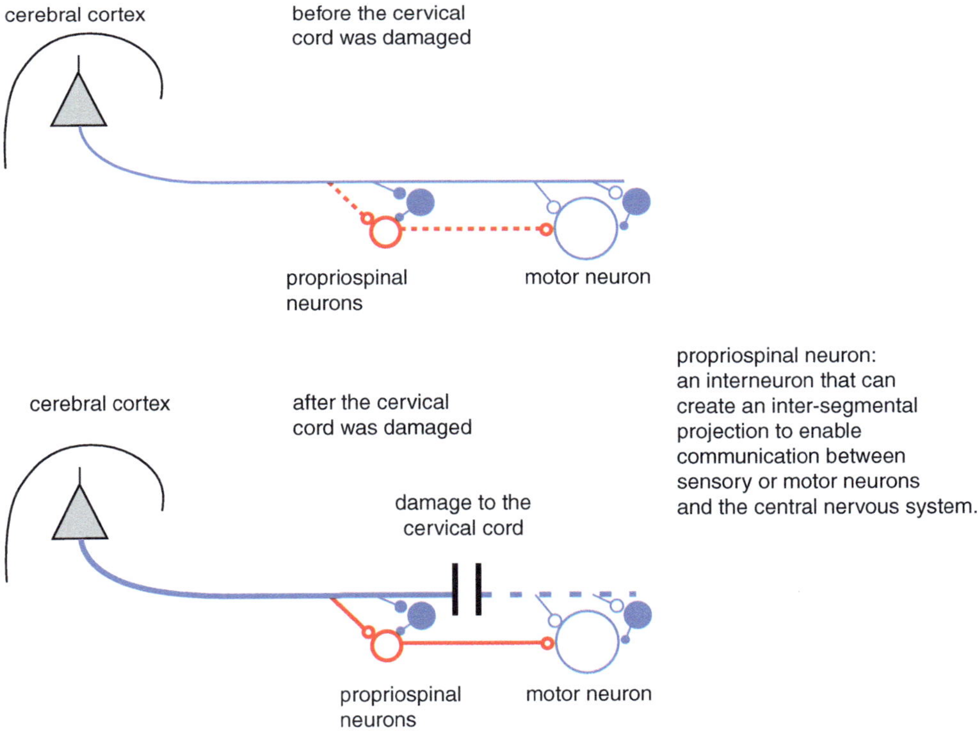

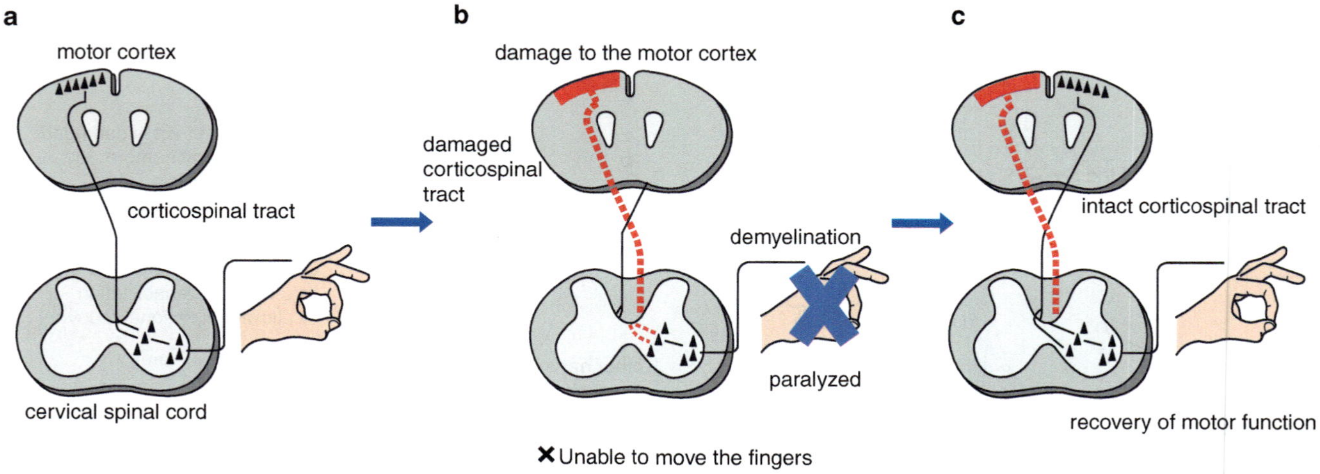

Fig. 1.14 Sprouting in the spinal cord [21]. Most of the corticospinal tracts starting from the motor area descend in the contralateral side of the spinal cord to connect to the spinal anterior horn cells (**a**). If a wider area of the motor cortex is damaged (**b**), collateral sprouting will extend from neurons originating from the motor cortex of the intact hemisphere that descend in the contralateral spinal cord to connect to the contralateral side of the spinal cord. This collateral sprouting connects to neurons that are no longer controlled by the cortex due to the damage bypassing the route and thereby facilitates the recovery of motor function (**c**)

1.6 Mechanism of Plasticity

See Fig. 1.15a. After a part of the motor area of the cortex that controls the monkey's finger movement is damaged, the monkey's finger function will not recover without rehabilitation. Furthermore, even the undamaged motor area that used to control the finger will start to shift its control to the shoulder and the elbow. However, if the monkey is trained to pick up raisins with the paralyzed fingers under the condition in which the movement of the unaffected forelimb is restrained (Fig. 1.16), the intact area of the cortex that controls movement of the finger is enhanced and recovery from paralysis occurs as shown in Fig. 1.15b [12, 22]. Similarly, in humans, during the course of Repetitive Facilitative Exercise for functional recovery of the affected upper limb, repeated voluntary movement improves the transmission efficiency of the neural circuit, resulting in the recovery of motor function. After that, the recovery of motor function through the strengthening of connections in specific neural circuits occurs (Fig. 1.17) [23].

Sprouting and unmasking are recognized as important mechanisms underlying plasticity (Fig. 1.18). The relationship between the timing of the emergence of these processes and rehabilitation is considered below [13, 24, 25].

1.6.1 Sprouting

When a neuron is damaged, nerve sprouts grow from the axon of the undamaged neuron to form a new neural pathway. New sprouts start growing about 2 weeks after injury, reaching a peak about one-month post-injury. Thereafter, sprouting will begin to decrease. This means that sprouts that do not receive inputs during this period will no longer be useful and will be eliminated from the numerous sprouts formed in the post-injury period.

1.6.2 Unmasking

Each neuron connects to many other neurons through synapses but many of these connections usually remain suppressed and the functions of these circuits are unclear. As GABA neurons reduce their activities upon neural damage, the synapses that were suppressed improve their transmission efficiency and start functioning as the transmission pathway within a few seconds to a few hours after injury. This is called "unmasking."

In clinical practice, the period of time showing improved plasticity—the acute and recovery phases—is the most effective time to begin rehabilitation for recovery from paralysis. The Repetitive Facilitative Exercise is the most effective rehabilitation technique in these phases because by performing it, the brain develops neural circuits that process information in place of lost neural circuits. In this approach, similar to "learning without error," excitation is selectively and frequently transmitted along the circuit to be strengthened, achieving a good result not only during the acute and recovery phases but also during the chronic phase 6 months or longer after a stroke.

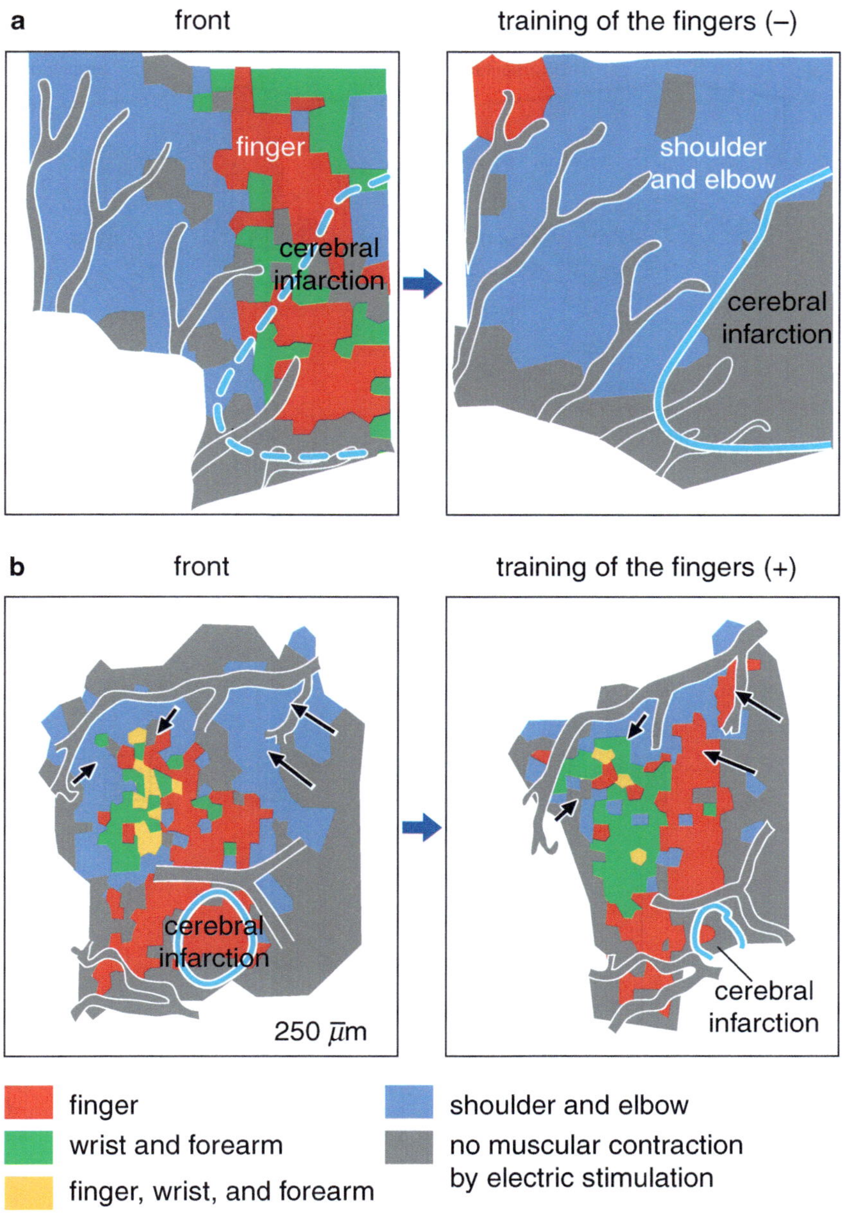

Fig. 1.15 Changes of the primary motor cortex by forced use training of the affected forelimb. (**a**) After electric stimulation of the motor cortex of a monkey to confirm the area controlling the finger, wrist, and shoulder, cerebral infarction was induced in the identified motor area. Thereafter, the monkey became unable to move the affected fingers. Even after a few months, paralysis of the fingers had not been improved and the motor cortex area that used to control the fingers and finger joints had shifted its control to the shoulder. (**b**) If movement of the unaffected forelimb is restrained, the monkey is forced to use the affected finger and paralysis of the finger improves. Furthermore, the motor cortex that controls the fingers and wrist extends to the area that had previously controlled the shoulder (*arrows*)

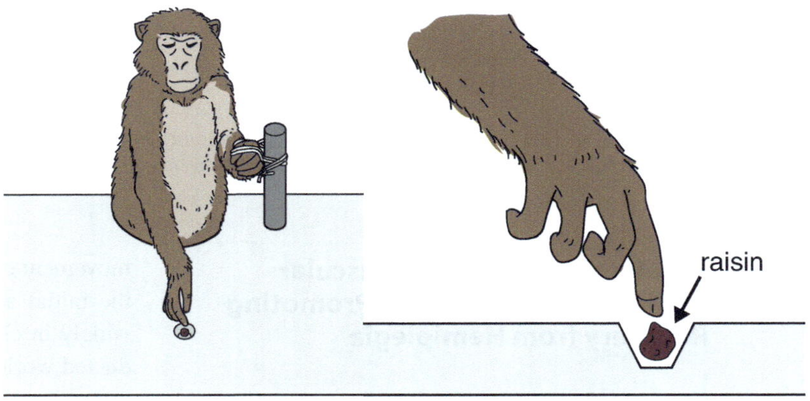

Fig. 1.16 Restraint of the unaffected forelimb and forced use of the affected finger. Restraining the unaffected forelimb forces the monkey to use the affected finger repeatedly to pick up the raisin placed in a hole

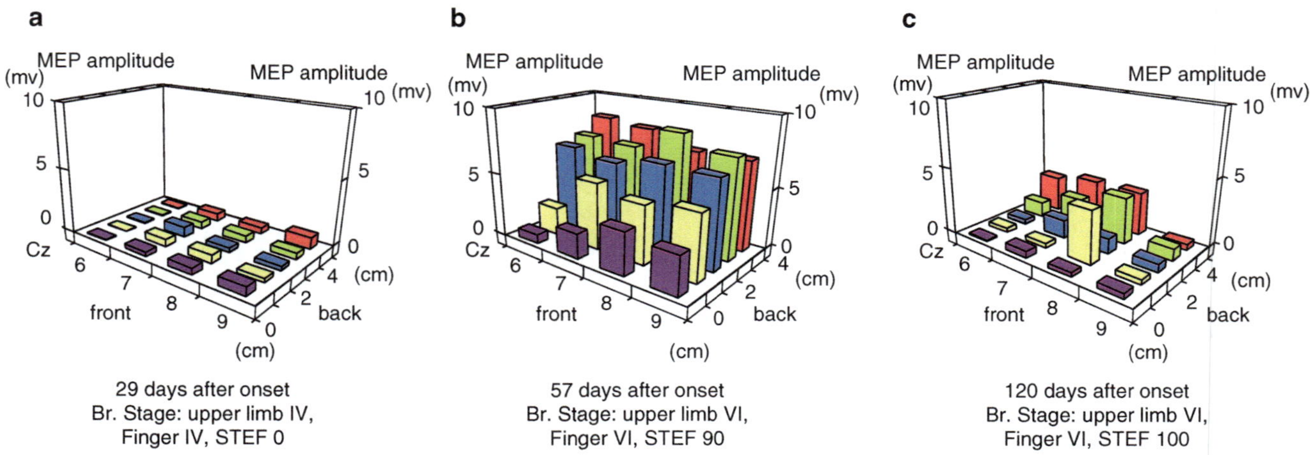

Fig. 1.17 MEP (motor evoked potential) and function of the upper limb affected by hemiplegia (56-year-old male, subcortical hemorrhage in the right frontal lobe) [23]. When the paralysis was severe, MEP (motor evoked potential) induced by TMS (transcranial magnetic stimulation) was low (**a**). As his condition improved, increased MEP appeared over a wider area (**b**). When paralysis improved further, MEP decreased and the area where MEP was recorded became smaller (**c**). In the early stage of recovery, many neurons and neural pathways were involved in finger movement. By using the Repetitive Facilitative Exercise, recovery advanced and a smaller number of neurons and neural pathways were involved in finger movement

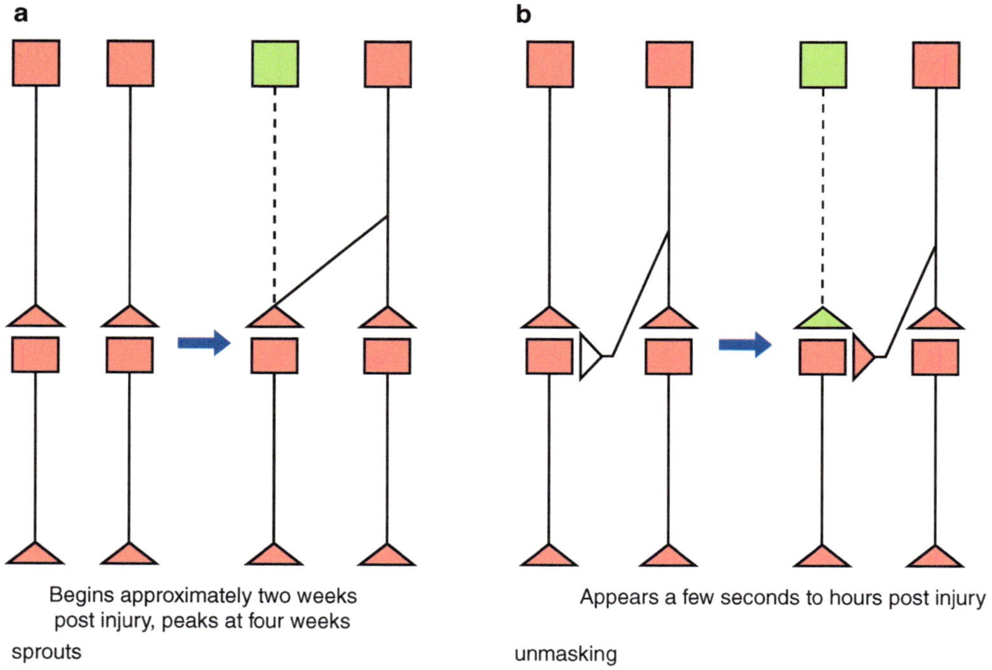

Fig. 1.18 Mechanism of emergence of plasticity [24]. Sprouting (**a**) and unmasking (**b**) are thought to be important mechanisms underlying plasticity. The green squares and triangles indicate necrotic neurons that no longer function

1.7 Problems of the Neuromuscular Facilitation Technique in Promoting Recovery from Hemiplegia

Neuromuscular facilitation techniques have been advocated as treatment methods to facilitate recovery from paralysis, that is, isolation of synergistic movement and recovery of voluntary movement control. Many of these techniques are named after the initial advocate of the method. Although they are used widely in clinical practice and many studies have been conducted worldwide, there is still little scientific evidence demonstrating that these techniques are more effective than conventional treatments, which are mainly comprised of joint range of motion training or basic movement training. Why is

Table 1.1 Characteristics and problems of neuromuscular facilitation techniques

(1) PNF (proprioceptive neuromuscular facilitation)
Good: Formation of practical movement patterns
Good: Focusing on movement development and body core ability
Neutral: Rather difficult to suppress spasms
Poor: For patients with hemiplegia, conventional treatments do not fully isolate the synergistic movements and do not provide sufficient upper/lower limbic movement patterns; furthermore, voluntary control is difficult when movements are quick, making the conventional treatment less effective.
(2) *Brunnstrom* approach
Good: Capable of realizing voluntary muscle activities
Neutral: The patient is not likely to realize and repeat useful movement patterns
(3) *Bobath* approach
Neutral: Tends to focus on motor development and trunk ability
Neutral: Does not emphasize voluntary suppression of spasm
Neutral: Poor understanding of hemiplegia and manual techniques for recovery
(a) Hemiplegia is not a disorder of postural reflex.
(b) The main purpose is relaxation, and there is no emphasis on realization and repetition of target movements. (This makes it difficult to reconstruct the pyramidal tract and form the necessary program.)
* Training without a brace, a cane or parallel bars will worsen abnormal gait patterns and promote spasms.
(c) The symmetrical approach does not recover functional symmetry.
* Lack of strengthening on the unaffected side will promote disuse syndrome

there no evidence that these neuromuscular facilitation techniques facilitate better recovery from paralysis? Why has the recovery of stroke patients from paralysis not been as good as the experimental result in animals? We must consider the problems of these neuromuscular facilitation techniques, which are summarized in Table 1.1.

1. *Does the human brain have less plasticity than the monkey brain?*

   In most researches involving experimental damage to the monkey brain, the cerebral cortex is removed. Greater improvement of paralysis can be achieved when the cerebral cortex is damaged, while a lesser degree of improvement can be observed when the internal capsule and basal ganglia is damaged.

   Poor recovery in human paralysis may be due to the following: (1) stroke lesions in humans most frequently occur in the internal capsule, corona radiata, and basal ganglia. Damage is rarely limited to the cerebral cortex; (2) humans have a larger and more complex central nervous system than other animals; and (3) not all neurons demonstrate the same level of plasticity.

   The cognitive therapeutic exercise approach does not include exercises to strengthen and reconstruct the transmission of stimulation to the pyramidal tract, which causes hemiplegia.

   The therapist moves the affected hand/foot of a patient, who closes his or her eyes, then asks the patient to say which part of the body is being touched or what is in the affected hand. It will provide training for the parietal association cortex.

2. *It is unclear whether the emergence of plasticity or improvement of the damaged region causes any degree of functional recovery in the early stages after brain damage.*

   Unmasking occurs within a few seconds to a few hours and sprouting occurs within a few weeks. For a few months after a stroke onset, there is some recovery from paralysis due to improvement in brain edema and perfusion. This phenomenon makes it impossible to determine whether the early stage of recovery from paralysis is due to plasticity or improvement from the disease itself.

3. *Functions cannot be fully recovered because the exercise and occupational therapy were started late or because the therapeutic regimen was not adequate.*

   Monkeys start moving immediately after waking up from anesthesia, but human patients do not start moving so soon after a stroke. Treatment restrictions related to the health insurance system can also limit the time and quality of exercise and occupational therapy for the affected upper limb/lower limb.

   Regarding rehabilitation contents, conventional neuromuscular facilitation techniques were insufficient to perform joint movement which is separated from the synergy movement. Moreover, it is difficult to fully perform and repeat the movement pattern of the upper limb necessary for manipulating items, as well as the movement pattern of the lower limb and trunk which is required for walking.

4. *Skillful finger movement must be done with intention.*

   Despite brain plasticity, paralysis of the fingers will not spontaneously improve to a practical level and most training programs do not fully eliminate the patient's trial and error, but rather rely solely on the patient's efforts. Such training should focus more on the movement of each finger and handling of objects as well as realizing and repeating the movements required for walking, and other actions.

1.8 Four Key Points to Facilitate Recovery from Hemiplegia

There are four points necessary for recovery from hemiplegia.

1. It is necessary to promote the restructuring and strengthening of new neural pathways from the cerebral cortex to the spinal anterior horn neurons. Furthermore, any treatment program that does not involve the patient's own effort to move will be ineffective.

Fig. 1.19 Guide the patient, utilizing their own effort, through the target nerve pathway to achieve the intended movement

2. The patient's attention should be focused on the movement during Repetitive Facilitative Exercise in order for the effort/intention of the patient to be conducted selectively to the target neural pathway. Then the intended movement can be realized, repeated, and strengthened effectively (Fig. 1.19). The therapist needs to monitor the patient closely for any differences between the target movement and patient's actual movement and then correct those differences efficiently.
3. During the acute and recovery phases, sprouting and unmasking cause dynamic alterations of the neural pathway. During these phases, the therapist should repeatedly provide as many strong stimulations and encourage as many intentional movements as possible instead of focusing on accurate sensory input or normal movements.
4. It is necessary to distinguish the repetition of movement patterns aimed at reconstruction and strengthening the damaged neural pathway from strength training exercises intended to enlarge the muscles. Since training for maximum muscle strength involves not only the target neural pathway but also other neural pathways, such exercises are not effective for selectively forming and strengthening the target neural pathway.

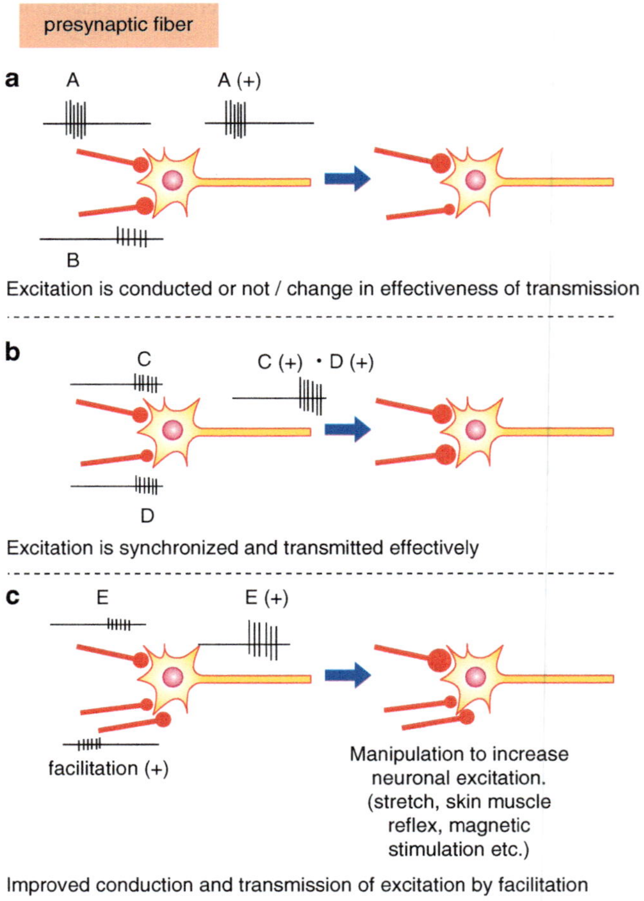

Fig. 1.20 Improved conduction of excitation and transmission efficiency by facilitation. (**a**) If excitement of the presynaptic fiber induces stimulation of postsynaptic neurons, transmission across the synapse becomes more effective and the neural pathway is strengthened. If the opposite occurs, transmission along the neural pathway is reduced. (**b**, **c**) If two excitatory inputs from presynaptic neurons are simultaneously transmitted leading to an excitation of the postsynaptic neuron, then connections in these two neural pathways are strengthened. In the treatment of hemiplegia, the facilitation procedure should be applied prior to the exercise to increase the level of excitation in the target pyramidal tract so that even a slight excitation from the cerebral cortex can achieve the intended movement. By repeating this process, the neural pathway related to the movement will be strengthened and the intended movement will become possible without the facilitation procedure

In order to improve paralysis, the neural pathway necessary for the intended movement must be formed and strengthened. First, transmit excitation to the target neural pathway to realize the target movement. Then repeat the movement to improve the transmission efficiency in the neural pathway [25], and strengthen the histological binding to develop a robust neural pathway. Finally, continue in order to improve the movements to a practical level.

If excitation in the presynaptic fiber is conducted to the postsynaptic neuron, synapse is strengthened, even when excitation is induced by electricity [26]. As shown in Figs. 1.20, 1.21, and 1.22, we should use positioning, stretch reflex, skin muscle reflex, electrical stimulation, vibratory stimuli, and other techniques to raise the level of excitation in the pyramidal tract, which connects the premotor cortex or primary motor cortex to the spinal anterior horn cells, to a level as close to the threshold as possible. Stimulation of the pyramidal tract should be strengthened prior to the conduction of excitation from the motor cortex via the prefrontal cortex (intention) or premotor cortex. When the level of excitation in the pyramidal tract is raised, excitation from the

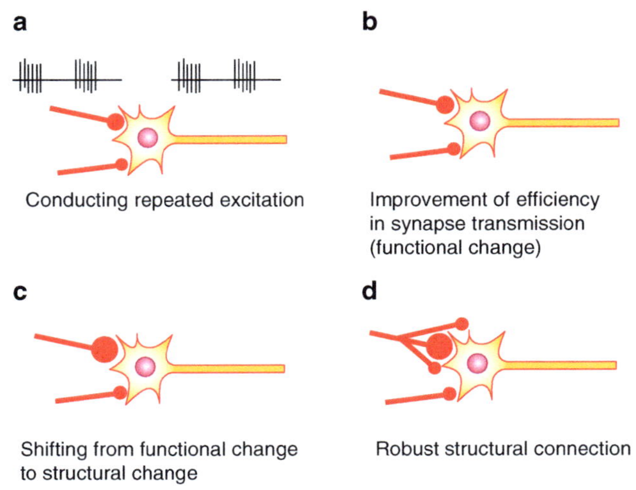

Fig. 1.21 Conducting excitation strengthens neurochemical binding at the synapse. The above shows the process of shifting the connectivity in the neural pathway from unstable functional strengthening to robust structural strengthening. Structural strengthening takes 4–6 h after training. In particular, this occurs most efficiently during sleep

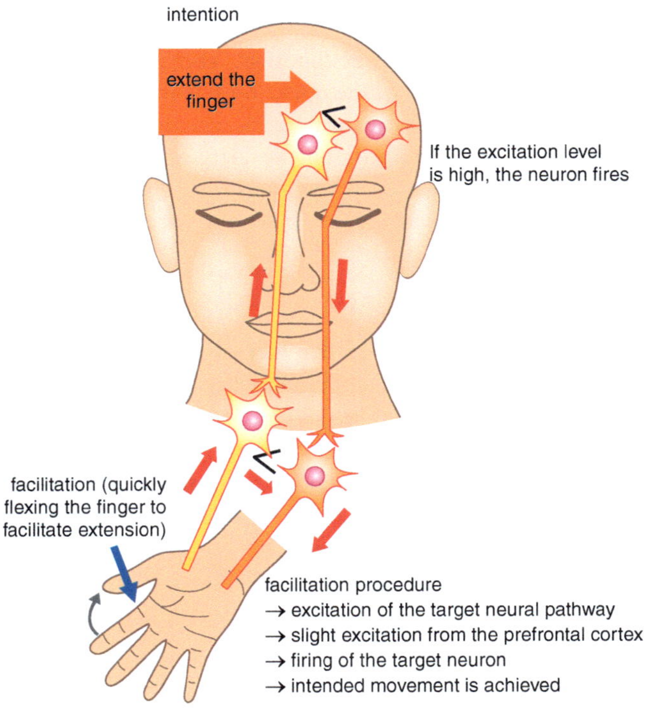

Fig. 1.22 Mechanism for facilitation procedures that induces finger movements

unaffected cerebral cortex depolarizes the spinal anterior horn cells and contracts the muscle. By repeating this procedure, connection in the pyramidal tract is strengthened and paralysis will be improved [24]. During the acute phase of cerebral infarction, Repetitive Facilitative Exercise under continuous electrical stimulation will improve paralysis more than other traditional therapies. Such improvements with Repetitive Facilitative Exercise have been confirmed in randomized control clinical trials and other studies [27–42].

For recovery from hemiplegia, inducing excitation along the neural pathway involved in the intended movement, that is, inducing the voluntary movement, is absolutely necessary. Any therapy that does not facilitate this process cannot be effective. Electrical stimulation of the muscle can be effective if it is combined with the patient's voluntary muscle contraction. As there is a strong link between the primary sensory cortex and primary motor cortex, passive movements guided by the therapist change the level of excitation in the neural pathway related to movement. However, the greatest excitation in the motor cortex occurs when the movement is achieved voluntarily.

A smart treatment strategy is needed to effectively promote recovery. Currently, numerous new therapies are being advocated. We need to classify these therapies into those that strengthen the target neural pathway and those that adjust the level of excitation in the neural pathway. By appropriately combining these therapies, we can set up an effective treatment plan with the least amount of trial and error by the patient (Fig. 1.23).

For neural regenerative medicine, research on methods of culture and injection of stem cells has advanced and some clinical trials have been launched. However, in order to achieve significant functional improvement that is in line with expectations, it is essential to establish the theory and technology of reconstructing and strengthening the neural pathways in which injected neurons are integrated to function. As shown in Fig. 1.24 [28], the mouse with brain injury showed greater recovery in its movements when treated with a combination of stem cell implant and rehabilitation (treadmill) than when treated with the implant alone.

In the scope of regenerative therapy for humans, rehabilitation techniques such as Repetitive Facilitative Exercise selectively conducting excitation to the target neural pathway to strengthen it are indispensable for reconstructing and strengthening the neural pathways connecting the body and brain, required for bipedal walking and manipulating each finger.

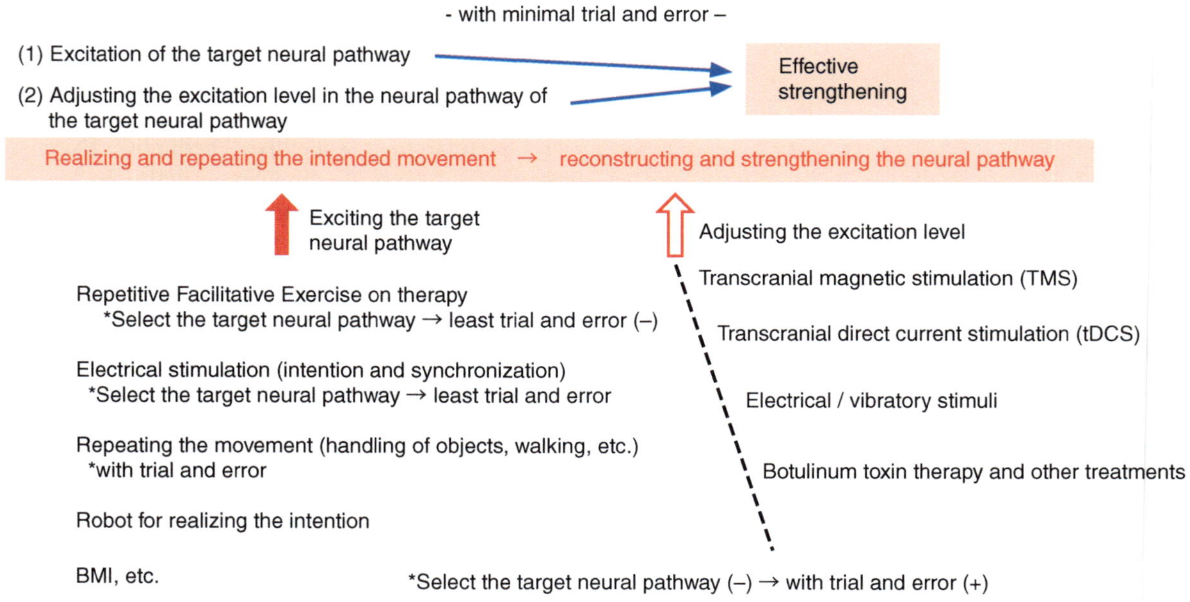

Fig. 1.23 Treatment strategy for recovery from paralysis

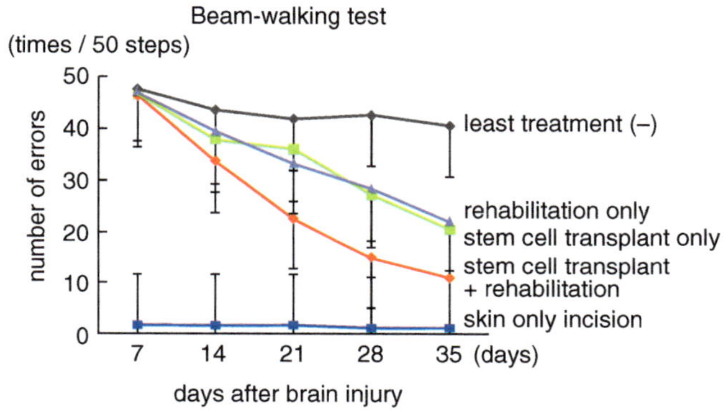

Fig. 1.24 Recovery of physical function of mice after brain injury: comparison between stem cell transplant with/without rehabilitation [28]. Stem cell transplant combined with rehabilitation shows better functional recovery

1.9 Outcome of Repetitive Facilitative Exercise

It has been reported that only 20–30% of patients with hemiplegia due to a stroke can recover their voluntary upper limb movements to a practical level that allows them to manipulate objects with their fingers. However, even if the improvement is under the practical level, as long as the patient can pinch an object, they can realize movements like holding a piece of paper and cutting it with scissors using the unaffected hand. Likewise, if the patient can support a buttonhole with one hand, they can fasten the button with the other. Even when the patient has persistent synergistic movement, if they can flex/extend the upper limbs, it will become easier to dress/undress. Therapy to improve voluntary movements of the upper limb is important. Even though the ceiling of level of the recovery from paralysis is determined by the location and severity of brain damage, more effective treatment regimens such as Repetitive Facilitative Exercise can promote recovery from paralysis.

1.9.1 Effectiveness of the Repetitive Facilitative Exercise for the Affected Upper Limb (Fig. 1.25)

If you add 20–30 min of Repetitive Facilitative Exercise–inducing and repeat each intended movement 100 times, then under a conventional occupational therapy regimen, the overall time for the entire treatment remains 40 min, but the patient will demonstrate greater improvement as shown in Fig. 1.25a (upper limb) and b (fingers) [29–32].

In a study comparing inpatients in the recovery phase from strokes who received either Repetitive Facilitative Exercise or conventional therapy, those receiving Repetitive Facilitative Exercise showed significantly greater functional recovery of upper limbs, fingers, and lower limbs [33]. Looking at ADL, they showed greater improvement of FIM scores (Fig. 1.26) [33]. In a multicenter randomized controlled trial using the international assessment index targeting stroke patients with upper hemiplegia rated III or higher

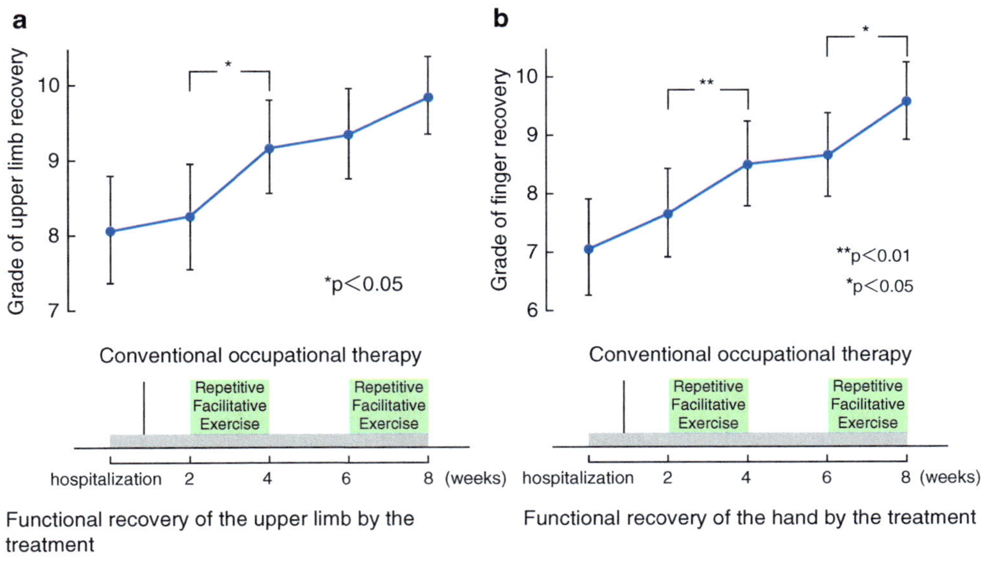

Fig. 1.25 Recovery of hemiplegic upper limb with Repetitive Facilitative Exercise. This figure (**a** and **b**) shows recovery from paralysis when Repetitive Facilitative Exercise is combined with conventional occupational therapy. In Repetitive Facilitative Exercise, a few movement patterns are selected, each of which are practiced and repeated 100 times. Although the treatment duration remains 40 min, the combination of Repetitive Facilitative Exercise and conventional occupational therapy induces greater improvement within 2 weeks

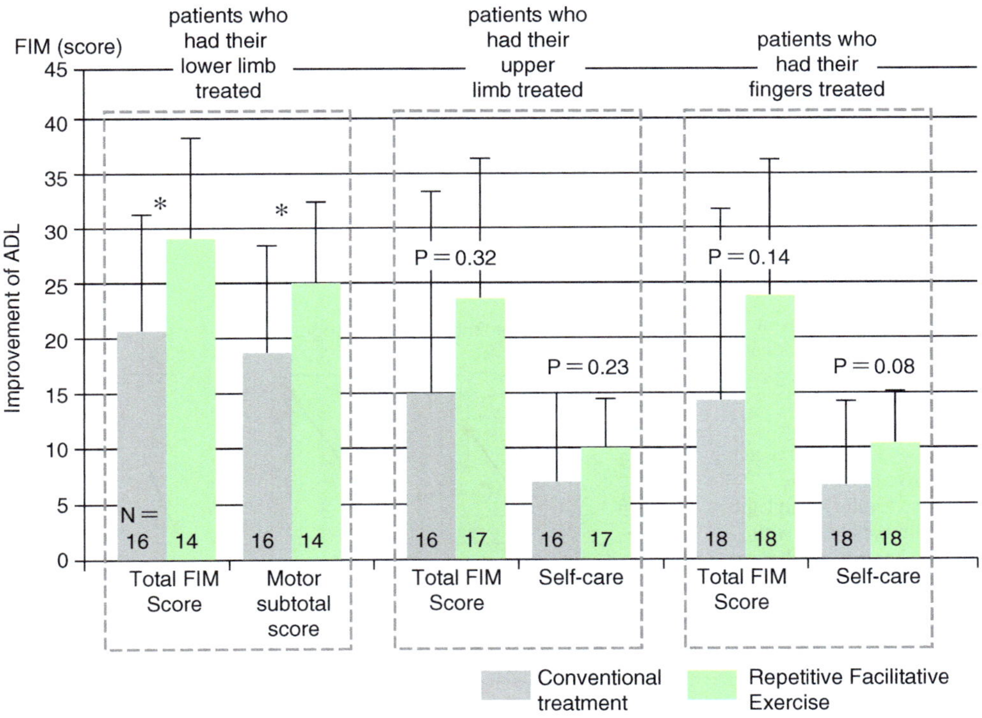

Fig. 1.26 Comparison of ADL improvement using Repetitive Facilitative Exercise and conventional treatment (in the recovery phase) [33]. Among patients receiving lower limb treatment, Total FIM and Motor FIM showed greater improvement of ADL (activities of daily living) following Repetitive Facilitative Exercise compared to that of patients receiving conventional treatment. Among patients receiving treatment of the upper limb and fingers, Total FIM and self-care FIM showed greater improvement following Repetitive Facilitative Exercise

on the *Brunnstrom Stage* of Stroke Recovery, some recovery from paralysis was observed (Fig. 1.27a), and the level of improvement in those with Repetitive Facilitative Exercise was much greater than in those receiving conventional therapy (Fig. 1.27b). The improvement was greater than MCID (minimal clinically important difference) [34, 35].

In a retrospective study on therapies during the acute phase of strokes, recovery from paralysis of the upper limbs and fingers was better in the group receiving a combination of Repetitive Facilitative Exercise and conventional therapy than the group which received only conventional therapy, even though the treatment times were the same, with 40 min/day rehabilitation treatment [36].

In a study of patients with chronic hemiplegia for 1 year or longer after a stroke, patients demonstrating isolated synergistic movement at hospitalization (duration after onset: 59 months) showed a significant improvement in upper limb grade, finger grade, and Simple Test for Evaluating Hand Function (STEF). However, those demonstrating synergistic movements (duration after onset: 31 months) showed a sig-

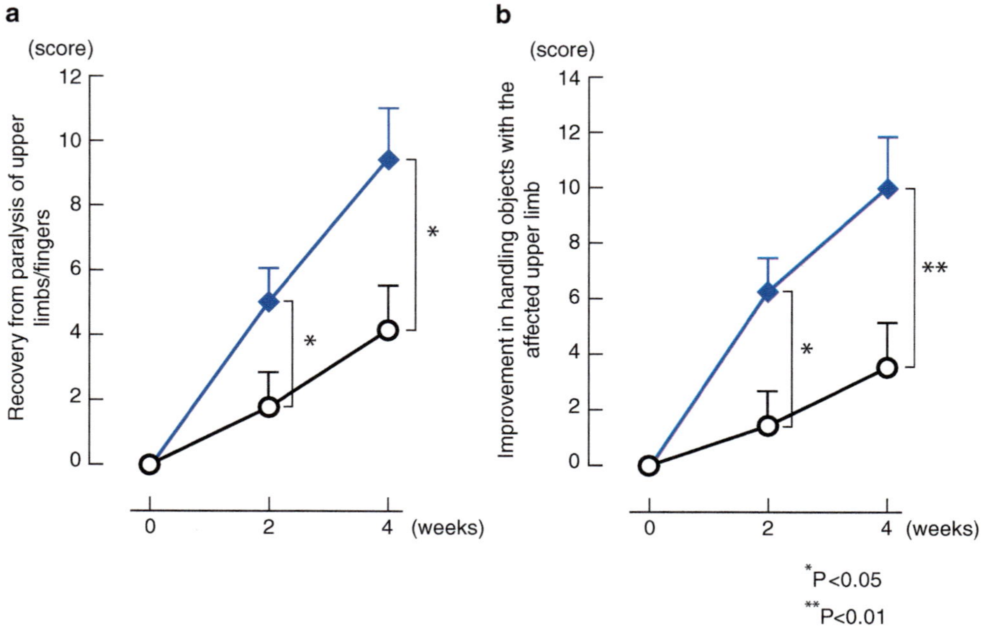

Fig. 1.27 Comparison of Repetitive Facilitative Exercise (*blue filled diamond*) and conventional treatment (*open circle*) in a multicenter randomized controlled trial (during the subacute phase). On both (**a**) recovery from hemiplegia [Fugl-Meyer Assessment for upper extremity (FMA), maximum score 66] and (**b**) upper limbs and fingers performance [ARAT: Action Research Arm Test, maximum score 57], patients receiving Repetitive Facilitative Exercise showed significantly greater improvement. The study included hemiplegic stroke patients in the subacute phase whose *Brunnstrom stage* was III or higher. Patients in both groups received a 40-min intervention plus 30 min of training in handling objects 5 days a week for 4 weeks. (The graph was based on data reported in reference [34].) Error bar indicates the average ± standard error

Fig. 1.28 Improvement of motor functions in patients with chronic hemiplegic upper limb following Repetitive Facilitative Exercise [37]. After receiving Repetitive Facilitative Exercise for the hemiplegic upper limb, patients with chronic hemiplegia showed significant functional improvement. Grade: median, 25–75%; STEF: average, standard deviation

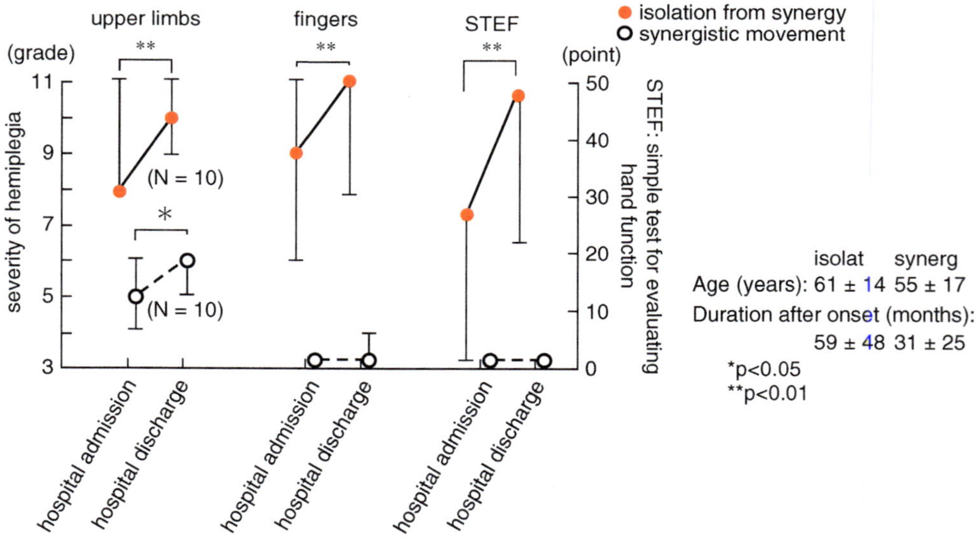

nificant improvement only in the upper limb (Fig. 1.28) [37]. The more repetitions of facilitative exercise were performed, the better the outcome. However, some case reports indicate that 50 repetitions are sufficiently effective considering the limitation of the treatment time [38]. The same effects can be seen in cases involving corticobasal degeneration [39] and spinal central injury [40].

1.9.2 Outcome of Repetitive Facilitative Exercise for a Hemiplegic Trunk

In a multicenter randomized controlled trial, stroke patients in the recovery phase were divided into two groups: one group received conventional physiotherapy only and the other group received conventional physiotherapy combined with Repetitive

Facilitative Exercise. The group receiving Repetitive Facilitative Exercise showed significantly greater improvement in the strength of the rotator muscles of the trunk as well as greater speed in the 10-m walk test (Fig. 1.29) [41].

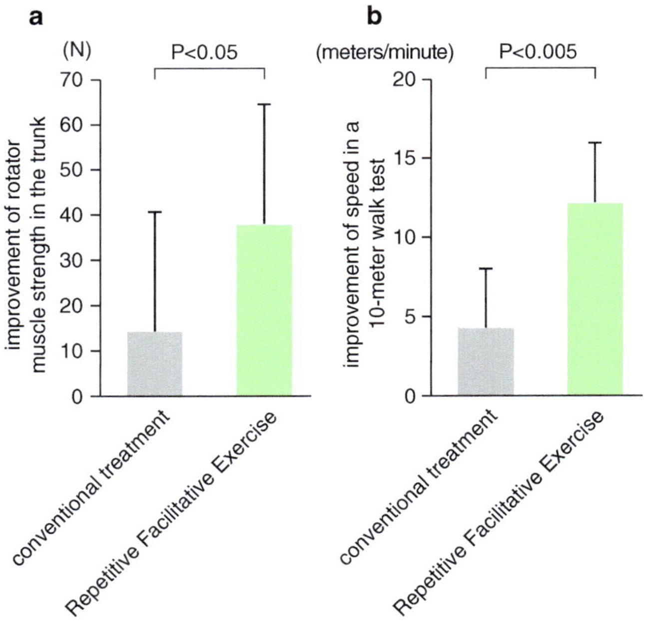

Fig. 1.29 Data from a multicenter randomized controlled trial comparing improvements in: (**a**) the strength of the rotator muscles in the trunk; and (**b**) a 10-m walk test between patients treated by conventional physiotherapy (*gray*) and those receiving Repetitive Facilitative Exercise (*green*) during the subacute phase. The study assessed patients in the subacute phase after a stroke who could sit upright independently or with assistance. The intervention was performed 6 days per week for 8 weeks. (The figure was based on data from reference [41]). *Error bar*: average ± standard deviation

1.9.3 Outcome of Repetitive Facilitative Exercise for Hemiplegic Lower Limbs

If the patient undergoes Repetitive Facilitative Exercise for the five movement patterns (100 times each) required for walking, Repetitive Facilitative Exercise improves recovery from hemiplegia (Figs. 1.30 and 1.31) [43, 44]. The effect of Repetitive Facilitative Exercise is not only due to the increase in exercise volume but also due to the strengthening of specific neural pathways as a result of repeating the movement induction. No significant recovery from hemiplegia can be expected if you simply prolong the duration of each treatment session.

(Megumi Shimodozono)

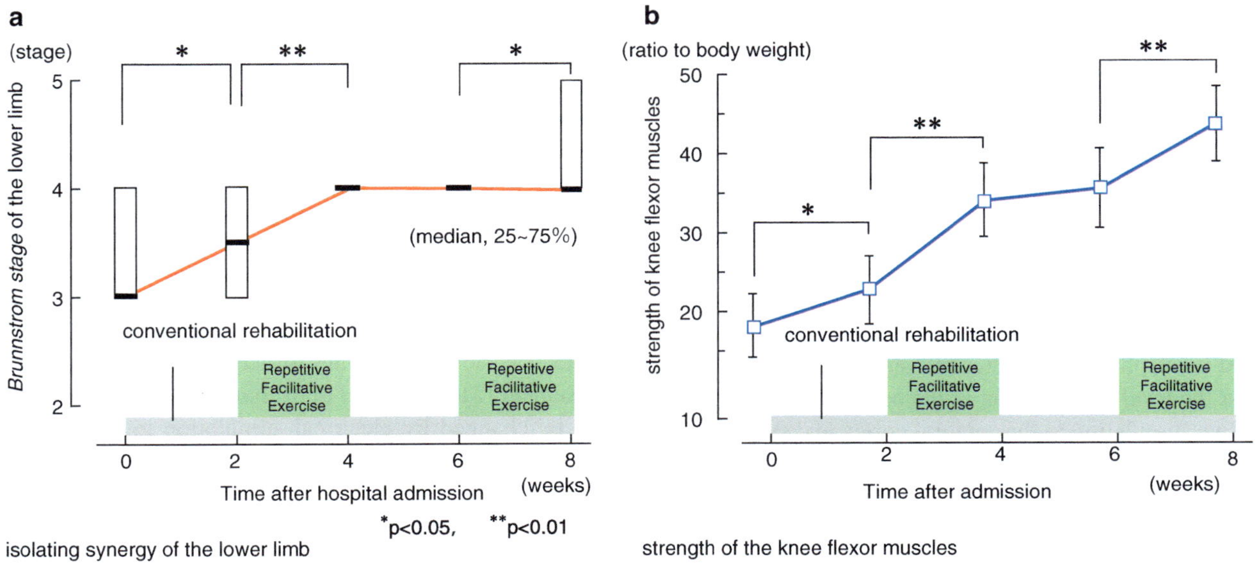

Fig. 1.30 Improvement of the hemiplegic lower limb following Repetitive Facilitative Exercise during the subacute phase [42, 43]. Two 2-week Repetitive Facilitative Exercise sessions (more than 100 repetitions a day for each of the five kinds of movement) were applied at 2-week intervals in patients with hemiplegia who were being treated with continuous conventional rehabilitation (**a** and **b**). The improvements after Repetitive Facilitative Exercise and conventional rehabilitation sessions were significantly greater than those after the preceding conventional rehabilitation sessions

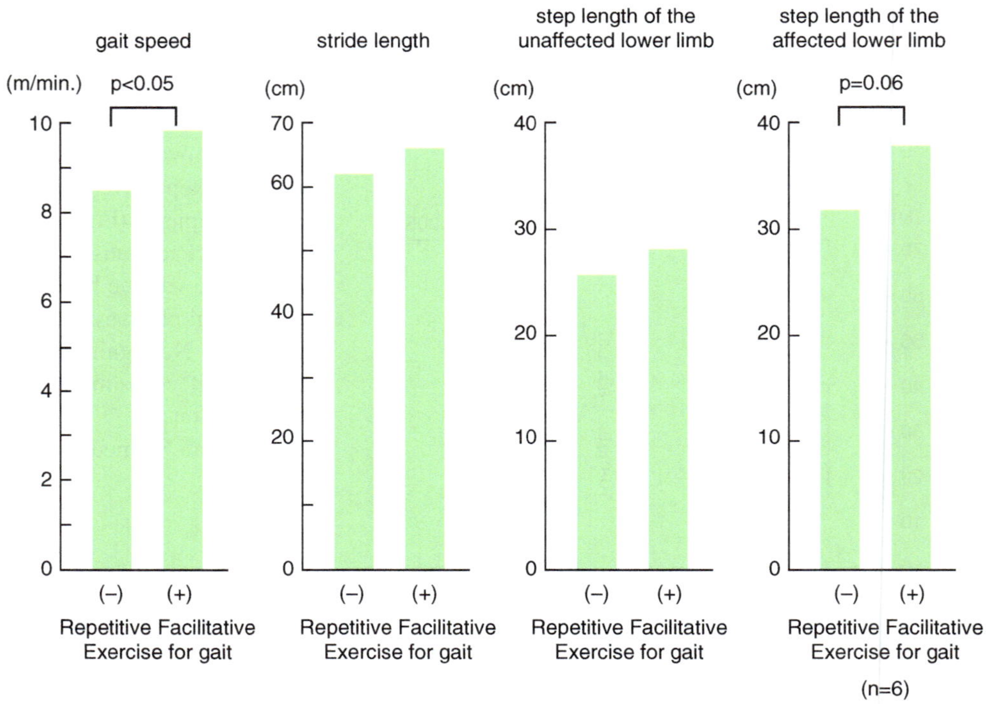

Fig. 1.31 Improvement of gait speed by Repetitive Facilitative Exercise [44]. Both gait speed and the step length of the affected lower limb were improved using Repetitive Facilitative Exercise

1.10 Repetitive Facilitative Exercise as the Primary Therapy in Combination with Other Supportive Therapies

1.10.1 Theory of Combination Therapy

Following a stroke or other damage to the neural pathway, motor function can be recovered with appropriate rehabilitation. This means that a functional structure is newly created in an area of the cerebral cortex that is adjacent to the damaged part and in the contralateral unaffected hemisphere. Rehabilitation to recover from motor impairments aims at reconstructing brain functions. Combining therapies that facilitate this process makes Repetitive Facilitative Exercise more effective. We expect that these combinations will be applied to an increasing number of stroke patients.

In general, damage to neurons and neural pathways after a stroke reduces the level of excitation in the affected brain hemisphere while causing excessive excitation in the unaffected hemisphere [45]. In recent years, techniques that can adjust the level of excitation in the cerebral cortex with less invasive stimulation to the cerebral cortex and stimulation to peripheral neural systems have been reported [46, 47]. Application of any of these therapies before or during Repetitive Facilitative Exercise enables patients to realize the intended movement and repeat it, thereby accelerating the reconstructing process of the brain functions.

Below, we will show techniques to adjust the neural pathways that are available in Japan and describe how to combine these techniques with Repetitive Facilitative Exercise.

1.10.2 NMES: Neuromuscular Electrical Stimulation

Electrical stimulation has long been widely applied as physical therapy to maintain and enhance muscle strength as well as reduce pain. In recent years, it has also attracted attention as a less-invasive technique that can adjust the neural pathways [48].

It is well known that multiple stimulations to the neurons can induce continuous improvement of transmission between the neurons, a phenomenon known as long-term potentiation (LTP) [49]. Researchers have shown that electrical stimulation in the sensory cortex or thalamus causes LTP in the motor cortex [50], and the combination of repetitive spontaneous movement with incoming inputs has been shown to make it easier for patients to acquire the ability for new movement [51]. The level of excitation in the cerebral cortex can be changed by electrical stimulation to the peripheral nerves, myoelectric potential produced when exercising, or functional electrical stimulation [52]. In the following section, we will show how to enhance the effect of Repetitive Facilitative Exercise by combining treatment with electrical stimulation.

1. Combination with continuous electrical stimulation

We use low frequency electricity (symmetrical waveform), which is inexpensive and easy to access. The frequency of the electrical current is 50 Hz, and the pulse width is adjusted between 150 and 250 μs depending on the degree of discomfort tolerated by the patient. Intensity of the electrical current is adjusted to produce slight contraction of the target muscle or slight joint movement. Bipolar electrodes are placed with the motor point located between the two electrodes. When providing Repetitive Facilitative Exercise under such conditions, the rubbing technique should not be applied. Tapping and pressing should be performed carefully to avoid causing the electrodes to peel off (Figs. 1.32 and 1.33).

Shimodozono et al. studied three groups of patients with severe upper limb hemiplegia in the subacute phase after a stroke: conventional treatment, Repetitive Facilitative Exercise, and Repetitive Facilitative Exercise with continuous electrical stimulation. Then they compared the improvement using Fugl-Meyer assessment scores after a 4-week intervention. The study showed that the effects of Repetitive Facilitative Exercise could be facilitated when Repetitive Facilitative Exercise is combined with continuous electrical stimulation (Fig. 1.34) [54]. Maesako et al. studied two groups of hemiplegic patients with hand edema in the acute phase after a stroke: one group was treated by a range of motion exercise and another group by Repetitive Facilitative Exercise with continuous electrical stimulation, both for 30 min per day, for 2 weeks. The study showed greater improvement in both motor paralysis and edema for patients treated by Repetitive Facilitative Exercise with continuous electrical stimulation (Fig. 1.34) [53].

2. Combination with electrical stimulation synchronized with movement

Electrical stimulation is given to the patient while synchronized with the patient's voluntary movement. An appropriate rest interval is allowed after each of the repetitive movements so the patient's muscle will not be fatigued, and the level of stimulation can be set above the motor threshold (Fig. 1.35). If the target muscle is rather big and the motor points are not close, it is possible to use the low frequency electricity with a symmetrical waveform described above. However, when many motor points are

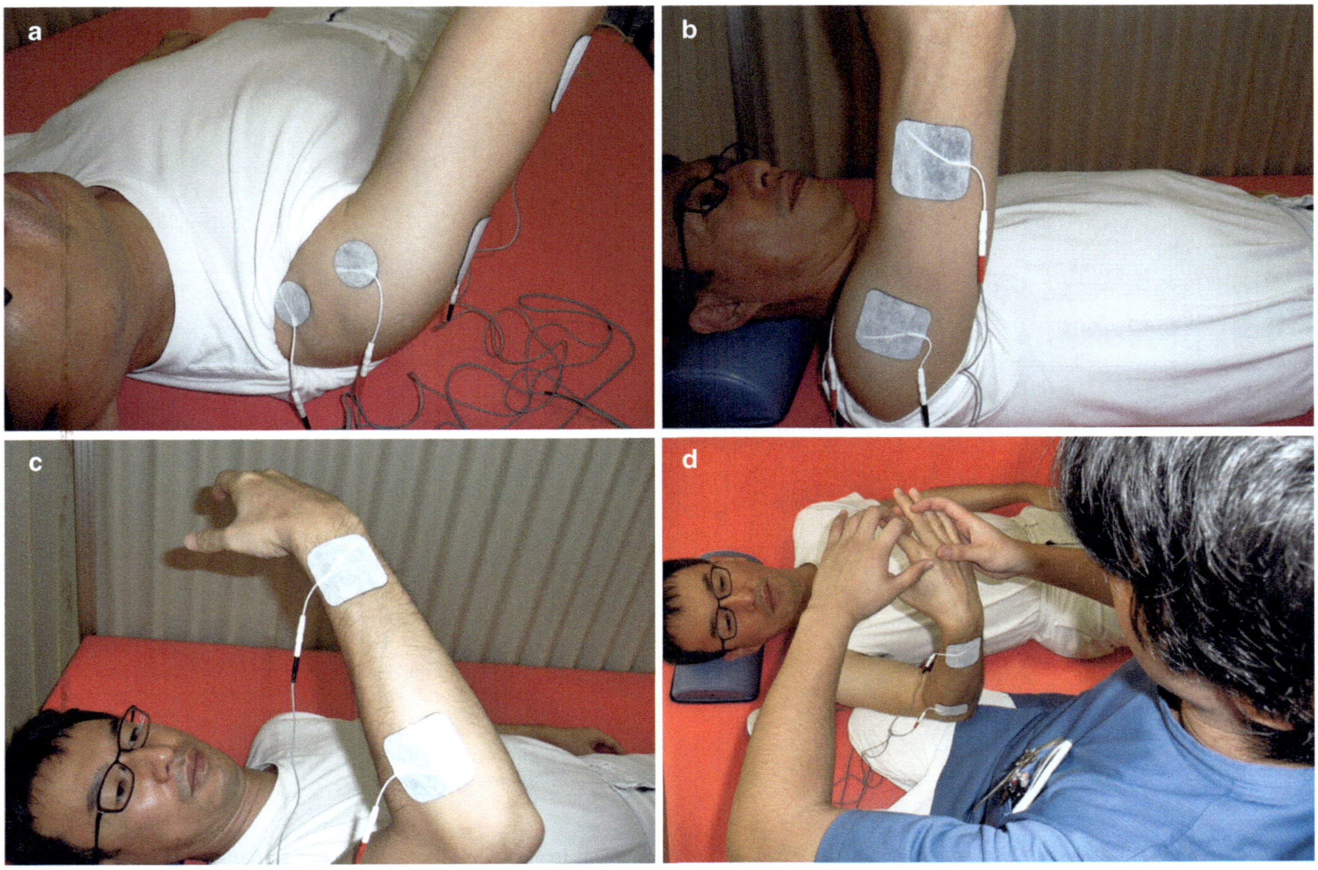

Fig. 1.32 Positions of the electrodes on the upper limb (Videos 1.1, 1.2, 1.3, and 1.4). (**a**) Deltoid front strand, (**b**) triceps, (**c**) hand and wrist extensor muscles. (**d**) Combination therapy

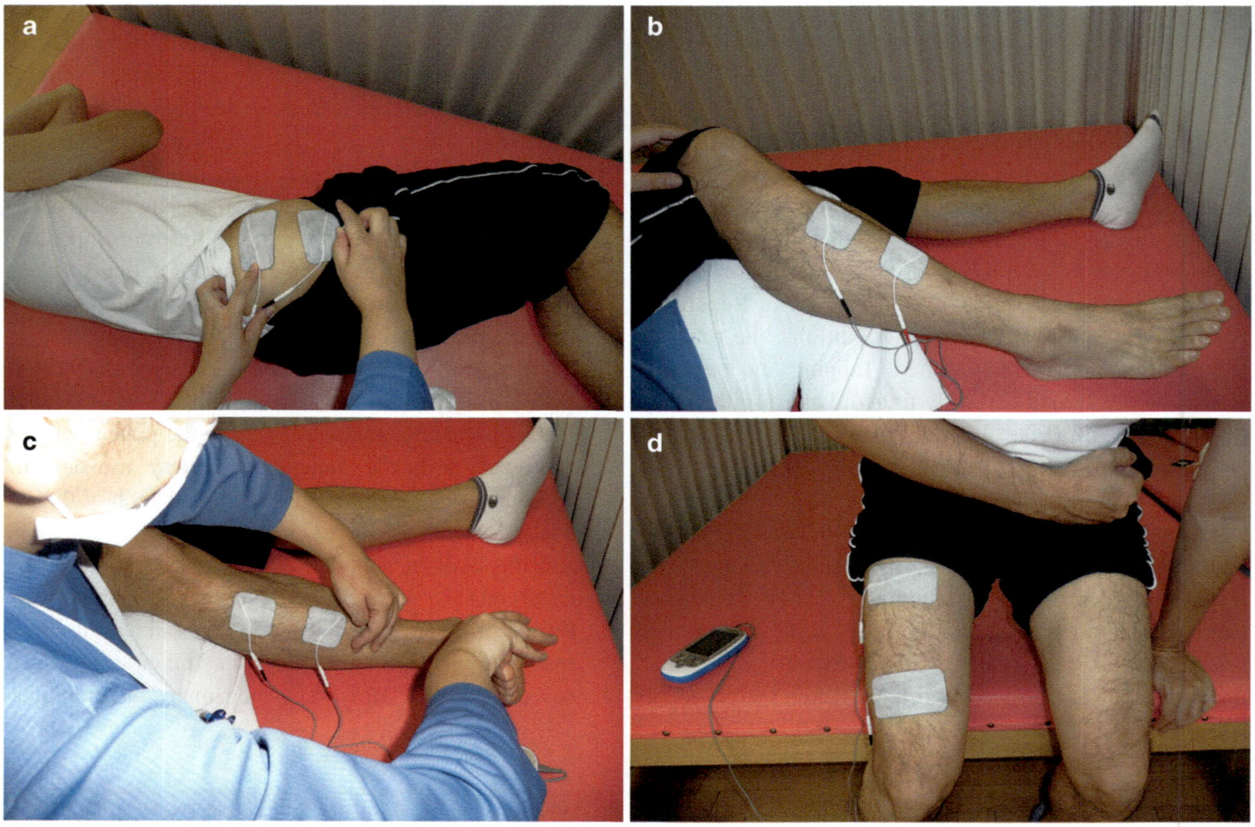

Fig. 1.33 Positions of electrodes on the lower limb (Videos 1.1, 1.2, 1.3, and 1.4). (**a**) Gluteus medius, (**b**) anterior tibial muscle, (**c**) combination therapy. (**d**) Quadriceps while training in an upright sitting position

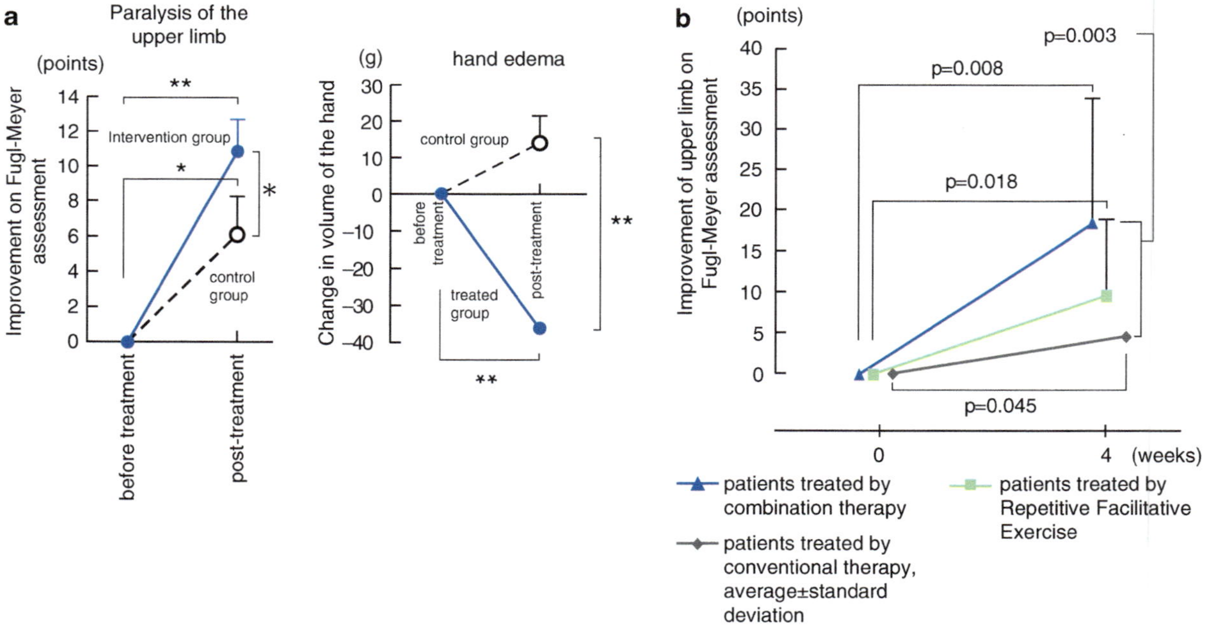

Fig. 1.34 Combined therapy of Repetitive Facilitative Exercise with continuous electrical stimulation. (**a**) Stroke in the acute phase [53]: In patients with a stroke in the acute phase, a combination of Repetitive Facilitative Exercise with continuous electrical stimulation showed a significant improvement of paralysis and edema compared to those in patients treated with conventional therapy (control patients). (**b**) Severe paralysis in the subacute phase [54]: Stroke patients in the subacute phase who received Repetitive Facilitative Exercise with continuous electrical stimulation showed significantly greater improvement than those receiving conventional therapy

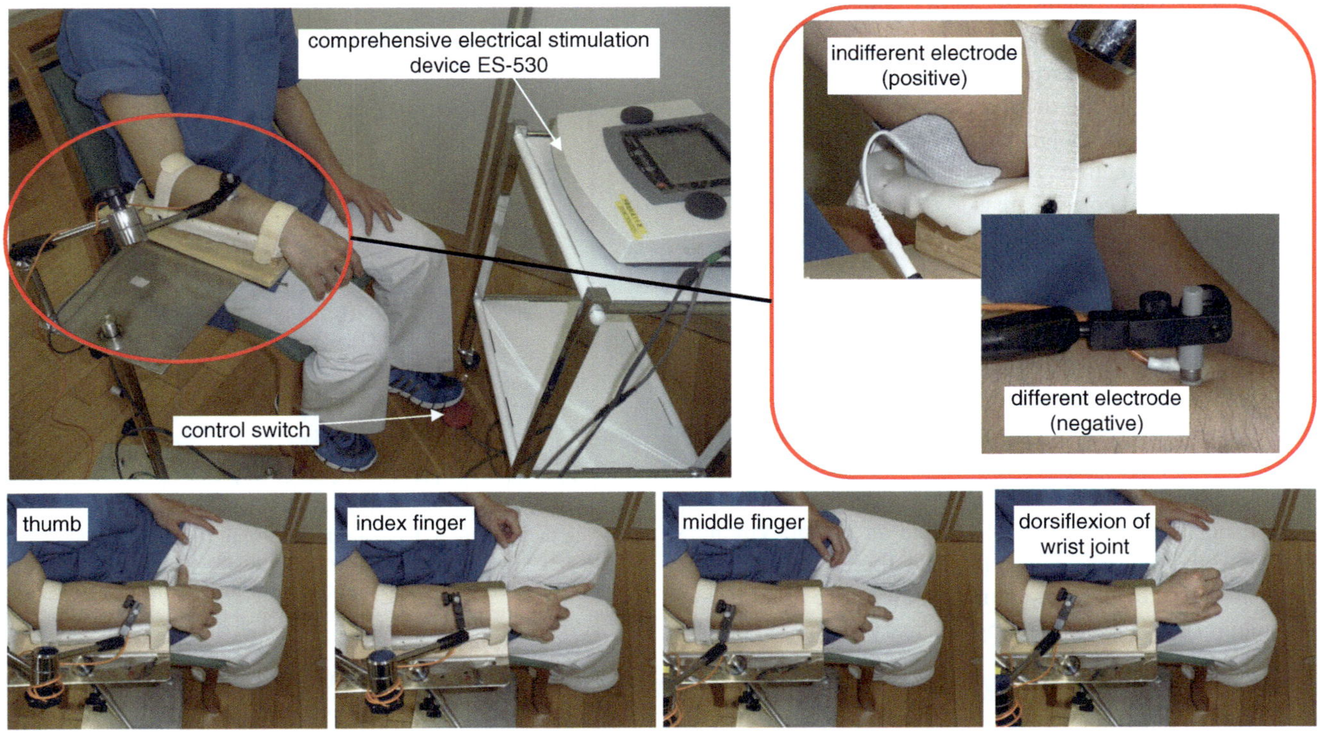

Fig. 1.35 Neuromuscular electrical stimulation synchronized to movement

located in close proximity, such as those in the finger muscles, low frequency electricity with a polar asymmetrical waveform or a high-voltage pulse should be used. The frequency of the electric current is 50 Hz, and the pulse width is adjusted between 150 and 250 μs depending on the degree of discomfort the patient can tolerate. When a monopolar electrode is used, the negative electrode is placed on the motor point or muscle belly, and the positive electrode is placed on the end of the muscle belly or tendon.

The level of stimulation is set higher than the motor threshold, though combination therapy will not be effective if electrical stimulation inducts strong muscle contraction throughout the entire range of motion. Fifty percent of the range of motion and 120% of the motor threshold may be the best fit.

Noma et al. reported the effectiveness of electrical stimulation to the hemiplegic upper limb at the chronic stage of a stroke (Fig. 1.36) [55], in which the electrical stimulation was synchronized using a high-volume pulse with voluntary movement and the patients could turn the electricity on or off using an external switch.

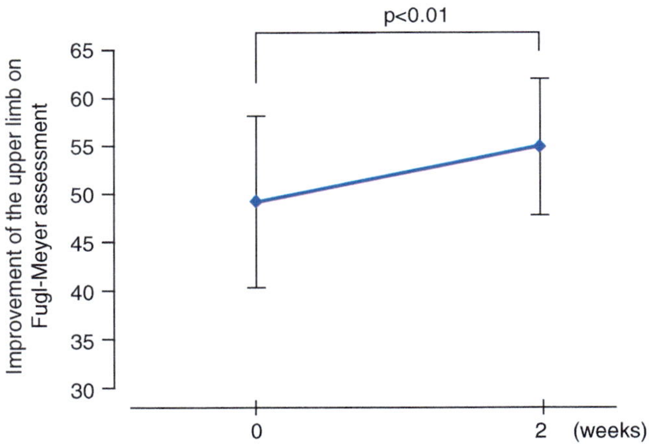

Fig. 1.36 Improvement of paralysis with high-voltage pulse stimulation [55]. Application of high-voltage pulse stimulation to muscles in the upper limb that is associated with handling objects (40 min/day, 5 days a week for 2 weeks) showed significant improvement from paralysis

1.10.3 Vibratory Stimulation

Vibratory stimuli induce the tonic vibration reflex (TVR), which effectively induces reflective contraction of the stimulated muscle, reciprocally inhibits the antagonistic muscle, and suppresses the monosynaptic reflex, that is, supresses the tendon reflex and stretch reflex, while stimulation is applied [56]. In stroke patients, removing suppression from the upper level of the cerebral cortex induces excessive activation of the stretch reflex and significantly increases muscle tone (spasm). Inducing these conditions disturbs rehabilitation of the upper limbs. Vibratory stimuli are effective in reducing spasms, and when combined with Repetitive Facilitative Exercise, they greatly contribute to the patient's ability to achieve the intended movement. In addition, vibratory stimuli are known to excite the muscle spindle and other receptors, transmitting an abundance of sensory information to the brain while affecting the sensory cortex and other movement-related cortical areas [57]. Vibratory stimuli have been shown to reduce suppression within the cortex, facilitate the learning of movements [58], and raise the level of excitement in the motor region of the cerebral cortex [59], although it should be mentioned that these reports are based on data from healthy subjects.

When the patient with forced grasping touches an object or other people's fingers with the patient's palm, the patient grabs due to tactile stimulus, and cannot open their fingers. In a training session, the therapist should ask the patient to clap their hands, making it easy to release an object. Vibratory stimuli to the fingers can also improve forced grasping. It is thought that the vibratory stimuli desensitize the sensory stimulation and activate the supplementary motor cortex [60].

1. Direct Application of Vibratory Stimuli: DAViS (Figs. 1.37 and 1.38)

   For DAViS, we devised an apparatus that does not indirectly stimulate the antagonistic muscles, but rather applies direct vibratory stimulation that simultaneously extends the spastic muscles [62]. At the beginning of the stimulation session, the stimulated muscle contracts intensely due to TVR (described above). However, if stimulation continues for a few minutes, TVR is suppressed. When the stimulation stops, there will be less spasm than before it was applied. Noma et al. demonstrated that 90 Hz DAViS for 5 min could suppress excessive excitation of spinal motor neurons for more than 30 min after the stimulation (Fig. 1.39) [61, 63]. Marconi et al. reported that direct 100 Hz vibratory stimuli to the spastic muscle induced suppression in the cerebral cortex, improved movement, and reduced spasms [64]. Miyara K et al. reported that direct 30 Hz vibratory stimuli of whole body vibration (WBV) to the spastic lower limb muscle improved walking speed and reduced spasms [65].

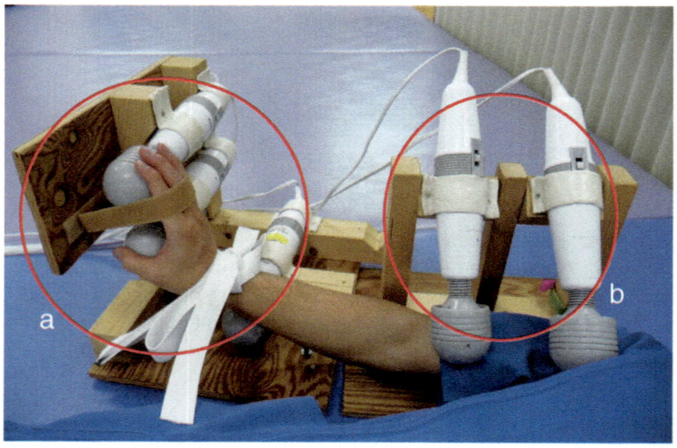

Fig. 1.37 Direct Application of Vibratory Stimuli (DAViS). (**a**) Device to stimulate the hand and forearm. (**b**) Device to stimulate the forearm. Vibration: 90 Hz

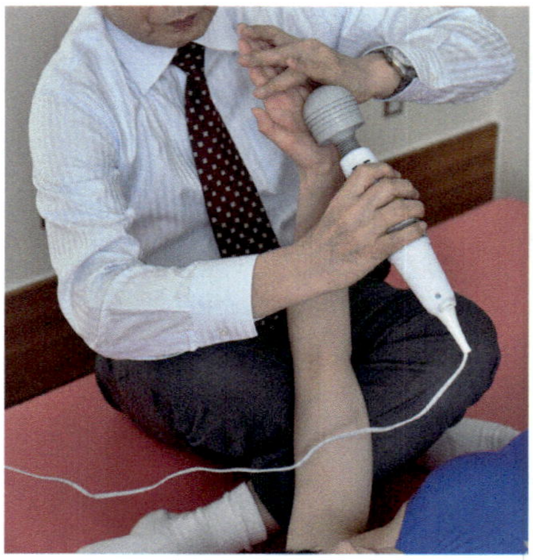

Fig. 1.38 Simplified DAViS. The therapist maintains the affected fingers in an extended position with the wrist joint dorsiflexed while applying the vibrator to the palm for 2–5 min. If flexion of the fingers due to the vibration causes discomfort to the patient, the therapist should hold the wrist/fingers in position with both hands and ask the patient, a family member or another therapist to apply the vibrator

   Noma et al. divided patients into two groups: 2-week Repetitive Facilitative Exercise only and 5-min DAViS followed by Repetitive Facilitative Exercise for 2 weeks, and then compared improvement between the two groups. Their results showed the potential to improve spasm and achieve good handling abilities by combining Repetitive Facilitative Exercise with DAViS [66].

2. Vibratory stimuli for sensory input

   Sensory disorder is a factor that prevents the recovery of motor functions. Shirahashi et al. reported that a patient whose condition was complicated by sensory disorder received vibratory stimuli applied to the palm and thereafter showed significant improvement in voluntary control of the upper limb [67].

   Kawahira et al. reported the effectiveness of functional vibratory stimulation applied to the muscle synchronized with the waking cycle (Fig. 1.40) [68]. In our clinical experiences, application of vibratory stimuli at a frequency of 110 Hz for a few minutes to patients complicated by sensory disorder improved fine motor control of

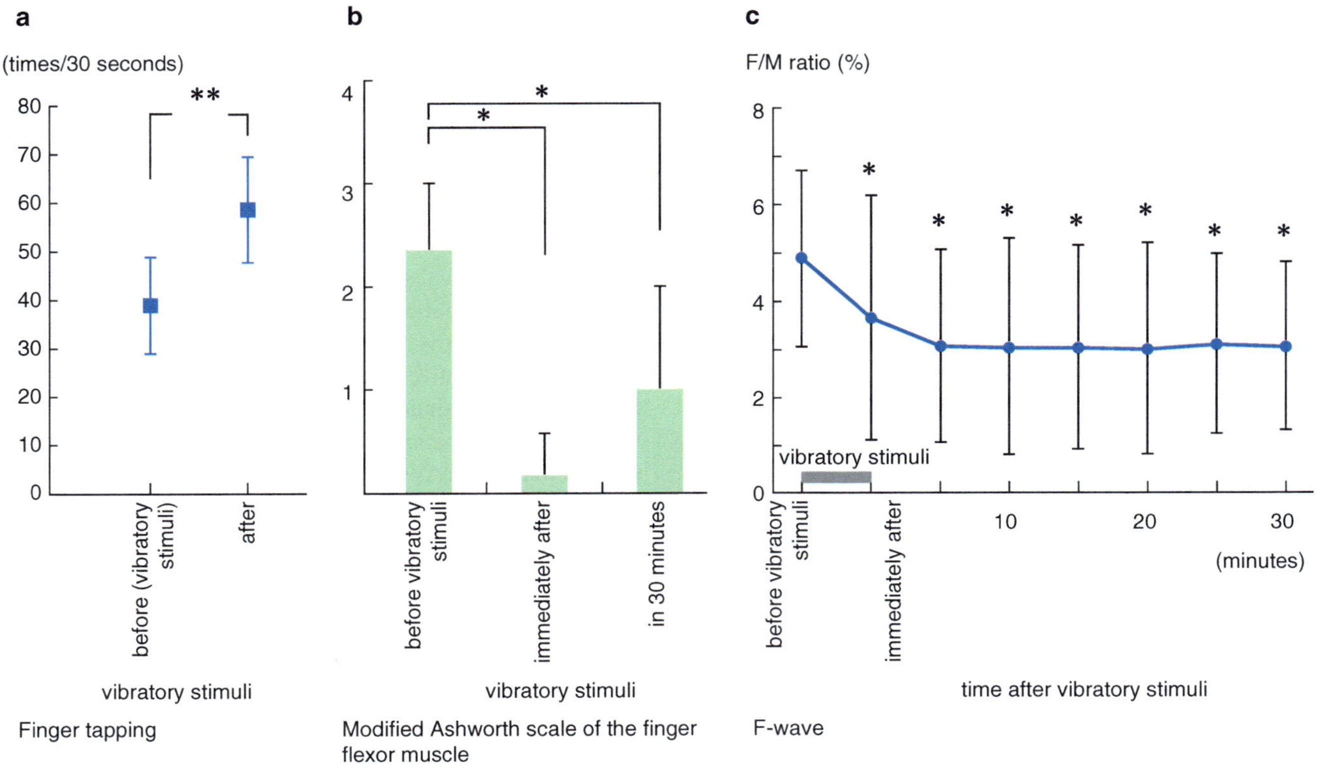

Fig. 1.39 Effect of Direct Application of Vibratory Stimuli (DAViS) [61]. When DAViS is applied, finger tapping speeds up (**a**). Spasticity of the finger flexor muscles measured using the Modified Ashworth scale (**b**) was significantly reduced. The level of excitation in the spinal motor neuron represented by F-waves (**c**) was also reduced

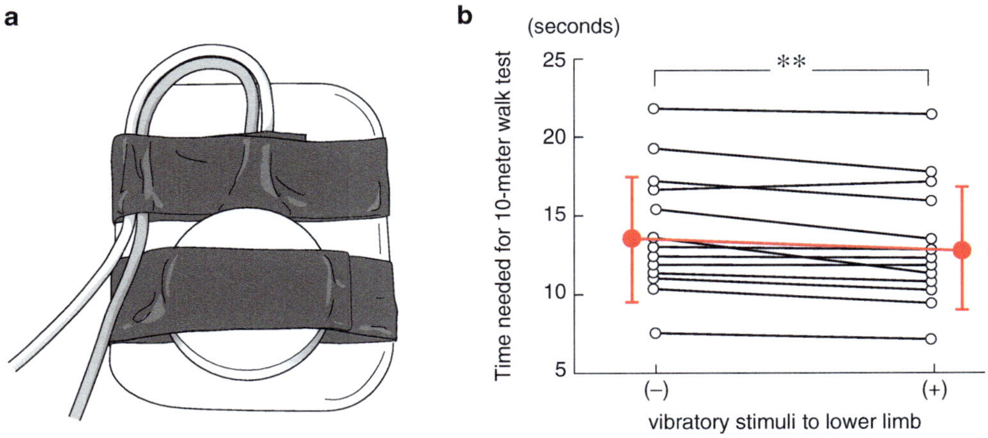

Fig. 1.40 Functional vibratory stimulation [68]. (**a**) Vibratory stimulation device. (**b**) Using this device makes the patient walk faster

finger movements as well as walking. We also found that sensory input using vibratory stimuli for patients complicated by both sensory disorder and neurological pain successfully reduced pain and numbness. In the future, we expect that more evidence about this phenomenon, including understanding of its mechanism, will be accumulated, and lead to further increase of applications of this method for clinical practice.

1.10.4 Combination Therapy with Continuous Electrical Stimulation/Vibratory Stimuli

As there are many limitations in the rehabilitation of hemiplegia during the chronic phase, it is necessary to have therapy that effectively improves paralysis but does not require prolonged application at a high frequency. Combining Repetitive Facilitative Exercise under continuous electrical stimulation and DAViS with sessions of 30 min/day, twice a week for only 12 weeks could achieve greater improvement of paralysis than other conventional treatments (Table 1.2) [69].

The mechanism underlying the beneficial outcome of this combination therapy may be considered as follows: the excitation levels of the sensory and motor pathways is increased by DAViS and the spastic muscle is suppressed while antagonistic muscle suppression is reduced. The motor threshold is also stimulated by continuous electrical stimulation and Repetitive Facilitative Exercise, making it easier for the patient to perform and repeat the intended movements.

1.10.5 Transcranial Magnetic Stimulation (TMS)

In TMS, an electric current is generated in a coil placed close to the patient's head. The variable magnetic field induces an eddy current that runs opposite to the electric current in the coil, resulting in stimulation of the cerebrum. Repetitive transcranial magnetic stimulation (rTMS) is commonly applied for rehabilitation from hemiplegia following a stroke. It has been demonstrated that stimulation at a frequency of 1 Hz reduces the excitation level of the brain while stimulation at 5 Hz or higher increases the excitation level of the brain [70]. To combine rTMS with the Repetitive Facilitative Exercise, rTMS at 1 Hz is applied immediately before Repetitive Facilitative Exercise. This reduces suppression of the affected hemisphere by the unaffected hemisphere through the corpus callosum, which is expected to promote the effect of Repetitive Facilitative Exercise.

The order of effectiveness when combined with Repetitive Facilitative Exercise is as follows: (1) continuous electrical stimulation + DAViS, (2) transcranial magnetic stimulation + continuous electrical stimulation + DAViS, and (3) transctanial magnetic stimulation.

Optimal effectiveness observed in combination with Repetitive Facilitative Exercise was observed in the following order: (1) Continuous electrical stimulation + DAViS, (2) transcranial magnetic stimulation + continuous electrical stimulation + DAViS, and (3) transcranial magnetic stimulation.

For patients with chronic hemiplegic upper limb, Etoh et al. compared the improvement of paralysis induced by 1 Hz rTMS applied to the unaffected motor cortex for 4 min or placebo stimulation before Repetitive Facilitative Exercise. Significant improvement was observed only for the period combined with rTMS [71]. Etoh et al. also reported another low-invasive treatment: Object handling was significantly improved by application of TMS to the affected motor area synchronized with movement during Repetitive Facilitative Exercise.

During treatment, therapists should combine electrical, vibratory, and magnetic stimuli in order to achieve greater effectiveness with combination therapy. A report by Etoh et al. indicated that a combination of Repetitive Facilitative Exercise under continuous electrical stimulation (peripheral stimulation) and DAViS is more effective than magnetic stimulation (central stimulation) [72]. The accumulation of more cases and evidence is expected (Fig. 1.41).

Table 1.2 Comparison of effects of treatment on hemiplegia in patients at an elderly care facility between a combination of Repetitive Facilitative Exercise, electrical/vibratory stimuli and conventional treatment (30 min/day, twice a week, for 12 weeks) [69]

		Before treatment		After treatment	Improvement		Effective rate	
Upper limb (grade)	Conventional treatment	6.7 ± 3.7		6.9 ± 3.5	0.2 ± 0.4		15(%)	
			$p = 0.0001$			$p = 0.005$		$p = 0.003$
	Repetitive facilitative exercise	6.1 ± 2.7		7.2 ± 2.5	1.0 ± 0.9		66(%)	
Finger (grade)	Conventional treatment	5.8 ± 4.3		6.0 ± 4.3	0.2 ± 0.5		10(%)	
			$p = 0.0035$			$p = 0.02$		$p = 0.015$
	Repetitive facilitative exercise	4.4 ± 3.2		5.0 ± 3.5	0.6 ± 0.9		44(%)	

Combination therapy (with both a shorter treatment time and with a lower frequency of treatment) showed significantly greater improvement from chronic paralysis

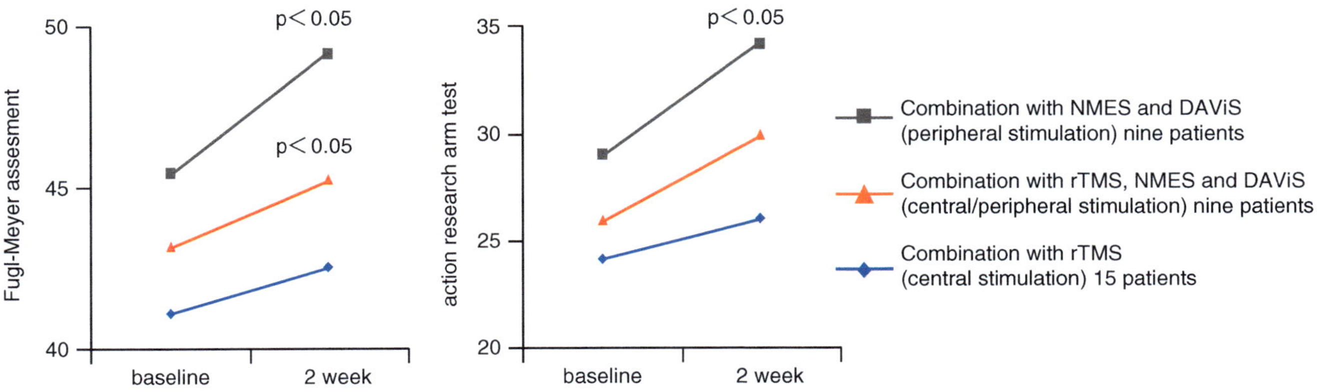

Fig. 1.41 The effect of combined Repetitive Facilitative Exercise with electrical, vibratory, and magnetic stimulation on the chronic hemiplegic upper limb of patients who suffered a stroke [71]

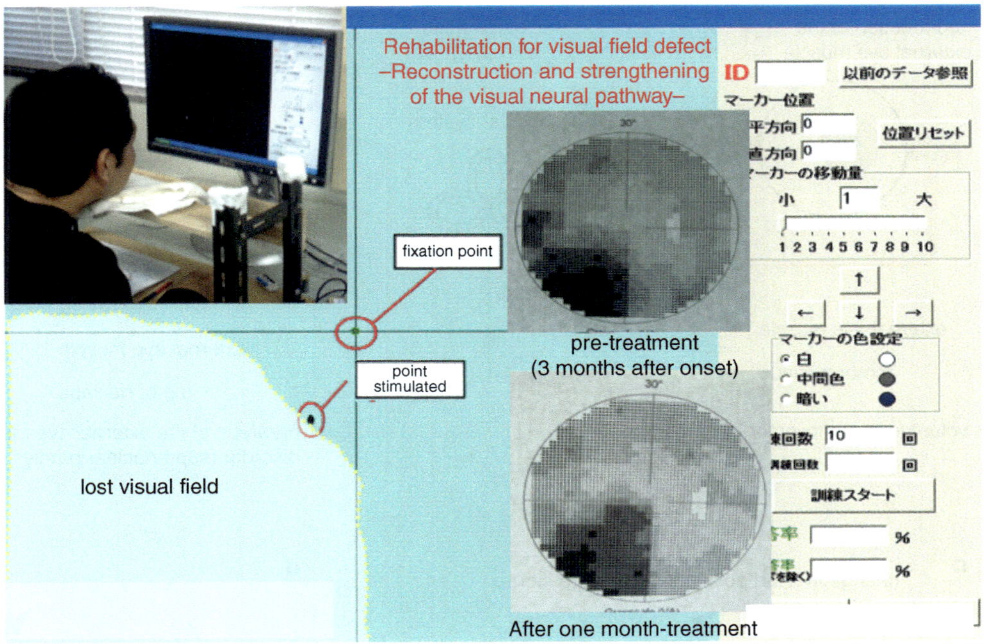

Fig. 1.42 Repetitive visual stimulation therapy for visual field defect [74]. By applying visual stimuli at the margin, the area of visual loss, reduction of the black area (where there is no perception of light stimulus) can be achieved

1.10.6 Botulinum Toxin Therapy

A study of chronic stroke patients by Hokazono et al. found that 4-week treatment with Repetitive Facilitative Exercise (RFE) alone led to an improvement in upper limb function with paralysis (FMA) of 3.0 points and article manipulation (ARAT) of 7.0 points. Also, a combination of RFE with botulinum toxin therapy showed an improvement in FMA of 11.0 points and ARAT of 12.5 points. Comparing these two procedures, the combination of botulinum toxin therapy with RFE proved significantly more effective than RFE alone [73].

1.10.7 Therapy for Visual Field Defect and External Ophthalamoplegia

Rehabilitation for visual field defects (hemianopsia, quadranopsia) and external ophthalmoplegia are still inadequate. However, visual field defect can be improved with repeated stimuli at the margin of the visual field loss (Fig. 1.42) [74].

There is no effective rehabilitation for external ophthalmoplegia and double vision, which are commonly caused by a disorder of the brainstem, although masking of one eye or surgical treatment of squint have been applied in some cases.

However, Repetitive Facilitative Exercise that combines the vestibulo-ocular reflex with voluntary eye movement to provide repeated eye movement training improves paralysis of the external eye muscle caused by brainstem disorder.

Excitation from the cerebral cortex is conducted to the oculomotor nerve nucleus in the brainstem, which contracts the external eye muscle to induce eye movement (Fig. 1.43a). Paralysis of the external eye muscle due to brain stem disorder can be classified into two groups: supranuclear paralysis, where the neural pathway from the cerebral cortex to the oculomotor nerve nucleus in the brainstem is damaged; and nuclear paralysis, where the oculomotor nerve nucleus itself is damaged (Fig. 1.43b). The vestibulo-ocular reflex involves labyrinthine input. If the therapist rotates the patient's head, the vestibulo-ocular reflex tries to maintain the patient's gaze on the original fixation point despite head movement (Fig. 1.43c).

Repetitive Facilitative Exercise using the vestibulo-ocular reflex conducts the excitation from the cerebral cortex and from the vestibulo-ocular reflex simultaneously to the oculo-

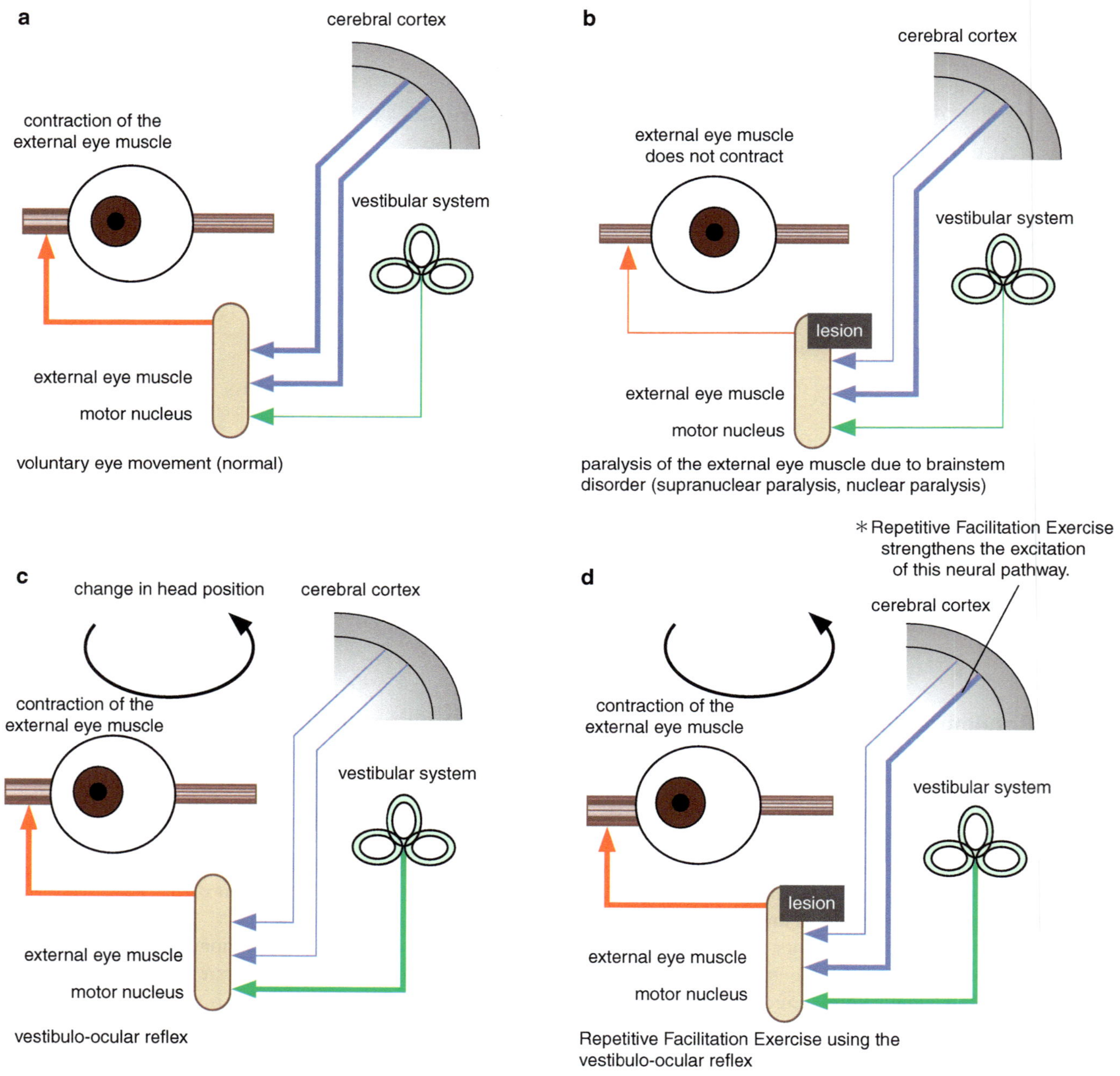

Fig. 1.43 Mechanism of Repetitive facilitation exercise using the vestibulo-ocular reflex

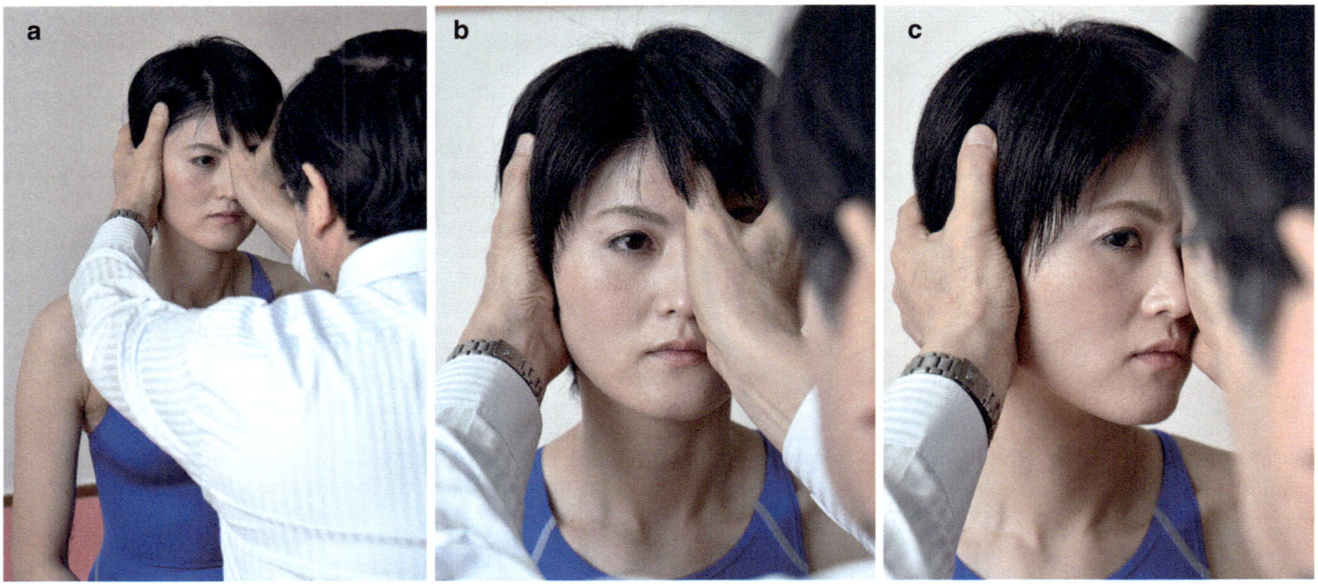

Fig. 1.44 Repetitive facilitation exercise using the vestibulo-ocular reflex (Video 1.6). When the patient tries to use both eyes to gaze, the damaged eye does not move adequately. (**a**) The therapist uses one hand to cover the intact eye, and the other hand to hold the patient's head. (**b**) The therapist says, "Look at my face." (**c**) The therapist turns the patient's head quickly in the direction opposite to the restricted eye movement. When the eye movement reaches its maximum, the therapist says again, "Look at my face." When the patient tries to look at the therapist's face, a strong resistance emerges against the therapist's hand, which is holding the patient's head, and tension is felt in the patient's neck. Even if the therapist tells the patient to relax her neck, some resistance remains. The therapist should not apply excessive force to turn the patient's head. At the beginning of treatment, involuntary rapid eye movement may occur, but as treatment continues, it will disappear

motor nerve nucleus in order to realize eye movement and strengthen the neural pathway connecting the cerebral cortex to the external eye muscle through the oculomotor nerve nucleus (Fig. 1.43d).

In Repetitive Facilitative Exercise using the vestibulo-ocular reflex, the therapist first asks the patient to look at the therapist's face (Fig. 1.44a). Then the therapist rotates the patient's head in the direction opposite to the paralyzed external eye muscle (Fig. 1.44b). When the eye movement caused by the vestibulo-ocular reflex reaches its maximum, the therapist asks the patient to look at the therapist's face again. We usually repeat this 100 times for each direction of restricted eye movement [75].

To treat the inability to focus the eyes, which is a common complication of external ophthalmoplegia, Repetitive Facilitation Exercise can be performed using the vestibulo-ocular reflex, and then the patient can wear a pair of glasses that block vision on the intact side for 1 h.

External ophthalmoplegia was clearly improved after a few weeks of treatment (Figs. 1.45 and 1.46) [75]. However, to improve the function of binocular vision, which requires coordinated focus of both eyes at the macular, the intact eye should be briefly but repeatedly covered and uncovered in order to help the patient gain binocular vision [76].

Some patients feel as if they are floating or dizzy, or have similar types of discomfort, when their head is rotated. The therapist should attempt vestibulo-ocular reflex facilitation therapy a few times but should not continue the treatment if the patient cannot tolerate it.

(Tomokazu Noma)

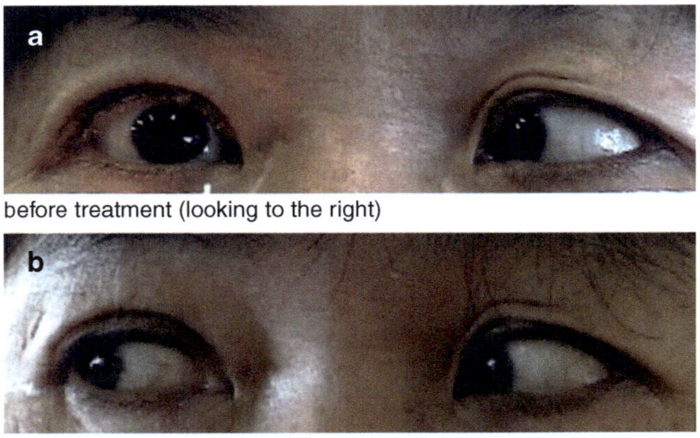

Fig. 1.45 Improvement from paralysis of external eye muscle with vestibulo-ocular reflex facilitation. Optimal effects of vestibulo-ocular reflex facilitation therapy are observed within a few weeks (**a** and **b**). If recovering the ability to focus is one of the goals, intermittent covering of the intact eye (for 1 h) should be added to the regimen and the treatment should be continued for a few months. Among patients who require some time before the vestibulo-ocular reflex is recovered, some show improvement of paralysis of the external eye muscle after a few months of facilitation therapy using the vestibulo-ocular reflex

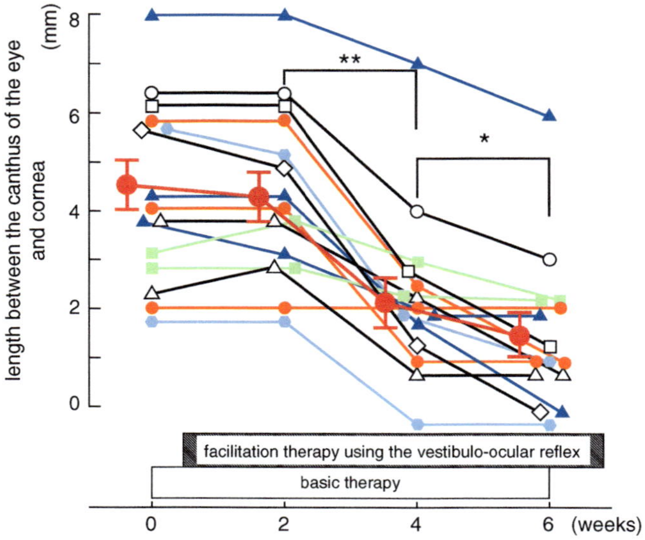

Fig. 1.46 Outcome of Repetitive Facilitative Exercise using the vestibulo-ocular reflex [75]. After 2 weeks of Repetitive Facilitative Exercise using the vestibulo-ocular reflex, the length between the corner of the eye and the cornea significantly decreased due to improvement of external eye muscle movement

References

1. Zeki (Japanese translation by Juro Kawachi). A vision of the brain. IGAKU-SHOIN Ltd.; 1995. p. 269–92. (in Japanese).
2. Sakata H, Lobe P, Iriki M, Toyama K. Physiology. BUNKODO CO, LTD; 1986. p. 396–414. (in Japanese).
3. Rizzolatti G. Functional organization of inferior area 6. In motor areas of the cerebral cortex. John & Wiley, Chichester (Ciba Found Symposium 132); 1987. p. 187–200.
4. Tanji J, Shima K. Role for supplementary motor area cells in planning several movements ahead. Nature. 1994;371:413–6.
5. Kubota K. Control of movement by the frontal lobe. Adv Neurol Sci. 1993;37:26–33. (in Japanese).
6. Kawato M. Model of the nervous structure involved in movement. Br Med. 1994;6:371–7. (in Japanese).
7. Tanji J. Motor-related areas and motor programming, brain and neuralnet. In Amari S, Sakata H, editors. Asakura Publishing Co., Ltd; 1994. p. 203–217. (in Japanese).
8. Baddeley A, Wilson BA. When implicit learning fails: amnesia and the problem of error elimination. Neuropsychologia. 1994;32:53–68.
9. Kawahira K. Effects of intensive exercise to facilitate the functional recovery of the hemiplegic lower limb and effects of an anabolic steroid and muscle strengthening exercise in patients with hemiplegia (current Technology in Rehabilitation for patients with stroke). Rehabil Med. 2002;39:86–90. (in Japanese).
10. Brashers-Krug T, Shadmehr R, Bizzi E. Consolidation in human motor memory. Nature. 1996;382:252–5.
11. Arima M, Shimodozono M, Etoh S, et al. Examining procedural consolidation with visuomotor learnig in the lower limb. Int J Neurosci. 2010;120(5):344–51.
12. Nudo RJ, Milliken GW, Jenkins WM, et al. Use-dependent alterations of movement representations in primary motor cortex of adult squirrel monkeys. J Neuroscience. 1996;16:785–807.
13. Kaas JH. Plasticity of sensory and motor maps in adult mammals. Ann Rev Neurosci. 1991;14:137–67.
14. Pons TP, Garraghty PE, Ommaya AK, et al. Massive cortical reorganization after sensory deafferentation in adult macaques. Science. 1991;252:1857–60.
15. Yang TT, Gallen C, Schwartz B, et al. Sensory maps in the human brain. Nature. 1994;368:592–3.
16. Jenkins WM, Merzenich MM. Reorganization of neocortical representations after brain injury; a neurophysiological model of the bases of recovery from stroke. Prog Brain Res. 1987;71:249–66.
17. Sadato N, Pascual-Leone A, Grafman J, et al. Activation of the primary visual cortex by braille reading in blind subjects. Nature. 1996;380:526–8.
18. Travis AM, Woolsey CN. Motor performance of monkeys after bilateral partial and total cerebral decortications. Amer J Phys Med. 1956;35:273–310.
19. Ishida A, Isa K, Umeda T, et al. Causal link between the corticorubral pathway and functional recovery through forced impaired limb use in rats with stroke. J Neurosci. 2016;36:455–67.
20. Nishimura Y, Isa T. Corticospinal tract involved in skilled hand movements and its recovery after injury. Exp Med. 2006;24:2410–7. (in Japanese).
21. Yamashita T. Significance of corticospinal tract sprouting after cortical injury in rodents. J Neurosci Neuroeng. 2016;4:21–5.
22. Nudo RJ, Wise BM, SiFuentes F, et al. Neural substrates for the effects of rehabilitative training on motor recovery after ischemic infarct. Science. 1996;272:1791–4.
23. Etoh S. Change of motor evoked potential induced by transcranial magnetic stimulation during movement training for paralysis due to cerebral stroke. Jpn J Rehabil Med. 2000;37:925–33. (in Japanese).
24. Kawahira K, Tanaka N. Neurophysiological background of information processing and plasticity in the brain. Jpn J Rehabil Med. 1995;32:670–87. (in Japanese).
25. Tsukada, M. Plastic neural circuit, brain and neutral net. Amari S, Sakata H, editors. Asakura Publishing Co., Ltd.; 1994, p. 268–329. (in Japanese).
26. Toyama K, Komatsu Y, et al. In vitro studies of visual cortical development and plasticity. Prog Neurobiol. 1993;41:543–63.
27. Maesako A, Nagase M. Tatsuo Nagado, et al: improvement of motor functions and edema of the upper limbs during the acute phase of cerebral stroke following repetitive facilitation therapy and continuous low-frequency electrical stimulation. Jpn J Rehabil Med. 2014;51(3):219–27. (in Japanese).
28. Imura T, Matsumotoa M, Fukazawa T, et al. Interactive effects of cell therapy and rehabilitation realize the full potential of neurogenesis in brain injury model. Neurosci Lett. 2013;555:73–8.
29. Kamada K, Noma T, Shimodozono M, et al. Functional recovery in the hemiplegic arm following occupational therapy and intensive exercise therapy focused on freedom from synergy. J Jap Occup Therapy Res. 2004;23:18–25. (in Japanese).
30. Hayashida K, Yorozuya W, Okudera E, et al. Functional improvements in the hemiplegic upper limb by repetition of facilitation exercise. Sogo Rehabilitation. 2007;35:809–13. (in Japanese).
31. Kawahira K, Shimodozono M, Etoh S, et al. Effects of intensive repetition of a new facilitation technique on motor functional recovery of the hemiplegic upper limb and hand. Brain Inj. 2010;24(10):1202–13.
32. Mitani T, Kisa T, Sakai Y, et al. Effects of repetitive facilitation exercise for hemiplegic stroke patients in the recovery stage. Sogo Rehabilitation. 2010;38:165–70. (in Japanese).
33. Kisa T, Sakai Y, Mitani T, et al. Effect of additive application of repetitive facilitation exercise to conventional rehabilitation in hemiplegic stroke patients in the recovery stage: a controlled clinical trial on motor functional recovery in hemiplegia and activities of daily living. Jpn J Rehabil Med. 2011;48:709–16. (in Japanese).
34. Shimodozono M, Noma T, Nomoto Y, et al. Benefits of a repetitive facilitative exercise program for the upper paretic extremity after

subacute stroke: a randomized controlled trial. Neurorehabil Neural Repair. 2013;27(4):296–305.
35. Miyasaka H, Ohnishi H, Hieda C, et al. A study of the training method of sub-acute stroke patients of the upper extremity: decision tree analysis. Jap J Compreh Rehab Sci. 2014;5:117–24.
36. Matoba Y, Kowa H, Sota T, et al. The efficacy of repetitive facilitation exercise on upper-extremity function recovery in hemiplegic patients with acute cerebral infarction. Sogo Rehab. 2015;43:563–6. (in Japanese).
37. Noma T, Kamada K, Kai Y, et al. Effect of repetitive facilitation therapy in the chronic phase on the hemiplegic arm following stroke. Sogo Rehab. 2008;36:695–9. (in Japanese).
38. Yorozuya W, Hayashida K, Abe Y, et al. Implementation of repetitive facilitation therapy for the hemiplegic upper extremity within the time allotted for occupational therapy and its effects. J Jap Occup Therapy Res. 2008;27:74–179. (in Japanese).
39. Kawahira K, Noma T. Junich Iiyama, et al: improvements in limb kinetic apraxia by repetition of a newly designed facilitation exercise in a patient with corticobasal degeneration. Int J Rehabil Res. 2009;32(2):178–83. (in Japanese).
40. Hujime K, Kawahira K, Kai Y. Effect of repetitive facilitation therapy for paralyzed arms following cervical cord injury. J Jap Med Soc Spinal Cord Lesion. 2008;21:64–5. (in Japanese).
41. Hirokawa T, Matsumoto S, Uema T, et al. Effects of intensive repetition of trunk muscle facilitation on motor functional recovery after stroke: study on a randomized clinical trial. Phys Therapy Jap. 2013;40:457–64. (in Japanese).
42. Kawahira K, Shimodozono M, Ogata A, et al. Addition of intensive repetition of facilitation exercise to multidisciplinary rehabilitation promotes motor functional recovery of the hemiplegic lower limb. J Rehabil Med. 2004;36(3):159–64.
43. Kawahira K, Ogata A, Tougou S, et al. Functional recovery of a hemiplegic leg by intensive exercise therapy focus on freedom from synergy. Jap J Rehab Med. 1997;34:598–604. (in Japanese).
44. Murayama M, Shimodozono M, Kawahira K. Effect of facilitation on gait for the hemiplegic patient. J Clin Rehab. 2007;16:1203–6. (in Japanese).
45. Takeuchi N, Oouchida Y, Izumi S. Motor control and neural plasticity through interhemispheric interactions. Neural Plast Article ID. 2012;823285
46. Simonetta-Moreau M. Non-invasive brain stimulation(NIBS) and motor recovery after stroke. Ann Phys Rehabil Med. 2014;57:530–42.
47. Ridding MC, McKay DR, Thompson PD, et al. Changes in corticomotor representations induced by prolonged peripheral nerve stimulation in humans. Clin Neurophysiol. 2001;112:1461–9.
48. Ikuno K. Current clinical indications for physical therapy. BUNKODO Co., Ltd; 2012. p. 190–207. (in Japanese).
49. Bliss TV. Lomo T: long-lasting potentiation of synaptic transmission in the dentate area of the anaesthetized rabbit following stimulation of the perforant path. J Physiol. 1973;232(232):331–56.
50. Asanuma H, Keller A. Neuronal mechanisms of motor learning in mammals. Neuroreport. 1991;2:217–24.
51. Asanuma H, Keller A. Neurobiological basis of motor learning and memory. Conc Neurosci. 1991;1:1–30.
52. Shin HK, Cho SH, Jeon HS, et al. Cortical effect and functional recovery by the electromyography-triggered neuromuscular stimulation in chronic stroke patients. Neurosci Lett. 2008;442:174–9.
53. Maesako A, Nagase M, Nagadou T, et al. Effects of repetitive facilitation exercise combined with continuous low-frequency electrical stimulation on upper-limb motor impairment and hand Edema during an acute phase of cerebral infarction. Jap J Rehab Med. 2014;51:219–27. (in Japanese).
54. Shimodozono M, Noma T, Matsumoto S, et al. Repetitive facilitative exercise under continuous electrical stimulation for severe arm impairment after subacute stroke: a randomized controlled pilot study. Brain Inj. 2014;28(2):203–10.
55. Noma T, Matsumoto S, Shimodozono M, et al. Novel neuromuscular electrical stimulation system for the upper limbs in chronic stroke patients: a feasibility study. Am J Phys Med Rehabil. 2014;93(6):503–10.
56. Hagbarth KE, Eklund G. Tonic vibration reflexes (VTR) in spasticity. Brain Res. 1966;2:201–3.
57. Naito E. Role of sensory input in acquiring a physical exercise model. J Soc Biomechan Jap. 2007;31:178–86. (in Japanese).
58. Rosenkranz K, Rothwell JC. Modulation of proprioceptive integration in the motor cortex shapes human motor learning. J Neurosci. 2012;32:9000–6.
59. Siggelkow S, Kossev A, Schubert M, et al. Modulation of motor evoked potentials by muscle vibration, the role of vibration frequency. Muscle Nerve. 1999;22:1544–8.
60. Noma T, Kamada K, Kai Y, et al. The effect of vibratory stimulation on forced grasping and a tonic plantar reflex in two patients with stroke. Sogo Rehab. 2007;35:605–9. (in Japanese).
61. Noma T, Matsumoto S, Etoh S, et al. Antispastic effects of the direct application of vibratory stimuli to the spastic muscles of hemiplegic limbs in post-stroke patients. Brain Inj. 2009;23(7–8):623–31.
62. Noma T, Etou S, Kamada K, et al. Inhibitory effects of the direct application of vibratory stimulus to spastic muscles of hemiplegic limbs in stroke patients. J Jap Occup Therapy Res. 2008;27:119–27. (in Japanese).
63. Noma T, Matsumoto S, Etoh S, et al. Anti-spastic effects of the direct application of vibratory stimuli to the spastic muscles of hemiplegic limbs in post-stroke patients: a proof-of-principle study. J Rehabil. 2012;44(4):325–30.
64. Marconi B, Filippi GM, Koch G, et al. Long-term effects on cortical excitability and motor recovery induced by repeated muscle vibration in chronic stroke patients. Neurorehabil Neural Repair. 2011;25:48–60.
65. Miyara K, Matsumoto S, Uema T, et al. Feasibility of using whole body vibration as a means for controlling spasticity in post-stroke patients: a pilot study. Complement Ther Clin Pract. 2014;20(1):70–3.
66. Noma T, Kamada K, Kai Y, et al. Effects of combined therapy with repetitive facilitation exercise and antispastic vibratory stimuli on motor functional recovery of the hemiplegic upper limb in stroke patients. Sogo Rehabilitation. 2009;37:137–43. (in Japanese).
67. Shirahashi I, Matsumoto S, Shimodozono M, et al. Functional vibratory stimulation on the hand facilitates voluntary movements of a hemiplegic upper limb in a patient with stroke. Int J Rehabil Res. 2007;30(3):227–30.
68. Kawahira K, Higashihara K, Matsumoto S, et al. New functional vibratory stimulation device for extremities in patients with stroke. Int J Rehabil Res. 2004;27(4):335–7.
69. Hayashi T, Ishikawa S, Kawamura T, et al. Effect of repetitive facilitation exercise with electrical and vibratory stimulation for chronic hemiplegia patients using day-care rehabilitation. J Phys Ther Sci. 2017;32:129–32. (in Japanese).
70. Etoh S. Clinical use of transcranial magnetic stimulation. J Jap Assoc Phys Med Balneol Climatol. 2013;76:293–6. (in Japanese).
71. Etoh S, Noma T, Ikeda K, et al. Effects of repetitive trascranial magnetic stimulation on repetitive facilitation exercises of the hemiplegic hand in chronic stroke patients. J Rehabil Med. 2013;45(9):843–7. https://doi.org/10.2340/16501977-1175.
72. Etoh S, Noma T, Takiyoshi Y, et al. Effects of repetitive facilitative exercise with neuromuscular electrical stimulation, vibratory stimulation and repetitive transcranial magnetic stimulation of the hemiplegic hand in chronic stroke patients. Int J Neurosci. 2016;126(11):1007–12.
73. Hokazono A, Etoh S, Yuiko J, et al. Combination therapy with repetitive facilitative exercise program and botulinum toxin type

a to improve motor function for the upper-limb spastic paresis in chronic stroke: a randomized controlled trial. J Hand Ther. 2021;S0894-1130(21):00024–7. https://doi.org/10.1016/j.jht.2021.01.005.
74. Shimozono Y, Arima M, Kai Y, et al. Quadranopsia improved by repetitive visual stimulation using a computerized visual field training machine in a stroke patient: a case report. Jap J Rehab Med. 2007;44:613–9. (in Japanese).
75. Kawahira K, Shimodozono M, Etoh S, et al. New facilitation exercise using the vetibulo-ocular reflex for ophthalmoplegia; preliminary report. Clin Rehabil. 2005;19(7):627–34.
76. Tamari I, Noma T. Yuiko Kai, et al: facilitation exercise using the vestibulo-ocular reflex and binocular-gaze training improved ophthalmoplegia in a patient with traumatic brain injury. Sogo Rehab. 2011;39(7):695–9. (in Japanese).

Part II

Practice

Principles and Basic Techniques for Repetitive Facilitative Exercise

2

Basic Principles Underlying the Exercise and Treatment Procedures

Kazumi Kawahira

2.1 Principles of the Exercise

Key Points
Enabling the intended movement of paralyzed limbs

- Postural reflex, stretch reflex, and skin-muscle reflex are used to maintain the excitement level of the target nerve path. Electrical stimulation and vibration stimulus are also applied as frequently as possible.
- Instead of a passive range of motion exercises, the joint movement as an active assistive movement using the facilitation technique should be repeated within the time allowed each day. Voluntary movements assisted by facilitation techniques are performed daily, starting from the proximal joints, which remain more capable of voluntary movement, then proceeding to the distal joints.
- The therapist should prioritize reacquisition of movement patterns (combinations of muscle movements) that are necessary for walking and activities for daily living.

1. Performing iterative reinforcement of the same patterns will strengthen the target neural pathway efficiently.
   - When switching to another exercise, allow a break of 20–30 s in order to avoid reducing concentration and disrupting organizational strengthening of the synapse.
   - Treating two patients at the same time makes it easier for each patient to have a sufficient rest interval between exercises.
2. Enhance the function of the unaffected side
   - Muscle strengthening on the unaffected side is a top priority. Walking and weight transfer depend on muscular strength and balance in the lower limbs and trunk on the unaffected side. For example, if the unaffected side is weak, the patient is forced to concentrate on maintaining their balance while standing, making it more difficult to perform voluntary movements with the affected lower limb. Repeated training in standing up, walking (exercises using an orthosis), and using a parallel bar needs to be started at an early stage after a stroke.

2.2 Basic Procedure for Repetitive Facilitative Exercise

Repetitive Facilitative Exercise is a treatment method that repeatedly increases the excitement level of the target neural pathways using a new facilitation technique, and thus makes it possible for the patient to move their extremities intentionally. Initially, there will inevitably be an overflow to other neural pathways that are not being targeted, but increasing the excitement level along the targeted nerve pathway using the facilitation technique will make the intended movements possible.

2.3 Enabling the Movement Exactly as Intended

In order to promote the intended movements in the extremities, the excitement level of the motor pathway needs to be raised from the anterior horn cells to the primary motor cortex. Moreover, synergy, stretch reflex (myotatic reflex), dermal reflex, postural reflexes (tonic labyrinthine reflex, tonic neck reflex) and associated reactions are also used to stimulate movement in the extremities. In order to obtain a sufficient therapeutic effect, the therapist must use both

Supplementary Information The online version contains supplementary material available at [https://doi.org/10.1007/978-981-19-0789-0_2].

K. Kawahira (✉)
Laboratory of Repetitive Facilitative Exercise, Kirameki Terrace Health Care Hospital, Kagoshima city, Kagoshima, Japan

K. Kawahira et al. (eds.), *Exercise Therapy for Recovery from Hemiplegia*, https://doi.org/10.1007/978-981-19-0789-0_2

hands skillfully. For example, while one hand is inducing a stretch reflex, the other performs a tapping action to multiply the stimulant transmissions. The cue for encouraging the patient to attempt voluntary movements needs to be considered in order to avoid inducing the patient to perform an unintended movement. For example, the cue should not be a simple order, such as "One! Two! One! Two!" because this would urge the patients to strain themselves excessively, causing unintended or inadequate movements. Instead, it would be better to say more rhythmically "One … then, Two …" so that the patient can adjust when starting the voluntary movement and perform the target movement appropriately.

Key Points

1. Synergistic movement

   After considering flexor muscle synergy (Figs. 2.1a, c) and extensor muscle synergy (Fig. 2.1b, d), choose the approach best suited to the targeted movement. When the movement is initiated, the therapist adds resistance to the proximal part of the extremity that is performing the target movement, thereby raising the excitement level of the nerve path related to the target muscle activity. For example, ankle dorsiflexion can be induced by having the patient perform hip joint flexion while the therapist adds resistance to the thigh, which increases stimulation of the neural pathway connected to the ankle dorsiflexor muscle group.
2. Stretch reflex (Figs. 2.2a)

   Immediately after increasing excitement of the nerve tract stimulating the muscle by quickly stretching the target muscle and inducing stretch reflex, instruct the patient to contract the muscle. When inducing stretch reflex, the therapist can quickly stretch the muscle of the affected extremities using their fingers while taking care to avoid excessive force or cause damage to the tissue. Combining electrical and vibration stimulations can enhance stretch reflexes.
3. Skin-muscle reflex (Fig. 2.2b)

   While encouraging the patient to move the extremities voluntarily, the therapist rubs or taps the skin overlying the target muscle with his fingertips in order to promote contraction.

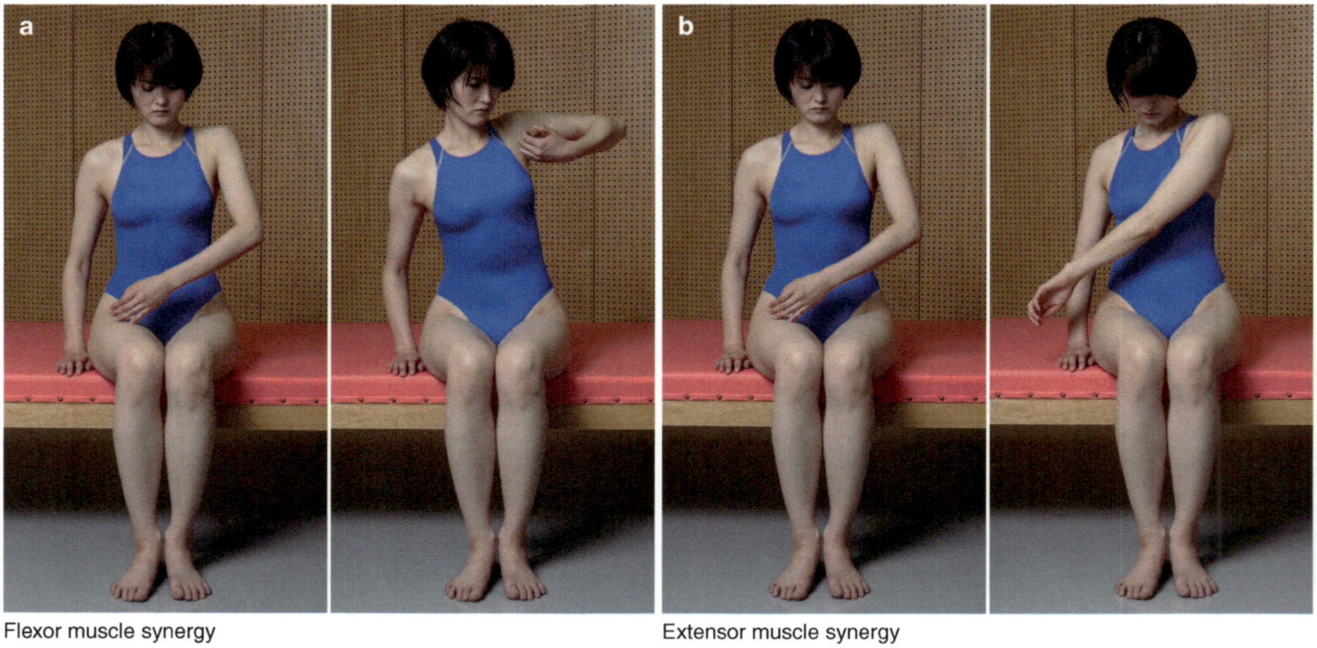

Fig. 2.1 Synergistic patterns in the upper and lower limbs (Video 2.1). There is a strong neural network between specific nerve tracts in the anterior horn cells of the spinal cord. In patients with a pyramidal tract injury, excitement transmitted from the cerebrum to the target nerve tract may also excite the anterior horn cells of non-targeted nerve tracts, causing synergistic movement. In the upper limbs, the effort (**a**) concerning construction of flexor synergy muscles (flexion, abduction, and external rotation of the shoulder, flexion of the elbow, and supination of the forearm) occurs in response to the instruction "Bend the elbow." The effort (**b**) concerning construction of extensor synergy muscles (extension, adduction, and internal rotation of the shoulder, extension of the elbow, pronation of the forearm) occurs in response to the instruction "Straighten the elbow." Synergistic flexion movement is stronger than extensor synergy in the upper limbs. In the lower extremity, the instruction "Bend your knee" will result in a flexor synergy movement

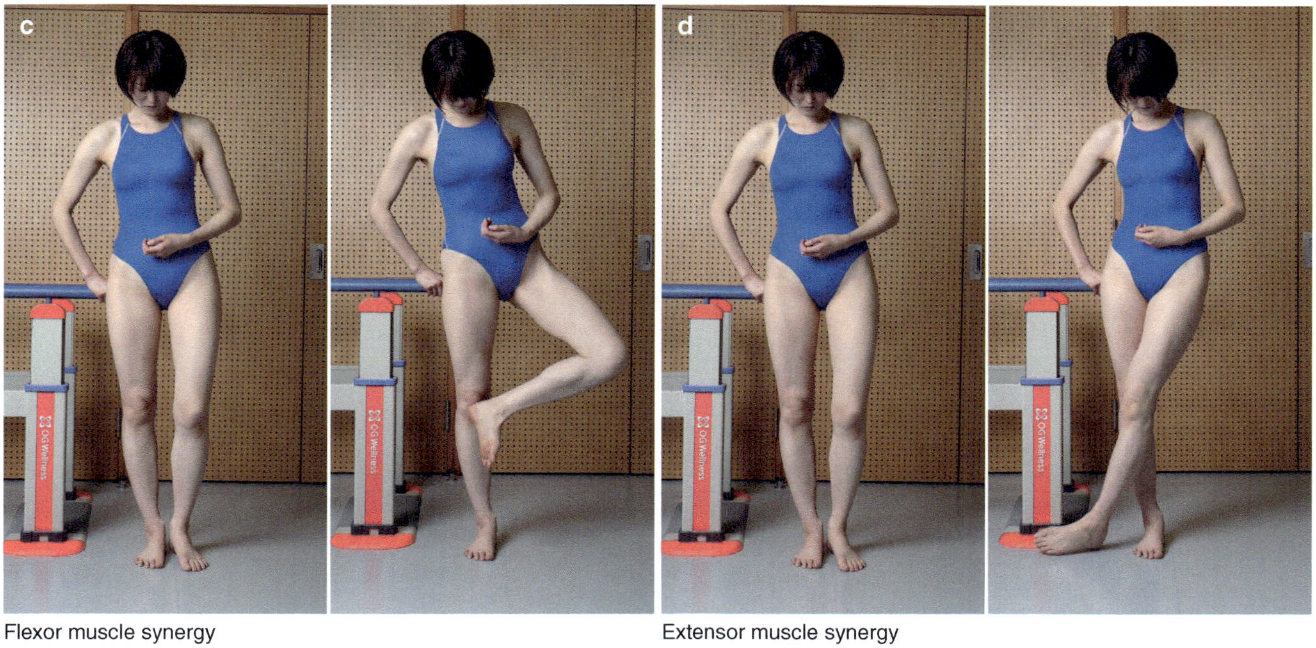

Fig. 2.1 (continued) (**c**) (flexion, abduction, and external rotation of the hip joint, knee flexion, and ankle dorsiflexion) while the instruction "Straighten your knee" will result in an extensor synergy movement (**d**) (extension, internal rotation, adduction of the hip joint, knee extension, and plantar flexion of the ankle). Extensor muscle synergy is stronger than flexor muscle synergy in the affected lower extremity

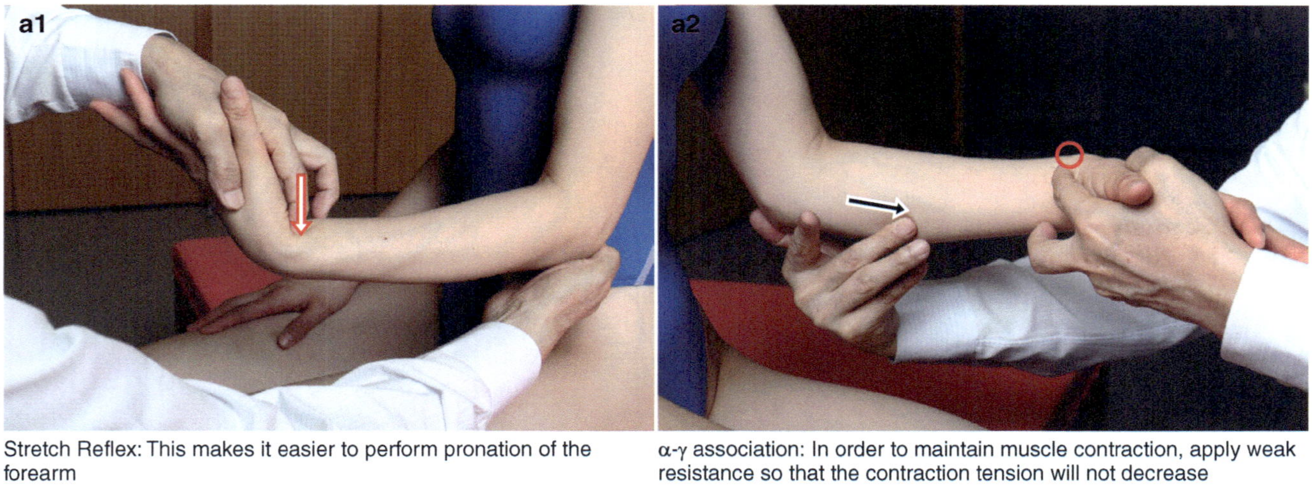

Fig. 2.2 Facilitation technique for the upper extremity. **Basic Knowledge**. When these procedures are performed effectively, the stretch reflex (**a1**) should achieve rapid muscle stretching and sustain the level of muscle tonus (**a2**), while the skin-muscle reflex

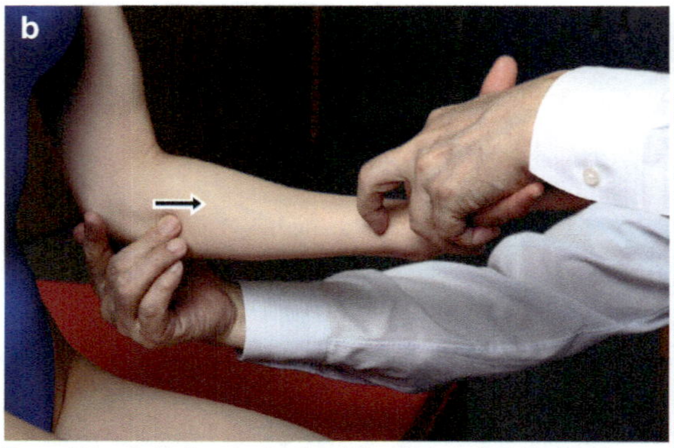

Dermal Muscle Reflex: Apply stimulation (external rubbing) to the pronator muscle (the muscles located below the elbow) while asking the patient to pronate their forearm. Maintain the forearm's pronation muscular activity by continuous stimulation (external rubbing or scratching) throughout the pronation movement

- ↓ Therapist moves the patient's body in this direction
- ○ Resistance to maintain muscle tone (just touch gently)
- ⇩ Stimulating (tapping or pressing)
- ⬇ Stimulating (rubbing)

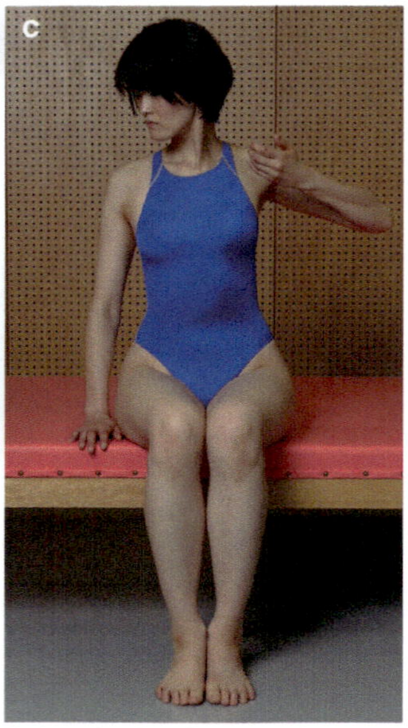

Tonic neck reflex: Facilitating the beginning of flexion in the shoulder and elbow

PNF pattern: shoulder extension, adduction, internal rotation

PNF pattern: shoulder flexion, abduction, external rotation

PNF pattern: shoulder extension, abduction, internal rotation

PNF pattern: shoulder flexion, adduction, external rotation

Fig. 2.2 (continued) (**b**) requires a longer duration of rubbing or tapping the skin. In order to increase extensor muscle tension in the upper extremity, it is important to manipulate the upper extremity in front of the face. Using movement patterns that involve the tonic neck reflex (**c**) and PNF pattern, stages (**d–g**) make it easier to induce voluntary movement. Limb position and the PNF movement pattern affect the level of neural activity. Experiments in monkeys showed that electrical stimulation to the primary motor cortex caused extension motion when the elbow was flexed and flexion motion when the elbow was extended, even though stimulation was applied to exactly the same site in the primary motor cortex. In other words, the excitement level in the motor neuron of the patient is determined, and this excitement from the cerebral cortex is conducted along the (highly excited) neural pathways. Therefore, it should be possible to adjust articulation of a hemiplegic limb by applying external stimulation

4. Postural reflex

Excitement of the nerve pathway to the extensor is increased in the supine or standing position, while excitement of the nerve pathway to the flexor muscle is increased in the prone and side-lying positions. Therefore, it is easier to flex the lower limbs when lying on the side or in a prone position (tonic labyrinthine reflex). It is easier to facilitate extension in the patient's upper limb when he or she turns their head in the direction of their affected limb. On the other hand, it is easier to facilitate flexion of the opposite limb when the back of the patient's head is aligned with the limb (tonic neck reflex, Fig. 2.2c). When applying the facilitation technique to the hand, instruct the patient to focus their attention on the hand being treated.

5. Patterns of movement and positions of the extremities

Certain movement patterns and limb positions involved in proprioceptive neuromuscular facilitation (PNF) make it easier for the patient to perform the movement of the hemiplegic limbs (Fig 2.2d, g). Raising the upper limbs in front of the face activates the nerve pathway of the extensor (Souques phenomenon). This makes it easier for the patient to extend their elbow and fingers. In this case, flexion of the lower limb becomes easier using the negative supporting reflex (flexion of the ankle and the toes), whereas extension of the lower limb becomes easier using the positive supporting reflex (dorsiflexion of the ankle and extending the toes) (Figs. 2.3 and 2.4).

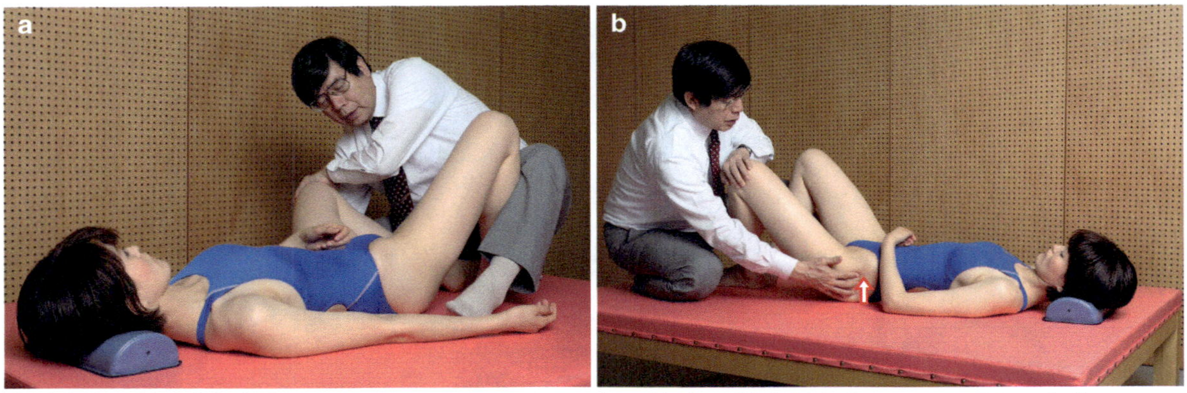

Hip abduction with *Raimiste's* phenomenon:
The therapist restricts abduction of the hip joint on the unaffected side by supporting the patient's lower limb. Abduction of the affected limb is then promoted using the skin muscle reflex.

A combination of *Raimiste's* phenomenon and dermal reflex (stimulation of the skin overlying the gluteus medius)

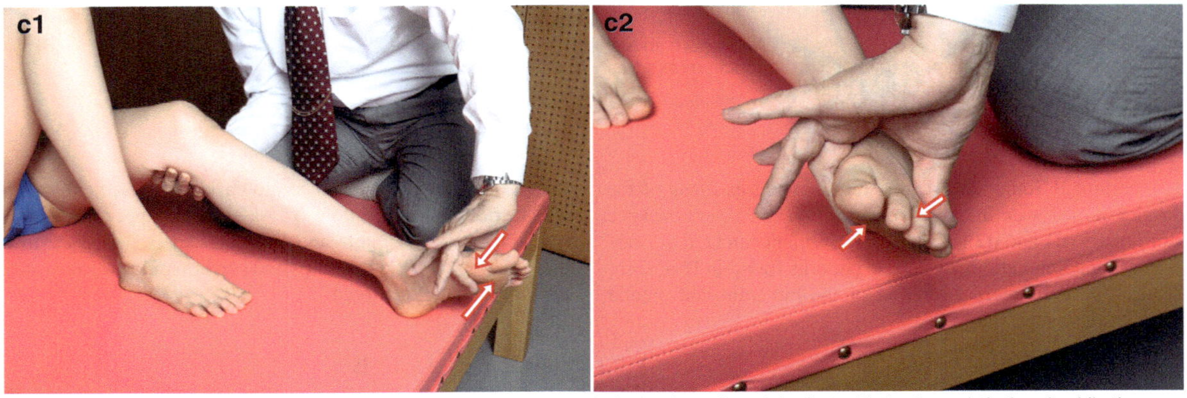

Negative supporting reflex:
The therapist holds the forefoot using his fingertips with the middle finger placed on the metatarsal as a fulcrum. Using the thumb, the therapist taps the outer dorsal region of the patient's foot. Inducing a quick inversion of the foot causes a negative supporting reflex, followed by a stretch reflex of the anterior tibialis muscle, thus making the lower limb flex easily.

Inducing inversion of the foot with the therapist's thumb while the therapist's middle finger acts as a fulcrum

Fig. 2.3 Facilitation technique for the lower extremity. *Raimiste's phenomenon* is an association reaction that causes abduction of the hip joint on the affected side when the unaffected hip joint is abducted (**a**, **b**). The same reaction occurs with hip joint adduction. A negative supporting reflex increases tonus of the flexor muscles by plantar flexion of the ankle joint and toe flexion (**c1** and **c2**), while positive supporting reflex increases tonus of the extensor muscle by dorsiflexion of the ankle joint and extension of the toes.

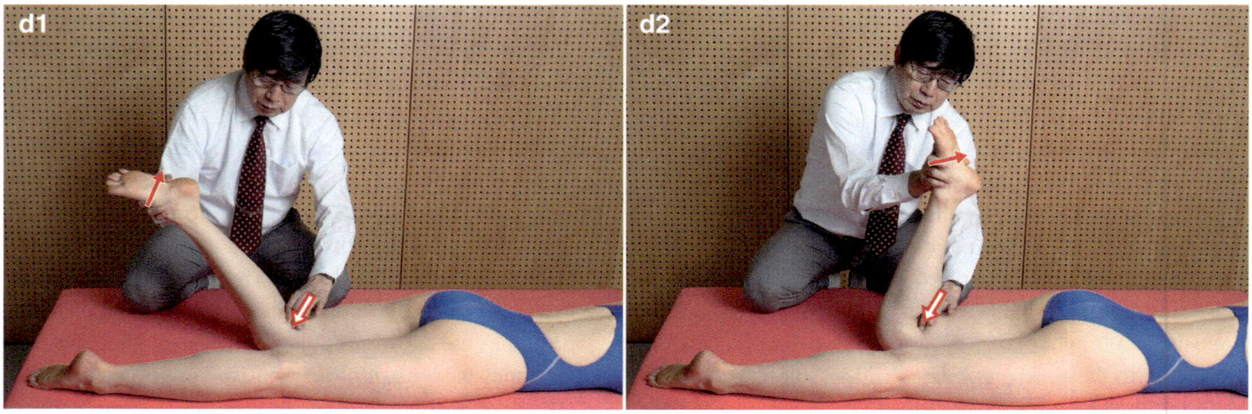

Knee flexion using negative supporting reflex: Tapping the hamstrings facilitates knee flexion.

The therapist supports the knee but does not apply pressure or interfere with the patient's efforts to voluntarily flex the knee.

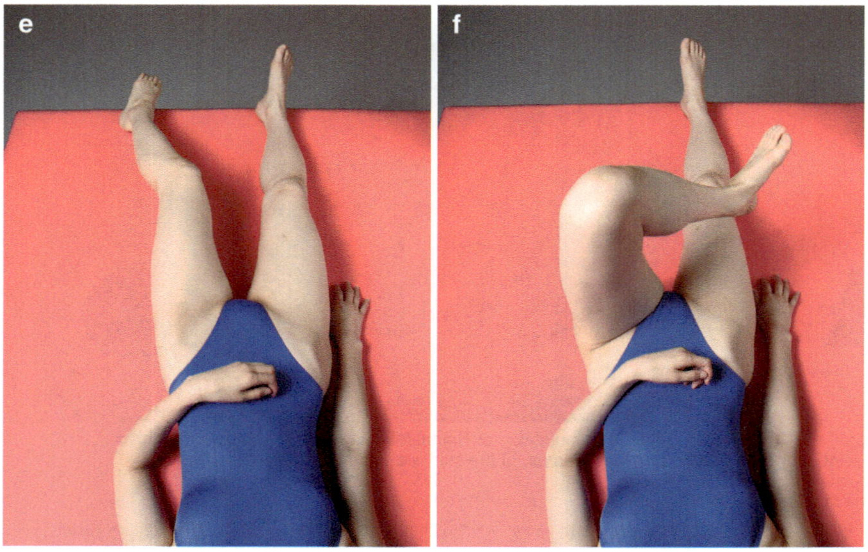

PNF pattern:
Hip flexion, adduction, and external rotation can be easily induced when internal rotation of the hip joint and plantar flexion of the ankle (in this position) occur quickly from a neutral hip position

PNF pattern: inducing hip joint flexion and adduction (externally rotated position)

Fig. 2.3 (continued) Tonic labyrinthine reflex increases tonus of the flexor muscles of the lower limb in a prone position and allows the knee to flex more easily (**d1** and **d2**). The target movement can be more easily induced using combinations of muscle patterns found in PNF techniques (**e**, **f**). By swift internal rotation of the hip joint and plantar flexion of the ankle (**e**), it becomes easier to induce external hip rotation, flexion, and abduction. (**f**) As with the upper limbs, in order to avoid injury to the joints and ligaments, it is necessary to hold the affected limb with the fingertips and perform quick manipulations without applying a strong force

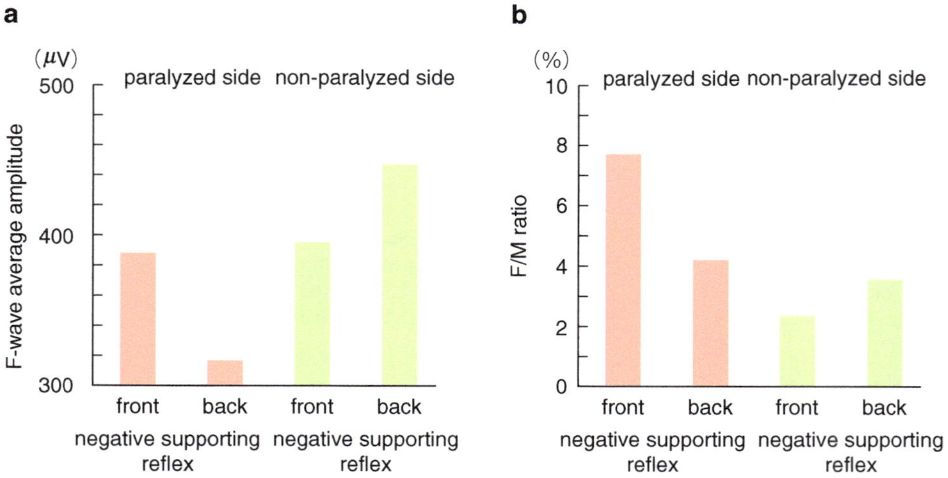

Fig. 2.4 Suppression of extensor spasticity by repeatedly inducing negative supporting reflex in the foot. The F wave (**a**) and the F/M ratio (**b**), which are indicators of extensor spasticity, both clearly decreased, indicating attenuation of extensor spasticity

2.4 Improvement of Paralysis (Isolation from Synergistic Movements)

Initially, even synergistic movement can be used to promote movement of the paralyzed limbs. Deviation from synergistic movement can be accelerated immediately after raising the excitement level of the nerve pathway for the target muscle involved in the intended movement by using an appropriate facilitation technique such as stretch reflex and instructing the patient to perform the target movement. Tell the patient to focus on matching their actual movement as closely as possible to their intended movement while they watch the affected limb move. Repeat the target movement to strengthen the nerve pathway formation and effectively "program" the target movement. Let the patient perform the movement smoothly from start to finish.

The PNF principle that movement starts in the distal extremities is applied to fully learned tool manipulation. In order to obtain the optimum movements that start from the distal limb, the parietal association area will send information about either the shape of the object or adjustment of the finger positions to the object; then the premotor and supplementary motor area will "program" an appropriate movement set. However, this response cannot be expected unless the underlying pyramidal tract from the motor cortex functions properly. Damage to the pyramidal tract causes hemiplegia, making it more difficult to voluntarily move the distal extremities than the proximal extremities.

Asking patients to move their distal joints voluntarily simply results in unnecessary and unproductive muscle tension. Thus, it is reasonable to start with voluntary movement involving the proximal joints, since the proximal muscles are innervated by bilateral pathways from the brain. Then distal joint movement can be added after voluntary movement of the proximal joint has started. In patients with only slight paralysis, movement starts from the distal joint. However, since patients do not have strong spasticity, there should be little difficulty starting with distal joint movement or inappropriate movement caused by synergy patterns.

2.5 Principles of Spasticity Control

Spasticity in the distal region can be controlled by highly voluntary movements of the proximal limbs. It follows that a patient can gain control over spasticity in distal muscles by initiating voluntary movement of the proximal region (i.e., the shoulder joint or hip joint). For example, it is relatively easy to suppress spasticity of the triceps surae muscle by keeping the hip joint in a slightly flexed position. Furthermore, it is useful for the patient to continuously watch their own affected limb's movement in order to facilitate their ability to gain control over spasticity of the affected limbs.

The instruction to "raise the face and expand the chest" during walking training often emphasizes spasticity of the lower limb's extensor muscles on the affected side. A passive approach to controling spasticity, such as sustained manual or mechanical stretching of the spastic muscle in the static state, is an effective procedure. Passively stretching the trunk muscle by rotating the trunk and passively stretching the triceps brachii muscle are also effective methods. In order to suppress increased spasticity during dynamic movements such as walking, it is necessary to improve balance on the unaffected side using strengthening exercises, correcting limb alignment, and providing a brace for walking assistance. While walking, increased spasticity of the extensor immediately before and after the affected limb touches the ground leads to scissors gait and foot drop. A lower limb brace can improve toe clearance, allowing the patient to walk without swinging the lower limb abnormally. It is also important to stabilize the standing balance using parallel bars or a cane for support during the swing phase before the foot touches the ground. These methods are necessary to suppress spasticity of the extensor muscles, which is increased during walking training. It is also easier to suppress spasticity when electrical stimulation is combined with these approaches. If the stretch reflex that has increased to the extent that clonus appears, or if it is impossible to suppress the stretch reflex during walking or other exercise, despite the efforts described above, for reasons such as foot orthosis and support, the patient will need a muscle relaxant, botulinum treatment, or nerve block. For details of such a procedure, please refer to a specialized text.

3 Planning Treatment Program

Kazumi Kawahira

The main points of a treatment program for hemiplegic patients are (1) prevention of disuse syndrome (strengthening muscles on the unaffected side, inducing frequent joint movement, providing strong sensory stimulation, and walking training in the early stages); (2) treatment of paralysis and sensory impairment as well as higher brain dysfunctions (such as aphasia, hemi-spatial neglect); and (3) training that includes gait training and ADL training. The actual components of treatment will differ depending on the patient's age, duration after onset, severity of paralysis, severity of sensory disorder, muscle strength on the unaffected side, and level of higher brain dysfunctions (Table 3.1). Physical therapy should place emphasis on the activities related to daily living and independent walking; therefore, it is essential to promote muscle strength in the unaffected limbs and trunk first. It will also be necessary to devote time to treatments that improve paralysis. Under the current medical insurance system in Japan, the time allotted for physical therapy and occupational therapy is about 20–60 min for each patient. In order to provide an effective treatment that improves the patient's paralysis, it is necessary to realize the intended movement within this limited treatment time and provide Repetitive Facilitative Exercise. Therefore, the technique for inducing excitement of the nerve tract should be completed within a few minutes. Instead of conventional passive range of motion exercises, the movements are performed as self-passive exercises with electrical and vibratory stimulation. This can be efficient as long as electrical and vibratory stimulation are not contraindicated, as described in the section on promotion methods in this book (Table 3.1).

K. Kawahira (✉)
Laboratory of Repetitive Facilitative Exercise, Kirameki Terrace Health Care Hospital, Kagoshima city, Kagoshima, Japan

K. Kawahira et al. (eds.), *Exercise Therapy for Recovery from Hemiplegia*, https://doi.org/10.1007/978-981-19-0789-0_3

Table 3.1 Treatment program

Level of paralysis	Treatment program	Points to remember
Upper limb (entire arm)	Increased voluntary movement of shoulders and fingers improves the spasticity of the Pectoralis major	Prevention of shoulder pain, combined use of electrical and vibration stimuli. (particulary for sensory disturbance)
Upper limbs Brunnstrom Stage I, II	Flexion of the shoulder, flexion/adduction Horizontal abduction/adduction Elbow flexion/extension, forearm pronation	Impingement prevention (restrained upward movement of the humeral head, shoulder external rotation), flexing shoulder sufficiently, prevention of contracture and pain Using an arm sling while in a sitting position and standing position
III, IV	Shoulder flexion Shoulder flexion/adduction PNF (shoulder flexion/adduction ⇆ extension/abduction) Pronation/supination of the forearm Reaching and pinching with self-assistive movement, lifting the hand from the knee to the chin	An example of flexion, abduction, internal rotation of the shoulder. If elevating the upper limb in a sitting position often hurts the shoulder, the exercise should be performed with assistance and reduced abduction of the shoulder by flexing the elbow accompanied by supination of the forearm.
V, VI	Flexion, adduction, external rotation of the shoulder, extension of the elbow PNF (flexion/adduction ⇆ extension/abduction) Dorsiflexion of the wrist joint Reaching and pinching with self-passive movement, lifting the hand from the knee to the forehead	Upper limb elevation while seated is mildly unloaded (during autonomous training the affected limb is lightly supported by the unaffected limb, avoiding flexion, abduction and internal rotation of the shoulder).
Fingers Brunnstrom Stage I, II	Flexion and extension of all fingers Dorsiflexion of the wrist joint Wrist dorsiflexion with fingers extended	While prompting the patient, perform a passive or nearly passive range of motion exercises with the fingers (edema alleviation, stretching the skin between thumb and index finger and between thumb and little finger by abducting the fingers).
III, IV	Extension of all fingers Individual finger extension Individual finger flexion	While counting slowly, flex and extend each individual finger. Do not simultaneously extend/flex all fingers together (do not open and close the hand into a fist).
V, VI	Individual finger flexion/extension Hand motion in the PNF pattern	Adjust the finger movement speed. Fast flexion/extension under full force can disrupt improvement in voluntary control.
Lower limb (entire leg)	Extension, abduction, external rotation of hip joint improves spasticity of the hip adductor muscles	Combined use of electrical and vibration stimuli, especially for sensory disturbance
Lower limbs Brunnstrom Stage I, II	Modified PNF pattern of flexion/adduction and extension/abduction of the hip joint ※, flexion and extension of the hip joint, extension of the lower limb	Even in cases of severe paralysis, bending the entire lower limb quickly using a flexor muscle synergy pattern (stretch reflex induction of hip adduction and triceps muscle) causes contraction of the extensor.
III, IV	Modified PNF pattern of flexion/adduction and extension/abduction of hip joint, flexion and extension of the hip joint, flexion of the knee joint, ankle dorsiflexion	When sitting with knees bent and legs tucked under or sitting cross-leged, which requires bending of the hip and knee joints of the affected lower limb with the ankle joint in plantar flexion, it is difficult to induce ankle dorsiflexion. This causes difficulties in the patient's recovery.
V, VI	Modified PNF pattern of flexion/adduction and extension/abduction of the hip joint, internal and external rotation of the hip joint, dorsiflexion of the ankle joint	During the modified PNF pattern of abduction, the hip joint is kept in external rotation but the ankle joint should not be in plantar flexion.
Trunk	Rotation and side bending	Useful for turning over and sitting up on the exercise mat

3.1 Points to Consider When Preparing Treatment Programs Including the Repetitive Facilitative Exercise

1. Rather than perform a passive range of motion exercise, Repetitive Facilitative Exercise is applied to a range of motion exercises as a combination technique during self-passive exercises. Then the range of motion using active movement is extended and the whole range of motion using passive exercise is performed. If the paralysis is severe, the therapist's instructions to the patient to "extend" and "bend" may not seem useful, but the therapist should always remember to give these instructions during the range of motion training with Repetitive Facilitative Exercise. In patients with severe spasticity, spasticity will be decreased by the facilitation method described here, making it easier to perform the range of motion exercises as a self-passive exercise. Particularly in patients with sensory impairments, ataxia, flaccid paralysis, and spasticity, vibratory stimulation and electrical stimulation should be used before and during training to adjust the excitement level of the motor neuron and sen-

sory pathways in order to facilitate smooth exercise. These stimuli prevent and alleviate neuropathic pain in the affected limb.

2. Patterns of Repetitive Facilitative Exercise aimed at restoring voluntary movement after paralysis may be limited to 4–6 subtypes according to the severity of the patient's paralysis and the skill level of the therapist. Since our institution is dedicated to investigating the effects of many different types of treatment, we often perform 8 different treatment patterns for the upper extremity and 5 for the lower extremity. For standard treatments, however, it would be more realistic to choose between 4 and 6 patterns with 50–100 repetitions for Repetitive Facilitative Exercise.
   - Essential movement patterns:

     In the upper extremity, (1) flexion of the shoulder joint; (2) elbow flexion/extension; (3) forearm supination; (4) Flexion and extension of the thumb and index finger.

     In the lower extremity, (1) flexion/adduction and extension/abduction (with the hip joint in external rotation/knee in a flexed position); (2) hip joint flexion/adduction-extension/abduction (hip joint in external rotation with knee flexion/extension); (3) flexion and extension of the knee joint; (4) dorsiflexion of the ankle joint; (5) trunk rotation and lateral bending.

     The time required to repeatedly carry out four kinds of repetitive exercise patterns 50 times per session depends on the severity of paralysis and whether the treatment target is proximal or distal. Excluding rest intervals, the required treatment time is 15–20 min for the lower extremity and about 20 min for the upper extremity.
3. Paralyzed limbs also need to undergo extensive training using voluntary active joint movements. The conventional method is to flex and extend the upper limbs on the desk by fixing the hand to a sanding board or a caster board. However, voluntary movement exercises should be performed while continuously holding the limb upward, supported by an unloading device. A patient who has become able to extend the fingers slightly can perform finger extension exercises at least 100 to 300 times per day with the wrist flexed and the finger dropping. During the exercise for wrist dorsiflexion with finger extension and for ankle dorsiflexion using low frequency electrical stimulation, encourage the patient to move voluntarily in synchronization with the electrical stimulation, and to attempt the exercise for 30–60 min.
4. Strengthen the unaffected side as much as possible using voluntary exercises. For patients needing some assistance, provide repeated instructions to a family member who can help the patient. To strengthen the trunk and lower extremity on the unaffected side, the patient should stand up from a sitting position 100–300 times/day (resting between every set of 20 repetitions) and get up from a supine position on a mat 20–40 times. For the upper extremity, pegboards, construction tasks using building blocks, script copying, etc. are utilized to treat impaired finger dexterity as well as higher brain dysfunction (mainly hemispatial neglect and constructive apraxia, grasping reflex). The current Japanese insurance system does not allow adequate treatment time, so family members should be encouraged to participate in treatment as much as possible in order to increase the therapeutic effect as well as to reduce the need for assistance in activities of daily living after discharge.
5. The patient's expectation for recovery from hemiplegia may be very high and often far exceeds the medical staff's objective predictions. At the start of treatment, the prognosis should be given as accurately as possible. It is recommended that the staff member tells the patient, "I will do my best to help your recovery, but from my experience there is usually only a slim chance that patients recover as expected."

   Regarding the upper limbs, it is necessary to tell the patient, "I will do my utmost for the affected side, but it is also important to train your unaffected hand in order to be able to perform certain activities for daily living in case the affected hand does not recover as much as you wish."

3.2 Methods of Maintaining Patient Concentration

In order to perform Repetitive Facilitative Exercise effectively and efficiently within the limited treatment time, it is necessary to devise a method that encourages the patient to maintain their concentration throughout the exercise.

1. Use the facilitation technique to quickly restrain the spasticity of the muscle group being treated. During self-passive exercises for the range of motion exercises using Repetitive Facilitative Exercise, encourage the patient using verbal cues for each movement, such as, "Bend" and "Extend."
2. The exercises should not be performed with explosive movements involving extreme stretching and recoiling but instead should attempt smooth movements adjusted intentionally by the patient by themselves.
3. For a patient with hemiplegia, moving the affected limb precisely takes a lot of concentration, so it is best to allow a 10- to 20-s rest before proceeding to another pattern.

Tips

When providing Repetitive Facilitative Exercise, it is better to treat two patients at the same time. Practice one pattern repeatedly 50 times for the first patient, then let the patient rest. During the rest period for the first patient, repeat one pattern 50 times with the second patient. Treating each patient alternately allows adequate rest periods for each patient. This intermittent rest is not only useful for maintaining the patient's mental concentration but also for improving the transmission efficiency of synapses following Repetitive Facilitative Exercise as well as enhancing the synapses after treatment.

4. Patients should be instructed to avoid the simultaneous contraction of numerous muscles during Repetitive Facilitative Exercise, as the therapeutic goal is to prevent synergistic movement using muscle combinations that differ from those involved in producing synergy. For example, if a patient repeatedly performs "show rock, show paper" immediately after repeatedly promoting the flexion/extension movement of each individual finger, the effect of the Repetitive Facilitative Exercise which flexed and extended each finger will be diminished.

1. Key points to keep in mind during treatment of the hemiplegic upper limb.

① Upper limb (shoulder/elbow) paralysis (*Brunnstrom stage* I, II)

- Repetitive Facilitative Exercise in the supine position repeatedly induces shoulder flexion, shoulder flexion and adduction, horizontal adduction and horizontal abduction of the shoulder, and elbow flexion and extension.
- Prevent shoulder pain. Use an arm sling while sitting and standing. Also, try to avoid impingement when flexing the shoulder joint.
- Combination therapy with electrical stimulation/vibratory stimuli

② Upper limb (shoulder/elbow) paralysis (*Brunnstrom stage* III–VI)

- Focus on the facilitation pattern that promotes the patient's ability to lift the upper limb by Repetitive Facilitative Exercise performing flexion, flexion and adduction of the shoulder, and flexion and extension of the elbow in a supine position as well as by modified PNF patterns. In the sitting position, the affected side requires the full support of the upper limb's weight with the shoulder relaxed in order to allow the fingers to demonstrate their fullest extent of voluntary movement. If the muscular strength for shoulder flexion is still weak, repeatedly lift the hand from the knee to the chin or forehead while performing external supination of the forearm.
- To prevent shoulder pain, do not hold the shoulder at 90 degrees abduction, which would cause impingement in the shoulder joints. Moreover, do not raise the hands in such a way that would cause a recoil response in the trunk.
- Combination therapy with electrical stimulation/vibratory stimuli

2. Key points to keep in mind for the treatment of hemiplegic fingers.

① Finger paralysis (*Brunnstrom stage* I, II)

- While repeatedly applying electrical and vibratory stimulus, frequent and repeated extension/flexion of the fingers provides a powerful sensory input to increase the excitement level of the nerve path and promote functional recovery as well as prevent contracture of the finger joints.

② Finger paralysis (*Brunnstrom stage* III, IV)

- Perform individual extension/flexion movements for each finger and simultaneous pronation of the forearm while the finger joints extend. Then, perform Repetitive Facilitative Exercise with finger extension and palmar flexion at the wrist. In order to prevent contracture of the finger joints, place the wrist joint in gentle dorsal flexion, then move the thumb and index finger in opposition as if trying to pinch some small object with the thumb and index finger.
- Combination therapy with electrical stimulation/vibratory stimuli

③ Finger paralysis (*Brunnstrom stage* V, VI)

- While performing extension/flexion movements for individual finger joints, simultaneously perform pronation of the forearm while extending the finger joints as described under stages III and IV. After that, move the thumb and other fingers in opposition to form a round shape, then repeatedly perform an opening and closing movement with the fingertips.
- Combination therapy with electrical stimulation/vibratory stimuli

3. Key points to keep in mind for treatment of the hemiplegic lower limb.

① Lower limb paralysis (*Brunnstrom stage* I, II)

- Perform flexion/adduction and extension/abduction of the hip joint at an externally rotated position of the hip while applying Repetitive Facilitative Exercise.
- Combination therapy with electrical stimulation/vibratory stimuli

② Lower limb paralysis (*Brunnstrom stage* III~VI)

- Perform flexion, flexion/adduction, extension/abduction and extension of the hip joint, modified PNF pattern for the lower limb, internal rotation and external rotation of the hip joint, and dorsiflexion of the ankle using the Repetitive Facilitative Exercise. If severe spasticity of the adductor muscle of the hip joint appears in the supine position, rapid internal rotation of the hip joint reduces adductor muscle spasticity by inducing outer rotation, flexion, and abduction of the hip joint.
- Even when dorsiflexion of the ankle joint in sitting and standing positions becomes possible, the patient will still be unable to dorsiflex the ankle quickly enough while walking. In many cases, plantar flexion of the ankle will occur due to spasticity of the extensor muscle. Therefore, a lower limb orthosis and cane will be needed for walking.
- In a patient whose stance phase of the unaffected lower limb has become stable, grounding on the affected side could be performed first with the cane followed by the affected foot. In addition, the patient who can perform walking exercises without using parallel bars or a lower limb orthosis should always contact the ground starting from the heel. In patients with a fast mid-stance phase right after the initial heel contact, walking exercises should be performed with small steps focusing on the heel contact while supported by parallel bars.
- Combination therapy with electrical stimulation/vibratory stimuli

3.3 The Purpose of Repetitive Facilitative Exercise

1. Recovery from paralysis

The upper limit for functional recovery of hemiplegic limbs is principally defined by the extent of brain damage and quality of rehabilitation. It is easier to reconstruct and strengthen new nerve tracts in a patient with less motor cortex damage and more of the pyramidal tract remaining. These new nerve tracts can then restore the function of affected limbs. Therefore, patients with mild paralysis can set relatively higher treatment goals. The extent of recovery from paralysis is better for patients with a shorter duration after onset and with milder paralysis and sensory impairment. Shoulder pain and contracture, spasticity, involuntary movements, and higher brain dysfunction (hemispatial neglect, dementia) are factors that inhibit progress. However, the effects of sensory impairment can be alleviated by the combined use of prompts during Repetitive Facilitative Exercise, and electrical and vibration stimuli. Brunnstrom stages at 7 days after onset of upper extremity paralysis show that full recovery within six months of onset is likely to cocur in about 10% of patients in Brunnstrom stages I or II, 55% of those in stage III, and 97% of those in stage IV.

The cumulative percentage of patients in whom hemiplegia recovery had ceased by 3 months after onset exceeds 90%. Furthermore, recovery from paralysis did not markedly differ between the upper and lower limbs [1]. On observation from the acute phase of 1,197 patients with cerebrovascular disease in Copenhagen, Denmark, neurological symptoms had stopped improving in 80% of patients at 4.5 weeks and in 95% at 11 weeks; in some patients, the improvement of neurological symptoms, including paralysis, is limited to the early weeks after onset [2].

However, there is no "six-month wall," a term used to suggest that the ability to improve paralysis ends at 6 months. Even after that six-month "limit," intense treatment will improve paralysis [3].

When the patient undergoes Repetitive Facilitative Exercise, recovery goals can be projected based on the duration after onset and the severity of paralysis as described below. However, these projections are not based on statistical data, which is currently insufficient because an adequate number of cases has not been accumulated.

① Duration after onset: 1 month

Targets for upper limbs (excluding fingers) and lower limbs for patients with a duration after onset up to 3 months are indicated by progressing through *Brunnstrom stage*. · Patients in stage I to II → III to IV; patients in stage IV to V → V to VI; and patients in stage V → VI.

Targets for hands with a duration after onset up to 3 months are indicated by advancement of each *Brunnstrom stage*. · Patients in stage I to II → III to IV; patients in stage IV to V → stage VI; patients in stage VI would likely remain in the same stage, but flexion and extension of the fingers will become smoother.

② Duration after onset: over 3 months

The goals for upper limbs (excluding fingers) and lower limbs following a duration after onset of several months are also defined by improvements of *Brunnstrom stage* · Patients in stage I to II progress to stage II to III; patients in stage III to IV progress to IV to VI; patients in stage V would likely remain in stage V.

Targets for fingers are similarly defined by *Brunnstrom stage* improvement· Patients in stage I to II → II to III; patients in stage III → IV; patients in stage IV to V → V to VI. Patients in stage VI would likely remain in stage VI, but flexion and extension of the fingers will become easier.

Thus, even though patients in *Brunnstrom stage* III to V with a duration after onset of over 3 months are supposed to have less possibility for improvement, such patients often demonstrate improvement following accelerated Repetitive Facilitative Exercise.

Therefore, by conducting self-passive exercise for a range of motion exercises using the Repetitive Facilitative Exercise for at least 3 months after onset, contracture and pain in finger joints and shoulders can be prevented. Moreover, depending on the degree of paralysis, combination therapy with Repetitive Facilitative Exercise and continuous electrical or vibratory stimulation should be applied so that voluntary muscle activity is increased. The goal for patients in *Brunnstrom Stage* IV or V is to allow each individual finger to be extended. Extension of individual fingers from a flexed position (gripping) is easier than flexion of the individual fingers.

2. Practical targets for upper limb use

Many patients have an impression of restoring the functionality of their upper limb that includes well-functioning shoulder and elbow joints, which allow the hand to reach an intended place using a high level of voluntary movement. In order for fingers and shoulder joints to perform skilled activities, it is necessary to achieve an advanced information processing ability in the motor cortex and pyramidal tract, which conducts the information required to perform the function. Therefore, the extent of recovery is greatly influenced by the severity of brain damage in the motor control area and pyramidal tract as well as the treatment plan for the affected limbs. In order to achieve functional usage of the affected upper limb for daily living, it is necessary to achieve both skillful finger function in order to manipulate objects and shoulder function to be capable of lifting the whole upper limb.

One of the goals is to enhance shoulder function without causing pain in the shoulder joint. However, the level of paralysis of the shoulder, elbow, and fingers in the upper limb greatly differs among individuals.

There are patients in whom shoulder function has deteriorated despite relatively good voluntary control remaining in the elbow and fingers. Conversely, the shoulder may show relatively good voluntary control while voluntary control of the elbow is limited and the thumb shows only slight movement. Accordingly, it is important for individual patients to set goals in which the upper limbs are as useful as possible for daily living. For example, even though the patient still has synergistic movements, dressing will be easier if the patient is able to extend the elbow. If it is possible to pinch with the thumb and index finger, the affected limb will be able to perform useful functions, such as fixing their medicine bag and cutting with scissors or holding one side when buttoning or unbuttoning clothes. For the patient who is only able to grip with the fingers through synergy even after several months, the goal is to pinch and hold objects with thumb and index fingers while bracing their hand against the desk or knee. A patient whose shoulder and elbow have sufficiently recovered from synergy, and who also has individual movement of the fingers (especially the thumb, index finger, and middle finger), should set a goal to achieve additional practical functions.

Patients who do not demonstrate improvement of individual finger movements and have severe spasticity will find it difficult to recover from paralysis. For patients with limited voluntary shoulder and elbow movement but good voluntary finger movement, it is better to keep the upper limb unloaded and use electrical stimulation on the shoulder muscles during shoulder flexing exercises in order to prevent shoulder joint pain while trying to improve voluntary shoulder movement.

3. Practical targets for recovery of the trunk and lower limb

If paralysis of the lower limb does not completely recover, walking without a brace or cane will be counterproductive.

In the presence of persistent paralysis and spasticity, walking without a prosthetic device will exacerbate spasticity and promote deformation of the knees and feet. This occurs when the patient lifts their affected lower limb excessively and strikes the floor forcefully during the landing stage in the swing sequence.

Balance while standing on the unaffected lower limb (in cases of double hemiplegia, the less-affected limb and trunk) determines whether walking is possible. Even though it does not seem to be related to strengthening the trunk and lower limb muscle on the unaffected side, increasing voluntary control of rotational and lateral bending on the affected side of the trunk will make it easier to slightly raise the pelvis on the affected side in order to lift the affected lower limb from the ground. As a result, it becomes easier to: swing the lower limbs forward smoothly (by flexion of the hip joint and rotation of the trunk), efficiently contract the abductor muscle of the hip joint and the quadriceps muscle during the stance phase, and suppress the triceps surae muscle during smooth movement from the center of gravity. Therefore, it is important to use the PNF modified pattern, extension/abduction exercise for the hip joint and the knee extension exercise with the ankle joint at 0°.

As part of balance training to improve standing, training to balance the weight on the affected lower limbs (by continuously weighing the other side and training to evenly load both lower limbs) is not effective because this fails to improve the affected lower limb's function and gait. Furthermore, postural control cannot be performed on the affected lower limb. It is important to balance the standing position with the unaffected lower limb while the affected lower limb pushes the received weight back to the unaffected lower limb. Repeating this lays the foundation for training.

With the patient's gait training, combinations of exercises necessary for walking should be repeated using parallel bars, lower limb orthosis, or a cane combined with the facilitation technique to induce walking so that the patient can learn these movement patterns quickly. In particular, repeating the pattern involved in swinging the unaffected upper limb forward followed by the affected lower limb as two walking movements (two-point gait) while supported by parallel bars promotes a firm standing balance on the unaffected side. It should be emphasized that placing your hands on the parallel bars, without gripping so the hands remain open, avoids exacerbating spasticity in the extensor muscles on the affected side right before landing on the affected lower limb. Even when wearing a short leg brace (ankle foot orthosis: AFO), promotion of extensor spasticity occurs when the heel strikes the ground with the knee extended. Walking should be taught using a pattern in which the heel of the sole touches the ground while the knees flex slightly. The more severe the paralysis, the more necessary it becomes to strengthen the trunk muscles by having the patient wear a long leg brace (knee ankle foot orthosis: KAFO) from the early stage of walking. This will enable the patient to lift the pelvis on the affected side. For hemiplegic patients without higher brain dysfunction, such as severe hemi-spatial neglect or joint disorder in the unaffected lower limb, independence will be the goal. In patients with severe paralysis and double hemiplegia (recurrent cases etc.), postural balance control on the unaffected side is limited. If the patient does not have sufficient strength even on the unaffected side, the possibility of walking independently decreases.

References

1. Ryu N. Nōsotchū kanja no shōgai no kōzō no kenkyū; henmahi to kikyo idō dōsa nōryoku no kaifuku katei no kenkyū [Structural disorder in patients with stroke; research on the process of recovery from hemiplegia and activities of daily living]. Sōgō Rehab. 1983;11:557–69.
2. Jørgensen HS, Nakayama H, Raaschou HO, et al. Outcome and time course of recovery in stroke. Part II: outcome. The Copenhagen stroke study. Arch Phys Med Rehabil. 1995;76:399–405.
3. Takuji H, Ishikawa S, Kawamura T, et al. Tsūsho rihabiritēshon ni okeru mansei-ki nōsotchū henmahi jōshi e no sokutsū hanpuku ryōhō to chiryō-teki denki shigeki shindō shigeki to no heiyō ni yoru mahi kaizen kōka [Effect of repetitive facilitation exercises with electrical and vibratory stimulation for chronic hemiplegia patients using day-care rehabilitation]. Physics Department. 2017;32:129–32.

Repetitive Facilitative Exercise for the Upper Limb

4

Kazumi Kawahira

Many stroke patients wish that their affected hand could function a little better, and rehabilitation staff wish they could make it happen quickly. Unfortunately, there are only a few effective treatment techniques. The purpose of Repetitive Facilitative Exercise for the upper limb, is to multiply the effect so that the affected upper limb will become more "useful." By repeatedly performing the intended movement to reconstruct and strengthen the target nerve tract, movement of individual fingers as well as movement patterns that involve the combined use of the upper limb and fingers will improve, enabling the patient to function better in daily living. Repetitive Facilitative Exercise can effectively promote functional recovery of the hemiplegic upper limb, including the fingers. For patients demonstrating poor cooperative movements because sensory impairment induces ataxia and excessive muscle contraction when the patient moves their limbs, it is especially effective to combine Repetitive Facilitative Exercises with vibration and electric stimulation.

4.1 The Principles Underlying Upper Limb Exercise Therapy

In order to improve voluntary control of the upper limb, it is necessary to begin with the shoulder first. Excessive pulling on the upper limb can damage the shoulder girdle and shoulder joint, so care must be taken to avoid injury. Even in patients demonstrating severe paralysis, the proximal part of the shoulder will have retained more voluntary control than the fingers. This point is important when eliciting voluntary control of the elbow and fingers.

Supplementary Information The online version contains supplementary material available at [https://doi.org/10.1007/978-981-19-0789-0_4].

K. Kawahira (✉)
Laboratory of Repetitive Facilitative Exercise, Kirameki Terrace Health Care Hospital, Kagoshima city, Kagoshima, Japan

The important aspects of upper limb therapy are summarized below.

Key Points

1. Thoroughly understand the functional anatomy of the upper limb and carefully manipulate the upper limb to avoid causing pain (Fig. 4.1).
2. Daily treatment begins from the part showing greater voluntary control and progresses to the areas with less voluntary control (in principle progressing from the proximal to distal areas).
3. Start with partial isolation of the synergistic patterns.
4. If exercise causes spasticity in the flexor muscles, first control the muscle tonus, then start the exercise.

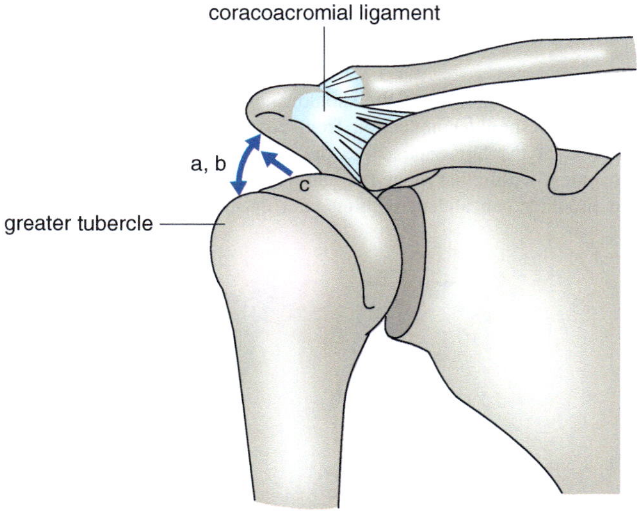

Fig. 4.1 Structure of the shoulder joint. (**a**) The positional relationship of the rotator cuff to the greater tubercle of the humerus causes the tubercle to hit the coracoacromial ligament during abduction. (**b**) When the shoulder is abducted while the shoulder joint is internally rotated, the greater tubercle of humerus strikes the coracoacromial ligament at 60° abduction (impingement). This impingement does not occur when the shoulder is externally rotated during abduction. (**c**) In patients with hemiplegia, impingement tends to occur because the muscles that stabilize the humeral head during abduction are not functioning well

K. Kawahira et al. (eds.), *Exercise Therapy for Recovery from Hemiplegia*, https://doi.org/10.1007/978-981-19-0789-0_4

5. Focus attention on the affected upper limb and encourage the patient to move the limb exactly as they intend to move it.

4.2 Basic Repetitive Facilitative Exercise for Shoulder and Prevention of Pain

The causes of shoulder pain in the hemiplegia patient are mainly categorized into two types: (1) strong mechanical stress caused by stretching the shortened joint capsule and stretching of the tendon which causes subluxation, muscle spasm, or ischemia; or (2) shoulder arthritis resulting from impingement and instability of the joint. For structural reasons, the rotator cuff or greater tubercle of the humeral is likely to hit the coracoacromial ligament during abduction of the shoulder, thus causing impingement (Fig. 4.2). In particular, flexor muscle synergy in hemiplegia patients is likely to damage the rotator cuff during shoulder abduction because the deltoid muscle lifts the humerus without the stabilizing effects of external shoulder rotation and engagement of the supraspinatus muscle.

4.2.1 Shoulder Joint Flexion and Manipulation

The therapist needs to keep the patient's humeral head below the shoulder joint in order to prevent the rotator cuff and greater tubercle of the humeral hitting the coracoacromial ligament. This can be achieved by using the fingers to push down on the humeral head (Fig. 4.3), then performing the flexion exercise combined with shoulder external rotation (Fig. 4.4).

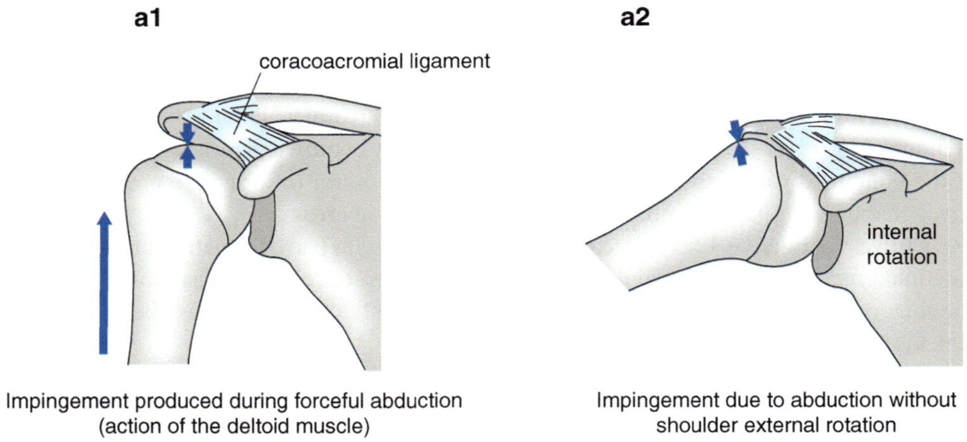

Fig. 4.2 Impingement of the shoulder joint. Resistance and pain are felt during shoulder abduction from 60° to 90° because the greater tubercle of the humerus strikes the coracoacromial ligament during movement within that range (**a1** and **a2**). Beyond that range, however, the shoulder can abduct without pain or resistance. If the therapist pushes the humeral head down with his fingers, it will be possible for the patient to abduct the shoulder without feeling resistance or pain. The Repetitive Facilitative Exercise involves frequent movements, so it is important to prevent impingement

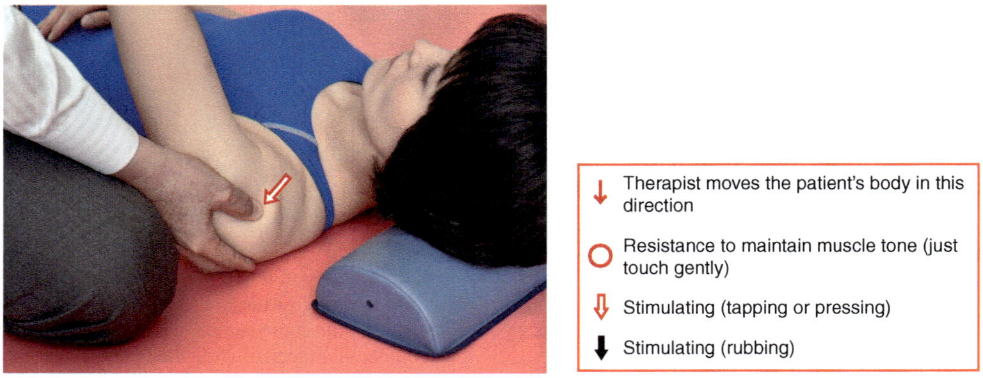

Fig. 4.3 Basic Technique: The position of the therapist's hand when holding the shoulder in order to suppress elevation of the humeral head. The therapist places his palm on the patient's lateral shoulder joint with his thumb on the edge of the acromioclavicular joint over the humeral head. Then the therapist places his index finger on the dorsal side of the patient's acromion near the humeral head. Using these fingers, the therapist should press the upper edge of the humeral head downward while flexing the patient's shoulder joint

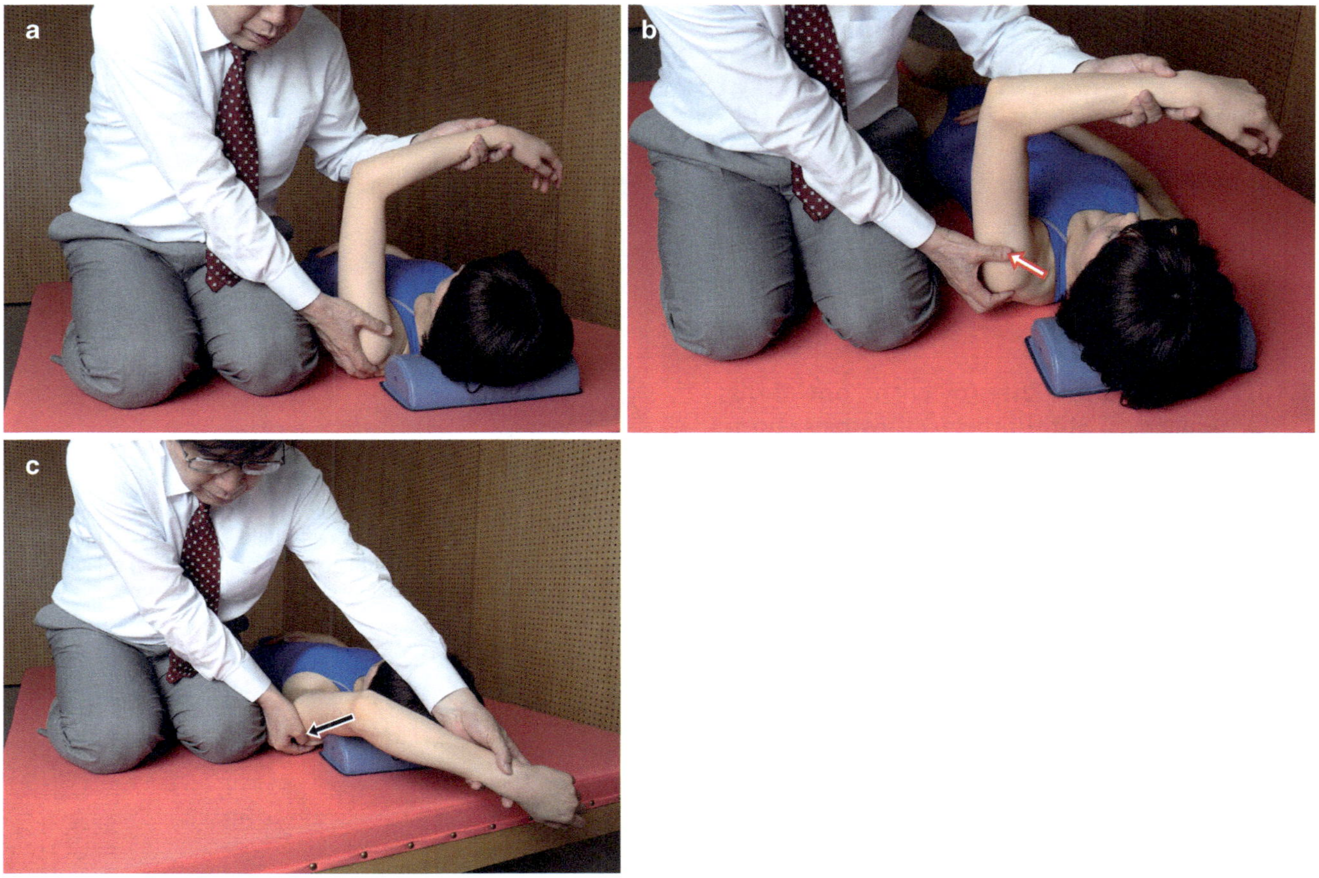

Fig. 4.4 Basic Technique: Shoulder joint flexion and manipulation technique. (**a**) Supporting the shoulder and the forearm. (**a**) The therapist places the patient's forearm on his palm and lightly holds the forearm slightly proximal to the patient's wrist joint. The therapist places the palm of his opposite hand on the patient's shoulder joint with his thumb on the inferior edge of the shoulder joint above the humeral head and the index finger on the dorsal side of the acromion and humerus head. The other fingers are placed under the shoulder. (**b**) Prevention of impingement. (**b**) The therapist flexes the shoulder joint while using his thumb to press the patient's humeral head downward in order to maintain the space between the humeral head and the acromion. (**c**) Prevention of impingement. (**c**) When the bent angle of the shoulder joint exceeds 60°, the therapist uses his thumb to press the patient's humeral head slightly downward in order to prevent humeral head thrust. The therapist supports the patient's forearm as the affected limb moves but does not push the limb in the direction toward bending

Key Points

When performing shoulder joint flexion and abduction, pay careful attention to the following points.

1. In order to prevent impingement, the tendency for the humeral head to push up is countered by the therapist's fingers pushing down (Fig. 4.3) and external rotation is induced during the manipulation.
2. When manipulating the shoulder joint, the therapist should hold the forearm lightly with his fingertips to avoid applying excessive force to the patient's shoulder joint (Fig. 4.5). If the therapist's technique during this procedure is poor, the patient will experience pain and the effect of the exercise will not be sufficient.

At the beginning of treatment, first flex the shoulder to confirm the range of motion, then carefully perform flexion and adduction movements of the shoulder. When flexing the shoulder, there is little chance of impingement. However, guiding the patient's movement is significant, so even when simply flexing the shoulder, try to perform the movement in combination with external rotation of the shoulder.

The therapist should also instruct the patient to keep the shoulder in external rotation by saying, "During this shoulder exercise, always keep your palm toward your face" and "Keep your hand higher than your elbow" (Fig. 4.6).

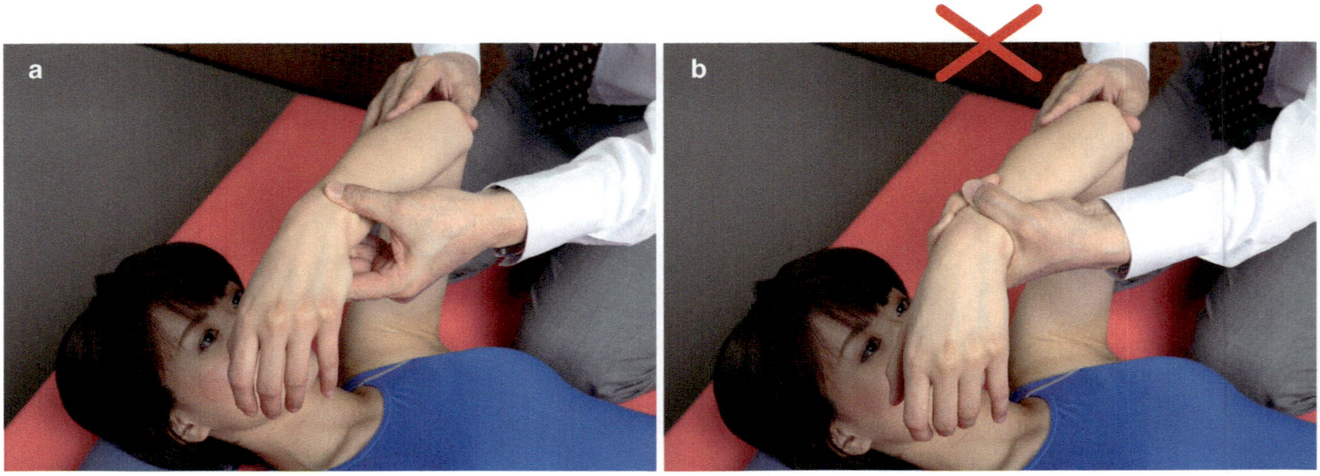

Fig. 4.5 Basic Technique: How to hold the arm without applying force to the shoulder. (**a**) The therapist holds the patient's forearm with his fingertips in order to support the patient's forearm lightly. Notice that there is a gap between the therapist's palm and the patient's forearm. (**b**) Poor technique. Clenching the patient's forearm makes it difficult to recognize active movement in the affected limb and disrupts the patient's ability to move. If the therapist holds the patient's hand, movement during the repetitive facilitative exercise becomes unstable, making it difficult to perform the movement rapidly

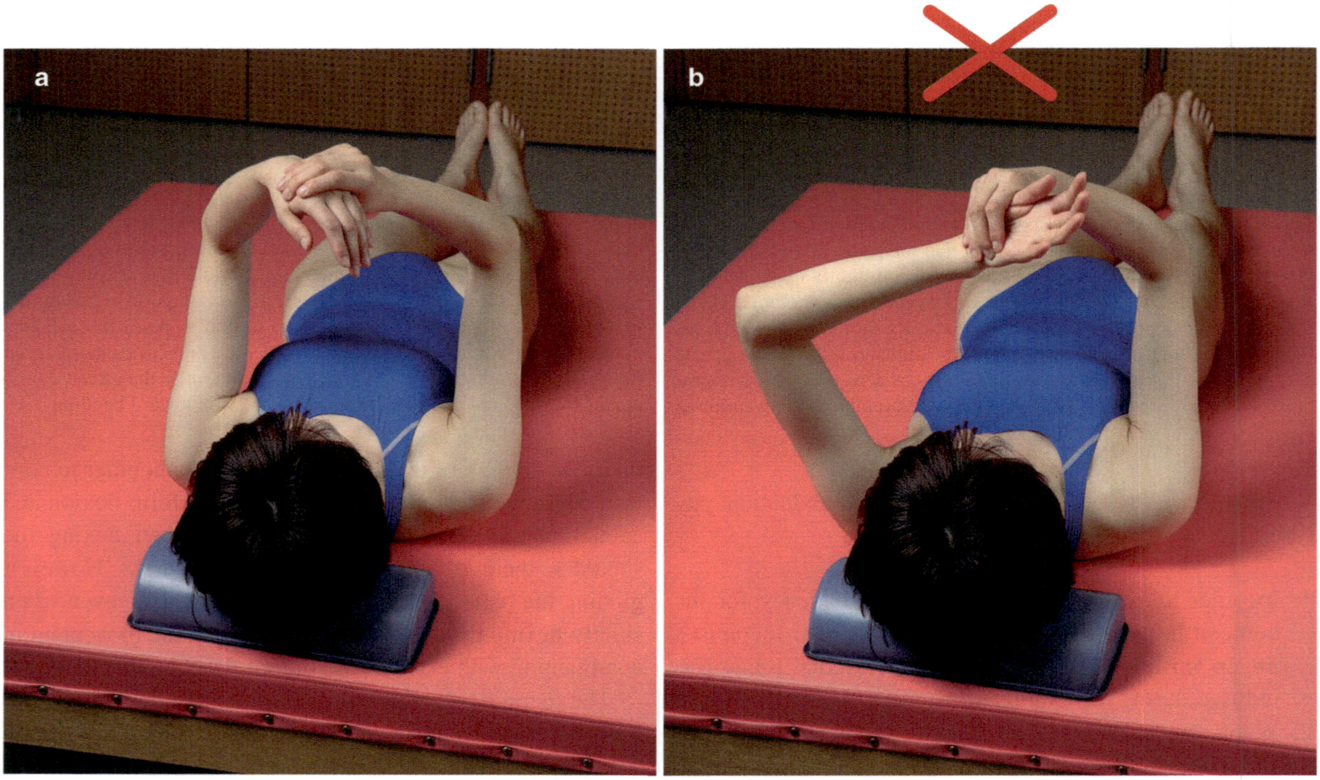

Fig. 4.6 Basic Technique: Active assistive exercise: Hold the shoulder joint in external rotation during shoulder flexion (turn the palm towards the patient's face). (**a**) In order to keep the shoulder externally rotated, hold the patient's hand so that the palm is continually facing the patient's face and instruct them: "Keep the hand position higher than the elbow (toward the head)." **Additional Tips**: If the wrist on the affected side is held and pulled during the exercise, the patient will likely experience shoulder pain. (**b**) Poor technique. When the back of the patient's hand is in front of his/her face, the shoulder can perform internal rotation, making impingement more likely to occur during shoulder joint movement. Holding the hand itself is acceptable

4.2.2 Procedure to Suppress Spasticity of the Flexor Muscles in the Fingers and Forearm (Supine Position)

During wrist flexion, gently grasp the fingers across the joint surface and pronate the forearm (Fig. 4.7). In order to avoid causing pain, it is important that the therapist performs dorsiflexion of the wrist joint slowly with the patient's fingers extended.

Pushing the patient's palm downward during dorsiflexion of the wrist can easily cause pain in the patient's shoulder (Fig. 4.8). Furthermore, pulling the patient's wrist when extending the elbow will also be painful for the patient (Fig. 4.9).

4.2.3 Repetitive Facilitative Exercise to Increase the Excitement Level of the Target Nerve Tract (Fig. 4.10)

Methods for increasing the excitement level only in the targeted nerve tract involved in the exercise include skin muscle reflex (rubbing or tapping the muscle with the therapist's finger) and stretch reflex (stretching the muscle quickly). In order to obtain the stretch reflex effectively, it is important for the therapist to use his fingertips in order to quickly manipulate the affected limb.

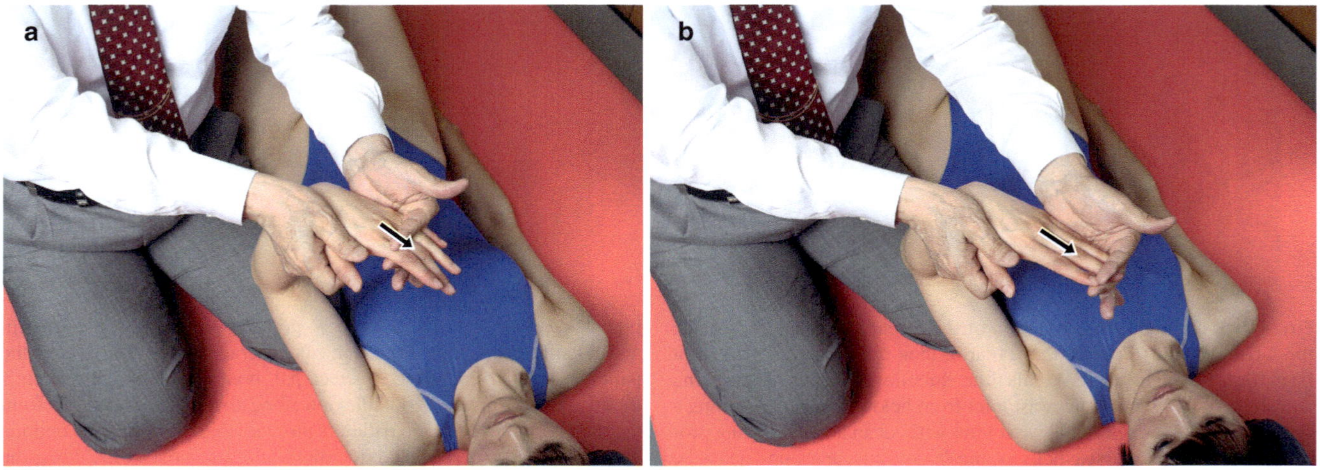

↓ Therapist moves the patient's body in this direction

○ Resistance to maintain muscle tone (just touch gently)

⇩ Stimulating (tapping or pressing)

⬇ Stimulating (rubbing)

Fig. 4.7 Basic Technique: Procedure to suppress spasticity of the flexors in the hands and fingers (in a supine position). Move the wrist joint slowly into dorsiflexion without exerting force on the fingers which are holding the patient's hand. How to hold the thumb and other fingers: (**a**) The therapist holds the patient's palm toward the patient's face, holding the thumb with one hand and the fingers with the other, then places the patient's elbows on the therapist's knee so that it reduces the load on the shoulders of the affected upper limb. (**a**) Gently pull on the finger joints so as to give space between joints. (**b**) Keeping the patient's wrist in flexion, the therapist pincers the patient's fingers between their index finger and the middle finger, then pulls them towards the patient's fingertips so as to give space between the joints.

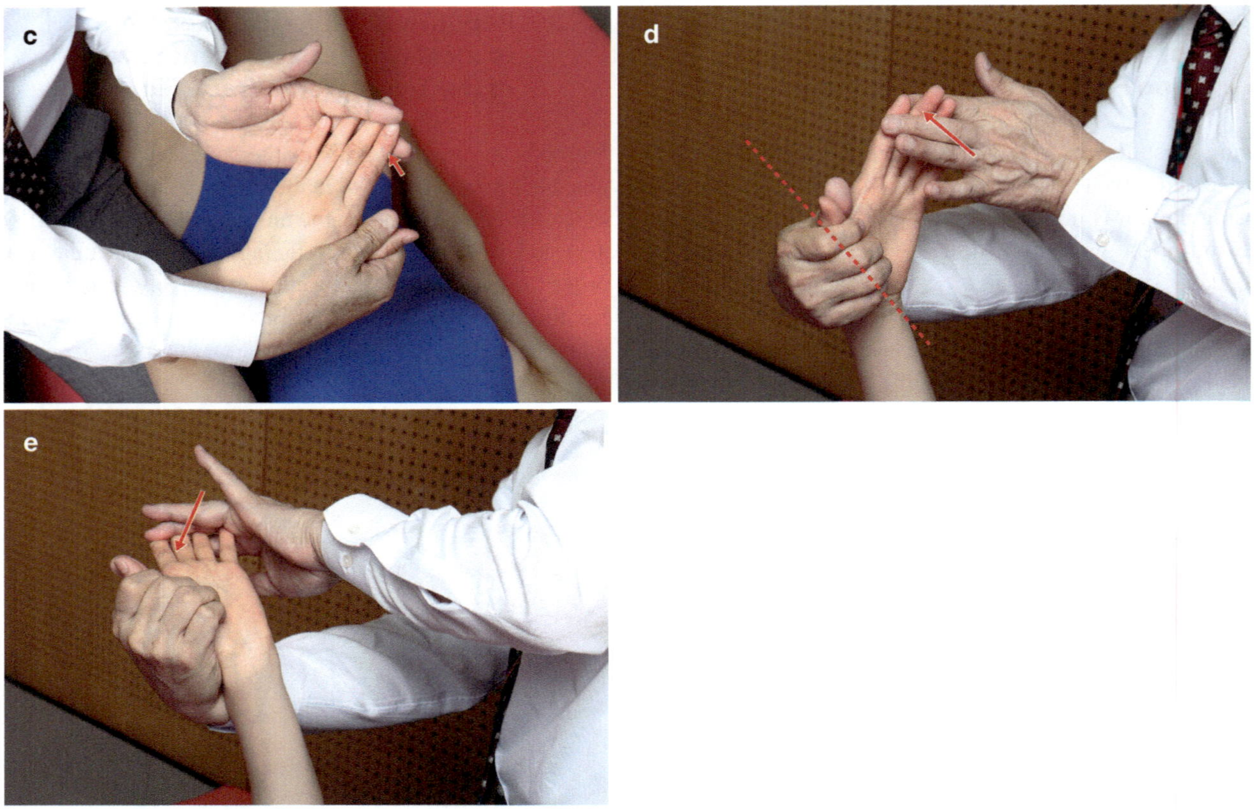

Fig. 4.7 (continued) (**c**) Pronation of the forearm and dorsiflexion of the wrist joint: (**c**) Pronate the forearm, then dorsiflex the wrist joint from the little finger's side. At this time, dorsiflexion is performed using a movement that scoops the wrist joint from below. When performing a dorsiflexion movement of the wrist joint without an accompanying pronation of the forearm, spasticity of the flexor muscle group cannot be suppressed, causing strong resistance. (**d**) The wrist is dorsiflexed, and an imaginary axis line should connect the thumb to the ulnar styloid process (*dashed line*). (**d**) In order to perform pronation of the forearm and dorsiflexion of the wrist joint, the therapist should first hold the patient's thenar eminence (metacarpal bones) below the patient's thumb, then position it in dorsiflexion along an imaginary axis line from the thenar eminence to the patient's ulnar styloid process. (**e**) Continue pronation of the forearm and dorsiflexion of the wrist. (**e**) Holding the thenar eminence (metacarpal bones), perform passive wrist dorsiflexion and finger extension exercises while pulling the patient's palm upward

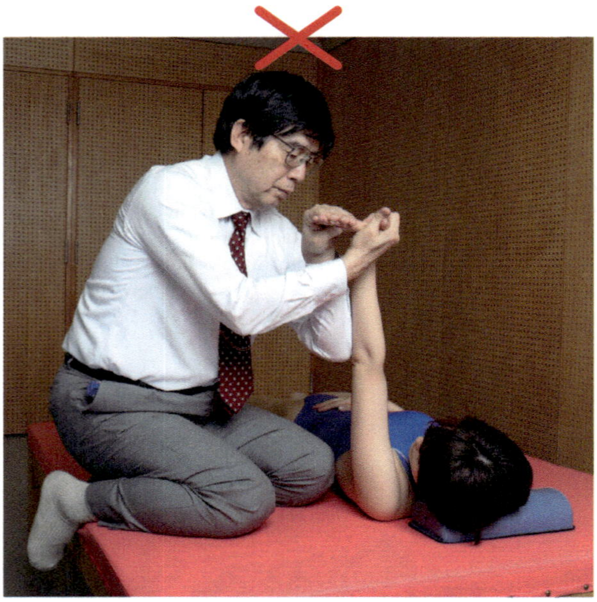

Fig. 4.8 Basic Technique: Poor technique: The procedure easily causes pain in the patient's shoulder joint (humeral head toward the back). If the palm is pushed downward while dorsiflexing the wrist joint, the humeral head will be pushed behind the shoulder joint and cause pain

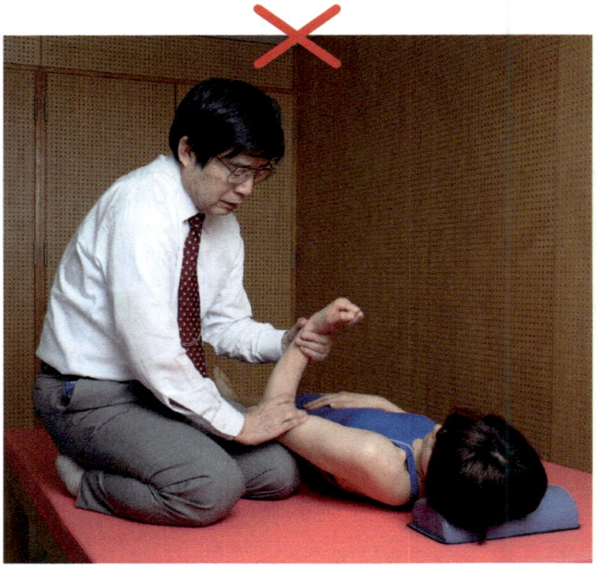

Fig. 4.9 Basic Technique: Poor technique: The procedure can easily causes pain in the patient's shoulder joint (the humeral head moves forward) (supine position). By forcing extension of the elbow joint. When trying to extend the elbow joint in a patient with spasticity in the flexor muscle, pulling the wrist joint while using the elbow as a fulcrum will cause the humeral head to protrude from the shoulder joint, which will be painful for the patient

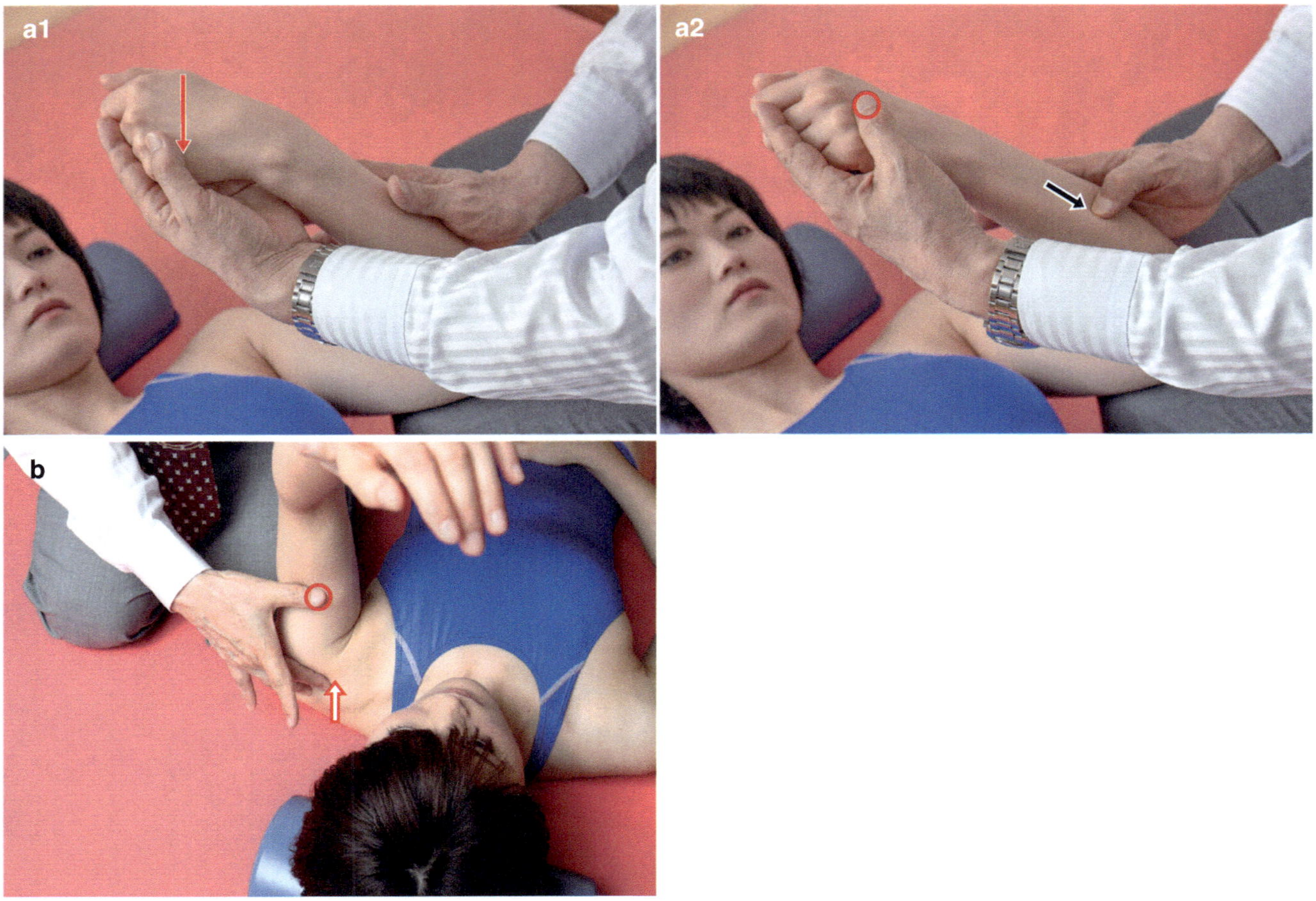

Fig. 4.10 Basic Technique: Repetitive Facilitative Exercise to increase the excitement level of the target neural pathway. In order to activate the stretching reflex, the therapist holds the affected limb lightly using his fingers so that excessive force is not applied (**a1**), and then applies light resistance (only touching) (**a2**) so that the target muscle does not lose its tension in order to maintain the effect. (**a1**) Rub in order to apply stretching stimulation to the wrist flexor muscles. (**a2**) The therapist continues by putting his thumb (dorsal), index (proximal), middle (media), and ring (distalis) finger on each patient's phalanx and flexing them to keep the tension on the extensor muscles. In order to facilitate pronation of the forearm and flexion of the wrist, the therapist quickly moves to spinate the forearm and flexes the wrist joint while scratching the forearm's extensor muscle. (**b**) Tapping: Stimulation of the anterior part of the deltoid muscle and suppression of the humeral head. (**b**) The therapist taps the anterior part of the deltoid muscle with his middle and ring fingers at the humeral head of the lower edge of the acromioclavicular joint and continues compression. In case the muscle contraction is still weak, tap firmly

4.3 How to Perform Upper Limb Exercise Therapy

Key Point

The individual treatment program depends on the severity of paralysis in each patient. However, from the stage at which synergic movement has started to appear until finger movement has recovered, the basic procedure starts with induction of shoulder flexion in the supine position. Then treatment proceeds in the order of flexion/extension of the elbow, flexion/extension of the fingers, and supination/pronation of the forearm while in a sitting position. Depending on the level of paralysis, several movements may be combined into a more complex pattern (Fig. 4.11). Instead of simply raising the affected upper limb, the movement is carried out by lifting the hand from the thigh to the forehead, which requires a modified PNF pattern involving flexion, adduction, external rotation of the shoulder, and elbow flexion (Fig. 4.12).

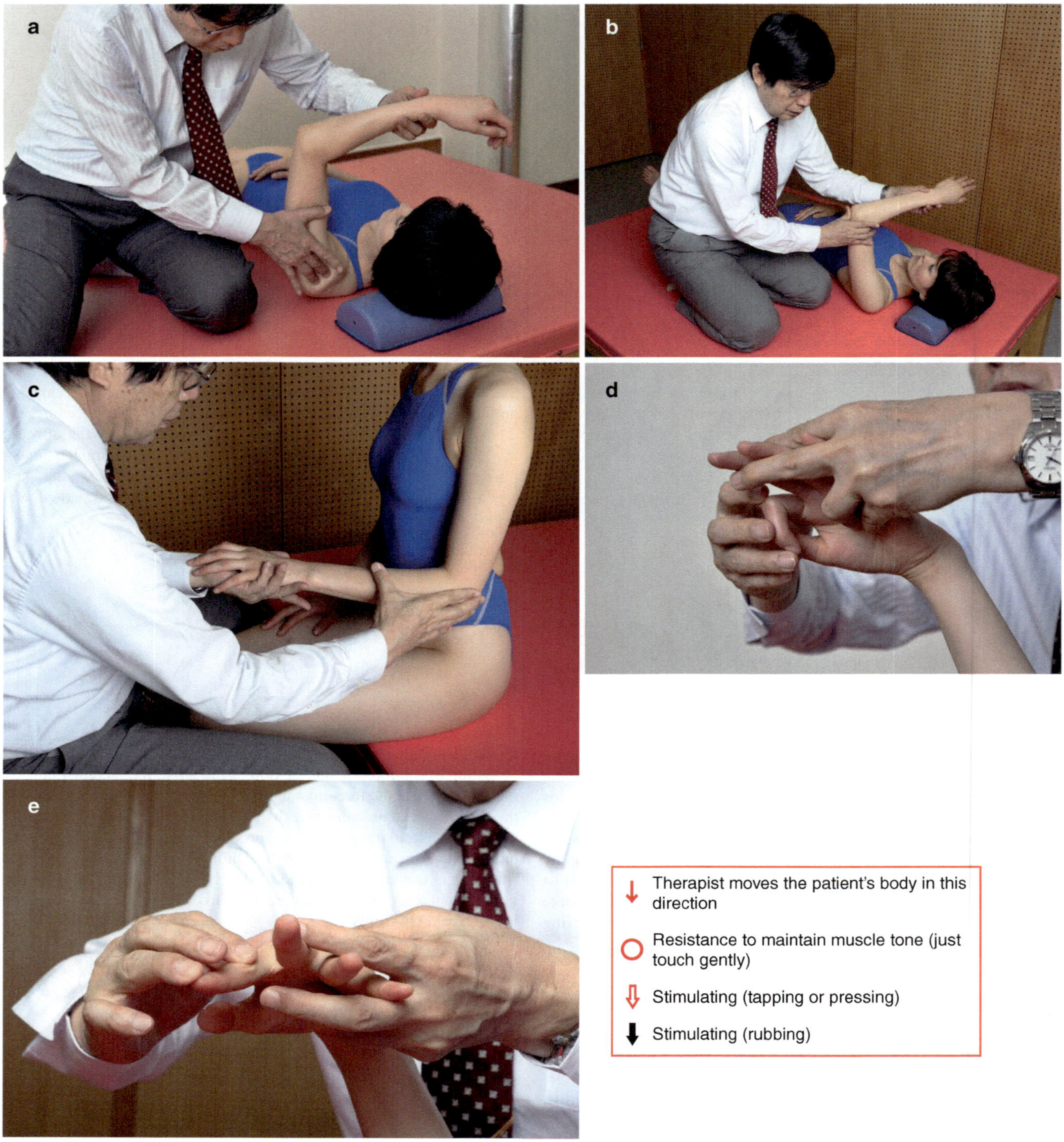

Fig. 4.11 Basic Technique: Procedure for upper extremity exercise therapy. The basic procedure for exercise therapy begins with flexion of the shoulder in the supine position, flexion/extension of the elbow, flexion/dorsiflexion of the wrist joint, flexion and extension of the fingers, and supination/pronation of the forearm in the sitting position. The figures above show (**a**) shoulder flexion, (**b**) elbow extesion, (**c**) forearm supination and (**d**) quick flexion of the fingers. In image (**d**) above, the therapist quickly performs flexion of the PIP and DIP on the patient's finger, followed by flexion of MP to induce excitement of the nerve tract when the fingers are extended. (**e**) Gentle resistance (light contact) is applied, continuing extension of the distal phalanx. The alpha-gamma linkage in the target muscle is engaged by applying gentle resistance with a light touch. Excessive resistance will make it difficult for the patient to continue extension, leading to unnecessary force on the patient

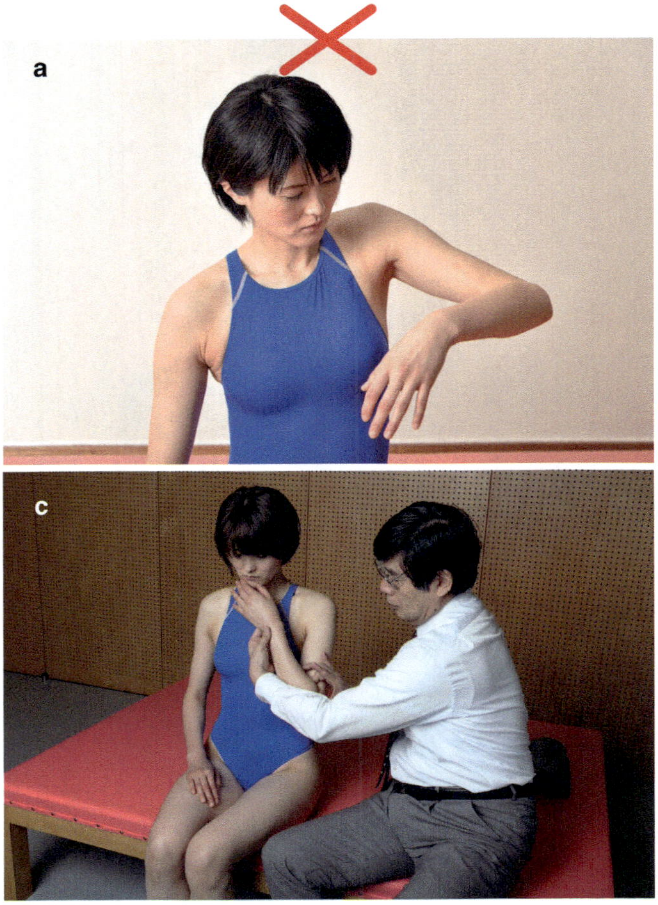

Fig. 4.12 Basic Technique: Shoulder abduction movement and shoulder pain while sitting. Poor technique: Shoulder abduction in the sitting position can often cause pain due to impingement because the deltoid muscle lifts the humeral head without sufficient contraction of the supraspinous muscle (**a**). Furthermore, the weight of the forearm will cause internal rotation of the shoulder during shoulder abduction in the sitting position (**b**). It is important to promote flexion/external rotation of the shoulder while avoiding abduction of the shoulder (**c**). (**a**) Poor technique: Shoulder abduction with the humerus lifted. (**b**) Poor technique: Shoulder abduction in an internally rotated position (**b**). This occurs during shoulder abduction in a seated position, when the muscle that externally rotates the shoulder joint is weak. This movement not only causes pain in the shoulder but also fails to improve the upper limbs useful in activities of daily living. (**c**) Shoulder flexion/external rotation (by Repetitive Facilitative Exercise). (**c**) One of the training objectives is to improve the patient's ability to lift her hand closer to her face while in a sitting position. If this is achieved, the upper limbs will be more useful in activities of daily living

4.4 Repetitive Facilitative Exercise for the Shoulder

The stimuli applied during the repetitive facilitative exercise are stretching the target muscle (quick extension) and stimulating the local skin area (tapping, rubbing). For example, for flexion of the shoulder joint, abduction of the scapula (pulling outward quickly) can be applied, and for flexion and extension of the shoulder joints, stimulation (tapping, compression) of the skin overlying the deltoid muscle can be used.

Key Points

1. Abduction/external rotation and adduction/internal rotation of the scapula, together with flexion and extension of the elbow (while lying on the side) (Fig. 4.13)

   In patients with severe paralysis, voluntary movement of the shoulder girdle is limited and facilitation for movement of the upper limb is difficult.

   For the first step, the therapist should apply a facilitation technique that focuses on improving voluntary movement of the shoulder girdle.

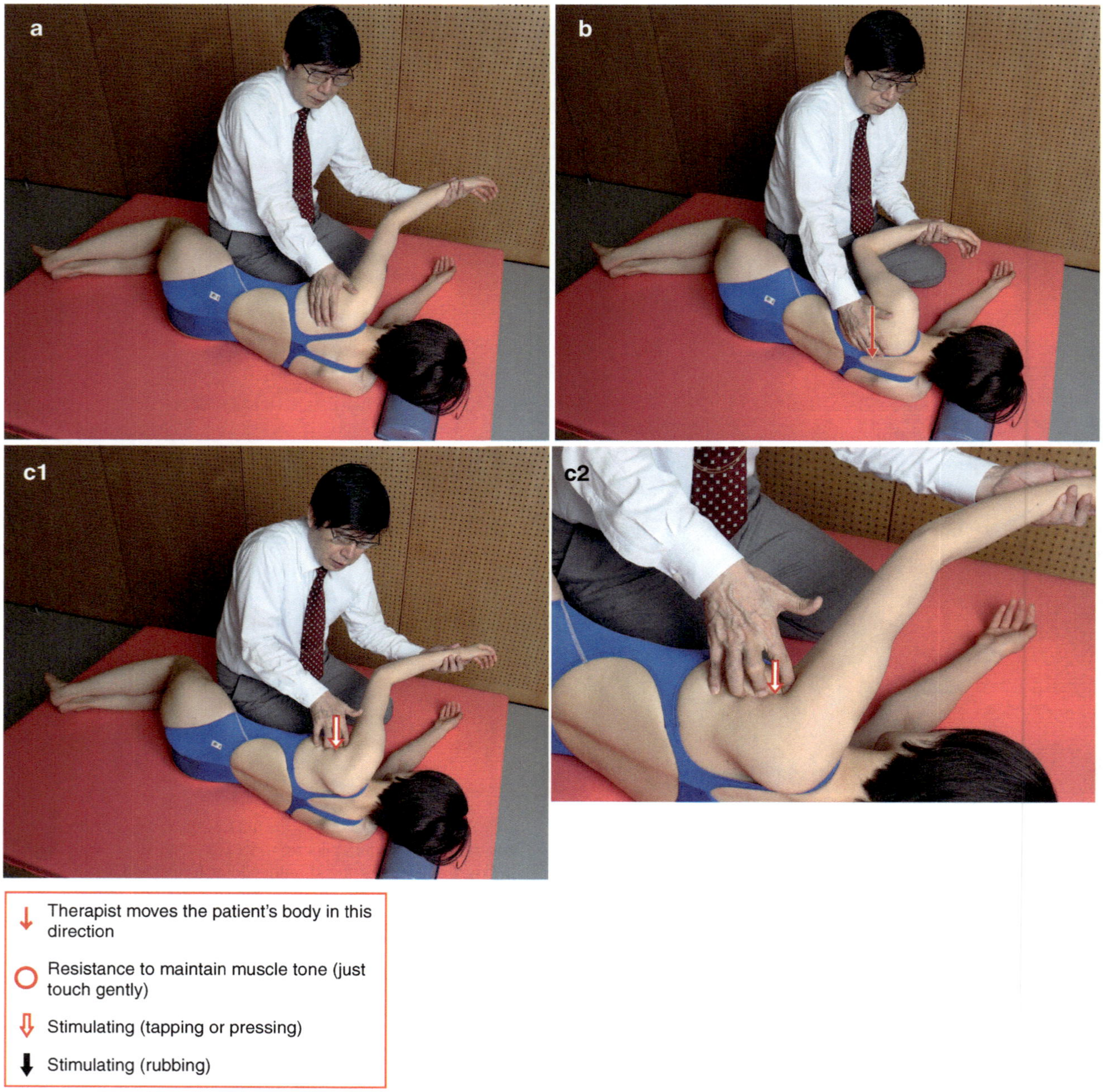

Fig. 4.13 Let's Try: Abduction/external rotation, adduction/internal rotation of the scapula, and flexion and extension of the elbow (while lying on the side) (Video 4.1). (**a**) Supporting the shoulder girdle and upper limb, passive abduction and external rotation of the shoulder girdle along with extension of the elbow. (**a**) The therapist should sit in front of the patient who is lying on the unaffected side with the affected side up. Place one hand on the patient's scapula on the affected side, gently hold the patient's forearm (near the wrist joint) with the other hand and turn the palm of the patient's affected hand towards the patient's face (the shoulder will be externally rotated). The therapist should instruct the patient: "Stretch your elbow out" or "Push your shoulder forward" and perform passive abduction and external rotation of the scapula along with elbow extension. (**b**) Passive adduction/internal rotation of shoulder girdle and flexion of the elbow. (**b**) The therapist should instruct the patient, saying, "Bend your elbow" or "Pull your shoulder backward," and then passively perform adduction/internal rotation of the scapula and flexion of the elbow. (**c1**) Facilitation of abduction, external rotation of shoulder girdle and elbow extension. (**c1**) After repeating (**a**) and (**b**), voluntary movement will appear. While the patient's elbow is flexed, the therapist should instruct the patient again, saying, "Extend your elbow" or "Push your shoulder forward." In order to induce extension of the elbow, tap the outer edge of the scapula using the middle finger, ring finger and little finger while simultaneously tapping the armpit with the index finger. While the patient's elbow is extended and the therapist is tapping the inner edge of the scapula with all four fingers, the therapist should instruct the patient: "Bend your elbow" or "Pull your shoulder backward." (**c2**) Focus on the hand. (**c2**) Stimulate the outer edge of the scapula with the tips of the little finger, ring finger and middle finger, while pressing the triceps brachii tendon at the axilla with the tip of the index finger.

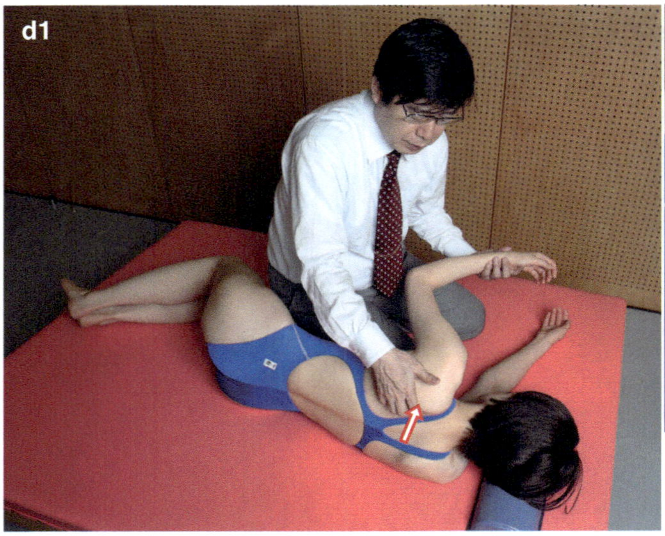

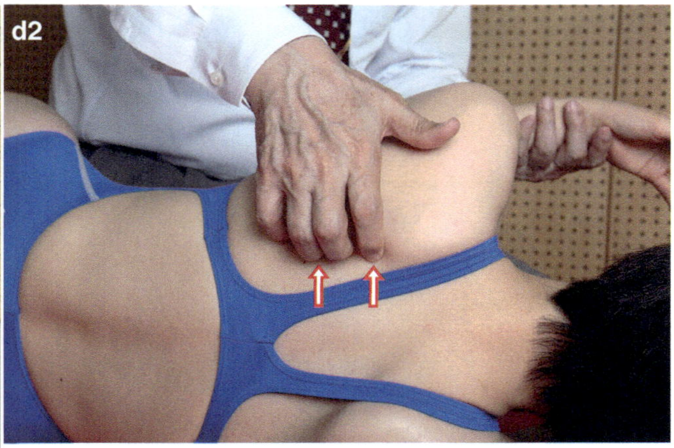

Fig. 4.13 (continued) (**d1**) Facilitating adduction, internal rotation of the shoulder girdle and elbow flexion. (**d1**) The therapist moves his hand, which is placed on the outer edge of the scapula, to the inner edge of the scapula. While keeping the elbow in an extended position, the therapist instructs the patient to "bend your elbows" or "pull your shoulders backwards." The therapist then taps the inner edge of the scapula with his index and little finger. (**d2**) Focus on the hand. (**d2**) Press against the inner edge of the scapula with the tips of the little finger and index finger

2. Flexion and abduction of the shoulder joint (supine position) (Fig. 4.14)

   It is easier to induce flexion/abduction of the shoulder joint in the supine position because this movement is similar to flexor muscle synergy. This exercise is the basic pattern used for patients with severe paralysis or spasticity of the large pectoral muscle in order to overcome flexor synergy.

   First, the therapist should tell the patient, "Gradually increase (your) effort, please. Don't try to move quickly" and "It will be difficult to control muscle contraction at the beginning, but gradually try to relax the muscles, please."

   After inducing flexion/abduction of the shoulder joint, the therapist should change his hand position so that the area to be stimulated is the anterior deltoid muscle; then flexion of the shoulder joint can be more easily induced.
3. Flexion of the shoulder joint (supine position) (Fig. 4.15)

   Since flexion of the shoulder joint is required in order to restore practical functionality of the upper limb, it is also necessary to use a light resistance method during the Repetitive Facilitative Exercise.

   Repeatedly perform joint flexion from 90 to 180° to improve the shoulder joint range (while rubbing the joint cartilage and stretching the joint capsule).
4. Flexion with resistance in the shoulder joint (supine position) (Fig. 4.16)

   When upper limb training is performed in the sitting position or when the patient is actually using the upper limbs, muscle strength must be sufficient to flex the shoulder joint while supporting the weight of the upper limb itself. The therapist should add light resistance to the upper arm to provide focused resistance exercise for the shoulder joint. The combined use of sustained electrical stimulation will enhance this effect.
5. Horizontal adduction and abduction of the shoulder joint (supine position) (Fig. 4.17)

   Horizontal adduction of the shoulder joint is relatively easy. If there is spasticity of the pectoral muscle during this movement and the elbow drops down, adjust the arm position and move the stimulated area to the anterior deltoid.

   Horizontal abduction of the shoulder joint is performed while stimulating the posterior deltoid.
6. Shoulder flexion/adduction, elbow flexion ⇆ Shoulder extension/abduction, elbow extension (supine position) (Fig. 4.18)

   After flexion and adduction of the shoulder joint have each been performed separately and synergistic movement has been controlled, proceed to a combined flexion and adduction movement using Repetitive Facilitative Exercise to promote flexion/adduction and extension/abduction movements while keeping the elbow of the affected limb at 90°.
7. External rotation of the shoulder joint (Fig. 4.19)

   When the patient is trying to move her hand to her face or forehead while in a sitting position, the shoulder will rotate inward and her forearm will drop in front of her chest if the movement is performed without external rotation of the shoulder joint. In order to restore functionality of the upper limb, it is important for the patient to gain the ability to maintain external rotation of the shoulder by contracting the infraspinatus muscle. This can be practiced in the supine position.
8. External rotation/flexion of the shoulder joint (supine position) (Fig. 4.20)

   Passing the hand in front of the face while lifting the arm places less burden on the shoulder. Shoulder pain can be prevented by performing the movement with sufficient repetitions in the supine position before carrying out training in lifting objects with the patient in a sitting position.

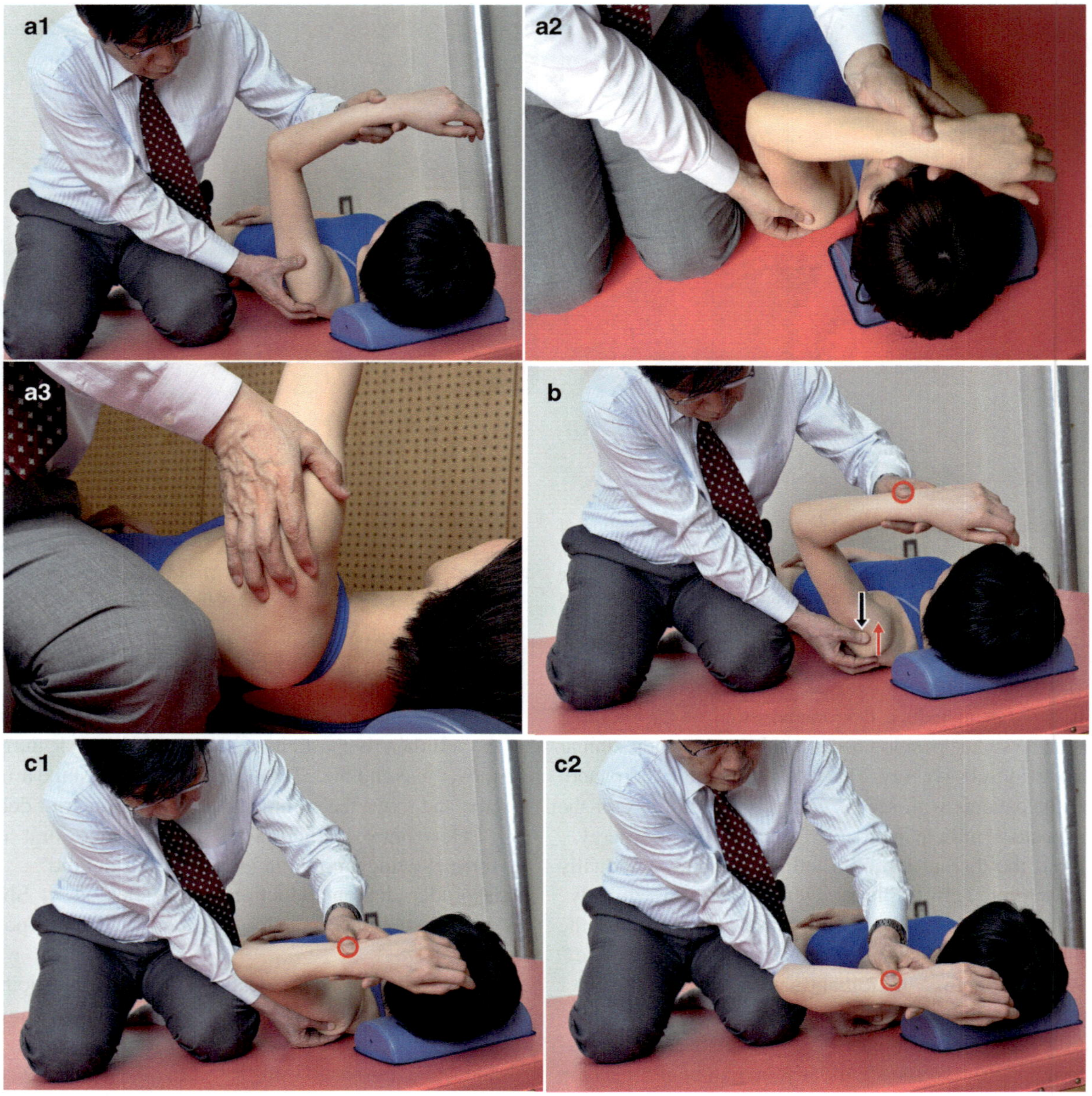

Fig. 4.14 Let's Try: Loosening the pectoralis major muscle by manipulating the shoulder girdle with passive flexion and abduction of the shoulder joint (supine position) (Video 4.2). While performing flexion/abduction exercise of the shoulder, instruct the patient in the initial direction of movement, saying "Let's try raising your arm diagonally upwards." The main causes of restricted shoulder joint range can be roughly divided into the following categories: (1) shortening of the joint capsule (contracture); (2) spasticity and shortening (contracture) of the pectoralis major muscle; and (3) pain due to impingement, etc. In many cases, more than one cause is involved. For patients with severe contracture of the pectoralis major muscle during active and passive shoulder flexion, try to manipulate (vibratory stimulation, flexion/abduction exercise of the shoulder) the muscle in order to relax the spasm. For patients in whom the shoulder flexion range does not improve after manipulation of the muscle, it should be considered whether or not the shoulder joint capsule has become markedly contracted. (**a1**) Supporting the shoulder girdle and the humerus. (**a1**) The therapist should lightly hold the patient's forearm with the fingertips of one hand and perform passive shoulder flexion (90°) with external rotation. With the other hand's thumb, the therapist supports the area between the center and posterior of the deltoid muscle, while the other four fingers support the shoulder girdle from below and the patient's forearm is stabilized against the therapist's thigh. (**a2**) Procedure seen from above. (**a2**) Externally rotate the shoulder joint and flex the shoulder joint so that the tension of the pectoralis major muscle relaxes a little. If the pectoralis major muscles are tense, the range of motion during Repetitive Facilitative Exercise will be limited. (**a3**) Procedure seen from a posterior position. (**a3**) Place the fingertips beneath the shoulder blade near the shoulder joint in order to quickly move the scapula outward. The therapist does not have to position his fingers at the inner edge of the scapula. (**b**) Lifting the shoulder blade and rubbing the deltoid muscle. (**b**) The therapist should brace his forearm against his own thigh and quickly raise the patient's shoulder blade with his fingertips. Then the therapist uses his thumb to rub the middle and posterior part of the deltoid muscle. (**c1**) Continuing flexion/abduction of the shoulder. (**c1**) Support the shoulder in an externally rotated position and continue the shoulder flexion/abduction movement. If the patient tries to move the shoulder while the upper limb is unsupported or if the therapist pushes against the shoulder joint, the patient will experience pain in the shoulder. (**c2**) Continuing shoulder flexion/abduction. (**c2**) Continue shoulder flexion/abduction with the shoulder in an externally rotated position while the therapist presses against the humeral head. If the pectoralis major muscle feels tight and the patient experiences pain, it would be better to limit further shoulder exercise during the current session

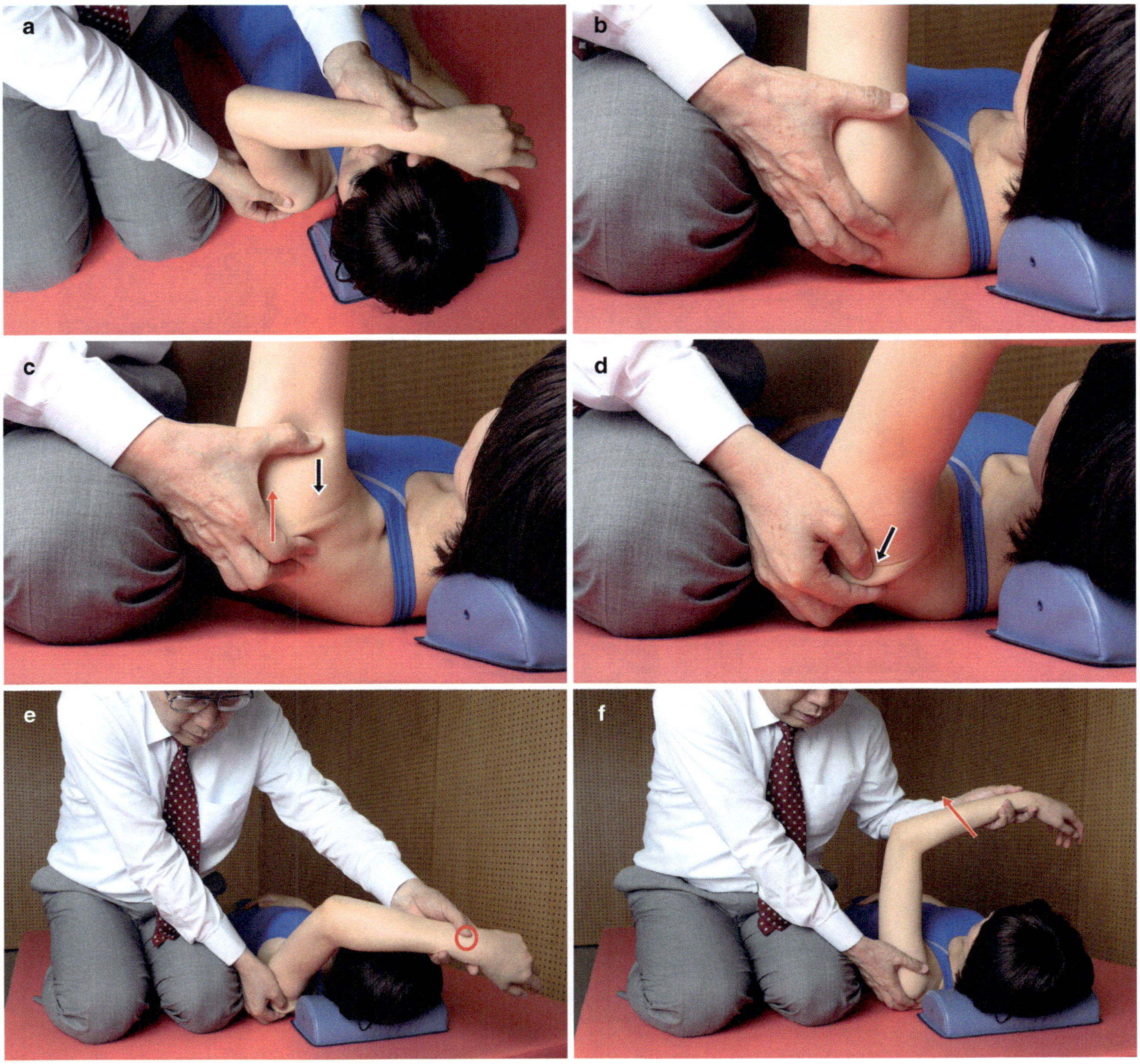

Fig. 4.15 Let's Try: Flexion of shoulder joint (supine position) (Video 4.3). (**a**) Procedure viewed from above. (**a**) The therapist should hold the patient's forearm gently with one hand and keep the shoulder joint in an externally rotated position. The therapist's other hand supports the shoulder and the scapula, and the therapist's forearm is braced against his own thigh. (**b**) Support of the scapula and upper arm. (**b**) The therapist positions his thumb so that he can stimulate the front part of the deltoid muscle, while the other four fingers are placed beneath the shoulder so as that they can be used for passive abduction (pulling up) of the scapula. (**c**) The therapist uses his four fingers (index finger to little finger) to quickly guide abduction of the scapula (pulling up). (**c**) The therapist quickly raises the shoulder blade 1 cm, while telling the patient, "Please raise your arm." If the therapist has not braced his forearm against his own thigh, it will be impossible to perform the movement rapidly and an adequate effect will not be obtained. (**d**) Rub the front part of the deltoid muscle, while pressing against the humeral head to maintain it in the correct position. (**d**) After raising the scapula, the therapist uses his thumb to rub the anterior part of the deltoid. At the same time, press against the humeral head to keep it aligned in the shoulder joint and prevent impingement. The therapist should quickly pull the scapula upward, then immediately stimulate the front part of the deltoid muscle in order to induce shoulder flexion. (**e**) Stimulate the front part of the deltoid muscle and continue flexion of the shoulder. (**e**) Rub the anterior of the deltoid muscle from the distal to proximal side. The thumb and thenar eminence should be placed between the shoulder joint and humeral head to help suppress the tendency of the humeral head should be pushed upward. When flexing the shoulder, the therapist should use his other hand to support the patient's forearm, thus preventing the affected upper limb from falling downward. (**f**) When flexion of the shoulder stops, return the upper arm to the starting position. (**f**) When the muscle contraction stops, quickly return from the flexed position to the 90° position, then return to (**b**). This method of rapidly returning the upper limbs back to the starting position of the exercise increases the excitement level in the nerve tract by using stretch reflex to stimulate the muscles, which facilitates shoulder flexion. Impingement tends to occur during active and passive movement of the shoulder when the shoulder angle is returning to 60°

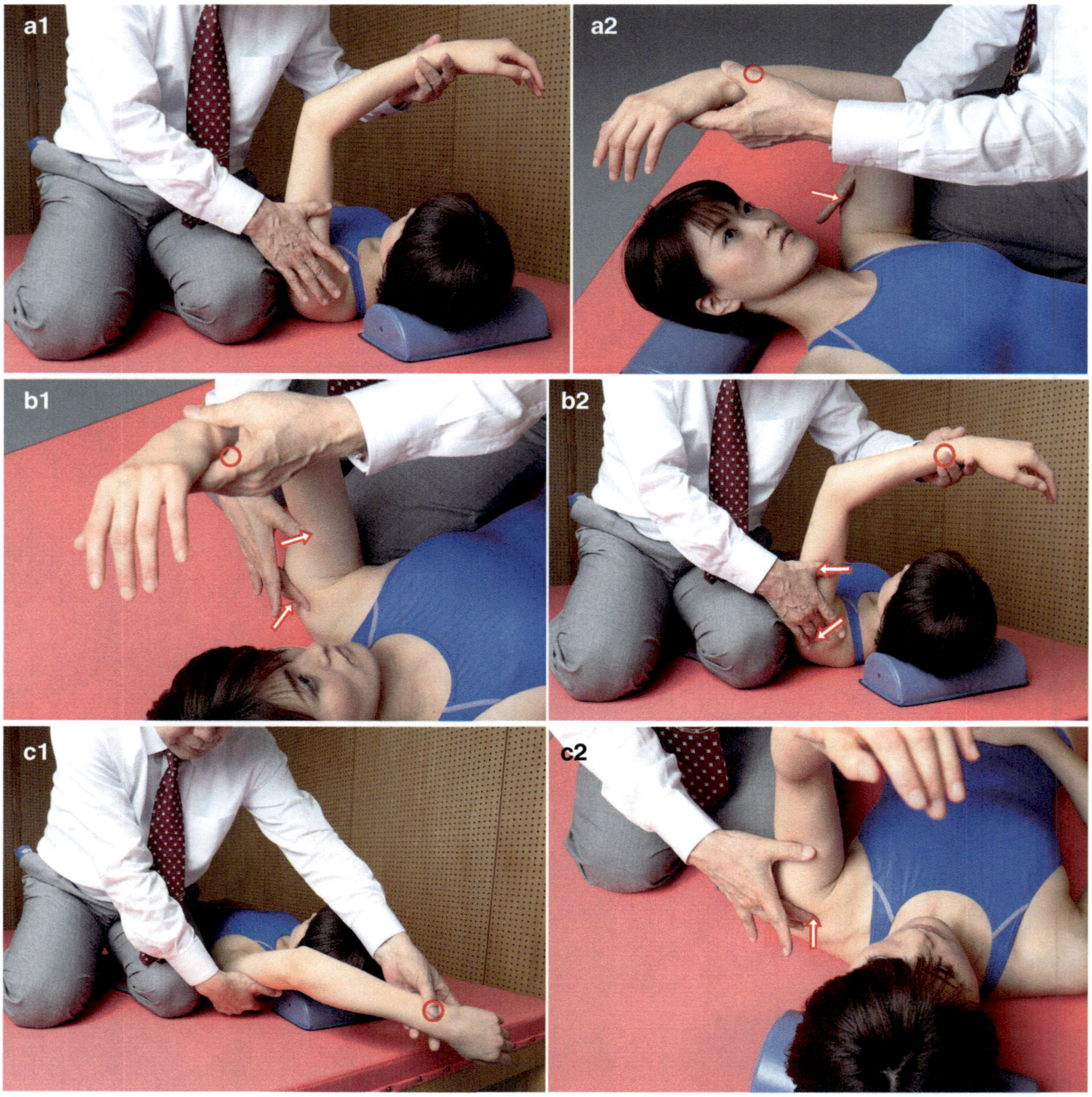

Fig. 4.16 Let's Try: Flexion of the shoulder with resistance (supine position). Do not typically apply resistance when flexing the shoulder. This movement is for a patient without sufficient muscle strength in the shoulder to manipulate objects even when able to lift their hand to the top of their head while sitting. The therapist should use his thumb to apply resistance to the patient's upper arm. Do not perform this process with the hand that is holding the patient's forearm. When resistance is applied with the hand that is holding the patient's forearm, the patient is likely to experience pain because the humeral head moves upward and pushes into the patient's shoulder joint. In patients with mild paralysis, however, there is unlikely to be a problem if resistance is applied to the forearm because the humerus will only push up slightly. (**a1**) Supporting the shoulder/upper arm, and the forearm. (**a1**) The therapist places the middle and ring fingers of one hand on the top of the humeral head and lower edge of the acromion, and places his thumb on the upper arm. The therapist should then grasp the forearm lightly with the fingertips of his other hand and hold the shoulder at 90° flexion in an externally rotated position. (**a2**) Procedure viewed from the opposite side. (**a2**) The photo shows how to use the middle and ring fingers to tap the front part of the deltoid muscle (above the humeral head at the lower margin of the shoulder). **Tip**: Tapping with the middle and ring fingers allows the fingers to touch the lower edge of the shoulder joint. Therefore, this technique is more effective than using the index and middle fingers to tap the deltoid. (**b1**) Stimulation is applied to the front part of the deltoid muscle. (**b1**) The therapist taps the anterior of the deltoid muscle while pressing the humeral head with the middle and ring fingers to form an axle. At the same time, the therapist should instruct the patient, "Please raise your upper limb." While tapping, the therapist presses his thumb against the upper arm. The patient's forearm is held in an externally rotated position. (**b2**) Another view of the procedure from the same side. (**b2**) The therapist continues stimulating the front part of the deltoid muscles while holding the humeral head with the middle and ring fingers, also adding light resistance with the thumb without disturbing the shoulder movement. (**c1**) Applying resistance with the thumb while supporting the upper limb. (**c1**) Even if the angle of the shoulder exceeds 90° at flexion, the therapist should use his thumb to add light resistance, and the range of motion for the shoulder flexion should be increased as much as possible. When flexion of the shoulder reaches its limit, quickly reverse the movement toward the extension direction (shoulder flexion at 90°). Return to (**a**). (**c2**) Another view of the procedure from above. (**c2**) The therapist should press against the humeral head with the middle and ring fingers while pressing the thumb against the upper limb

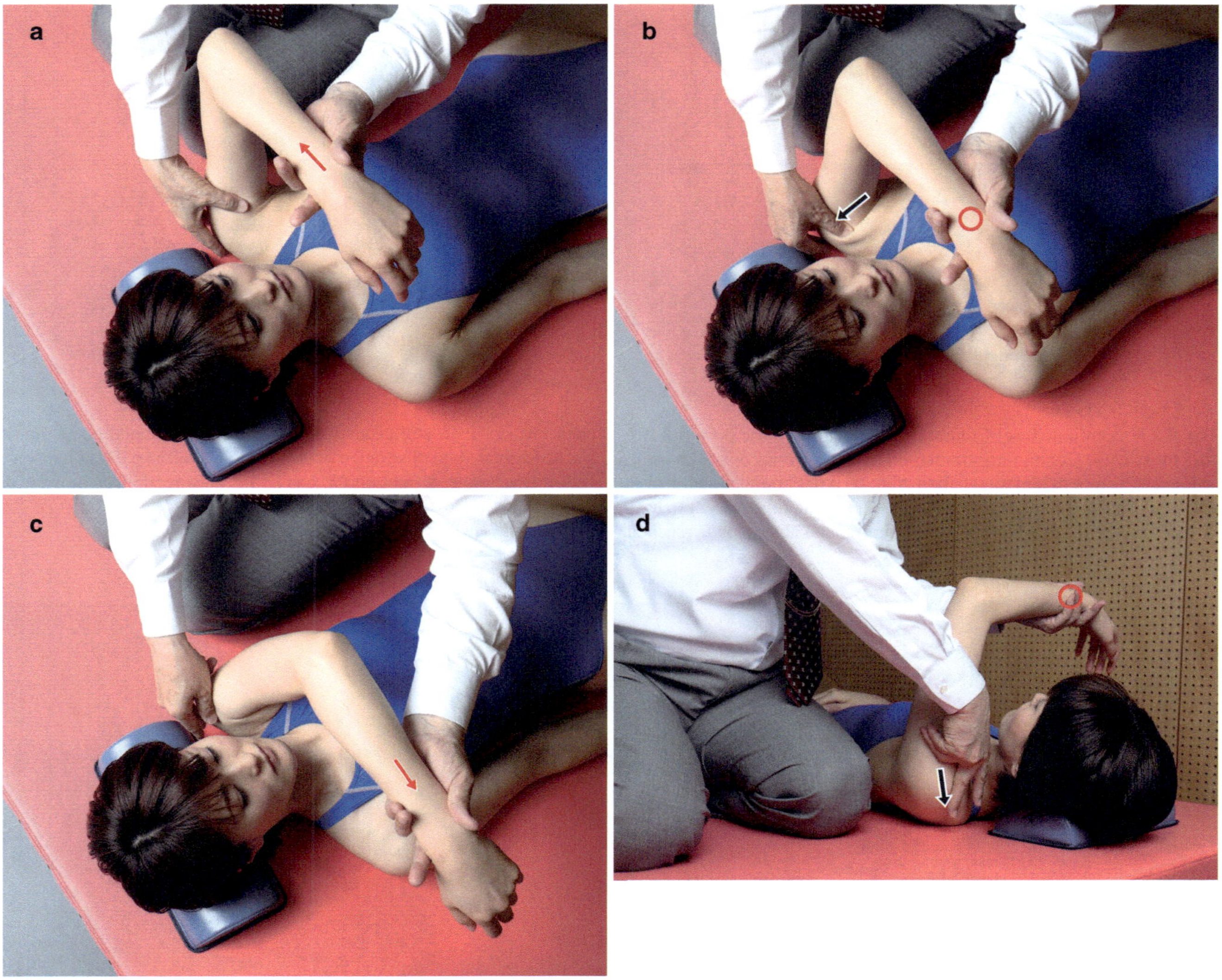

Fig. 4.17 Let's Try: Horizontal adduction and abduction of shoulder joint (supine position). This approach is used to treat patients with joint pain caused by restricted shoulder flexion and severe contracture of the shoulder joint. Repeated horizontal adduction/abduction of the shoulder joint will loosen the muscle group around the shoulder joint without causing shoulder pain, then Repetitive Facilitative Exercises becomes easier to perform. In patients with shoulder pain, this movement will also reduce stiffness in the shoulders. A description of the procedure for horizontal adduction (**a**–**c**) and horizontal abduction (**d**–**f**) of the shoulder joint is as follows: (**a**) Support the shoulder girdle and stretch the front part of the deltoid muscle. (**a**) The therapist's thumb should be placed on the anterior deltoid muscle with the other fingers on the posterior of the deltoid muscle (as if holding the patient's shoulder). Lightly support the patient's forearm with the fingertips of the other hand while holding the elbow at flexion 90° across the patient's chest. Using the hand to support the forearm, the therapist should quickly pull about 10° in the direction of shoulder abduction. Since passive adduction is easier to induce, this movement is performed first, and then passive abduction is performed subsequently. Thereafter, Repetitive Facilitative Exercise is repeated with a constant rhythm while repeating the verbal instructions: "To the inside"/"To the outside." (**b**) Stimulating the anterior deltoid muscle. (**b**) While instructing the patient to move "To the inside," the therapist should rub the patient's anterior deltoid muscle with his thumb. (**c**) Stretching the posterior deltoid muscle. (**c**) When the limit of horizontal adduction of the shoulder joint is reached, the therapist should quickly pull the forearm 10° in the direction of horizontal adduction while instructing the patient to move: "To the outside." (**d**) Stimulation to the posterior deltoid muscle. (**d**) Using the ring finger, the therapist should rub the patient's posterior deltoid muscle while simultaneously instructing the patient to move: "To the outside."

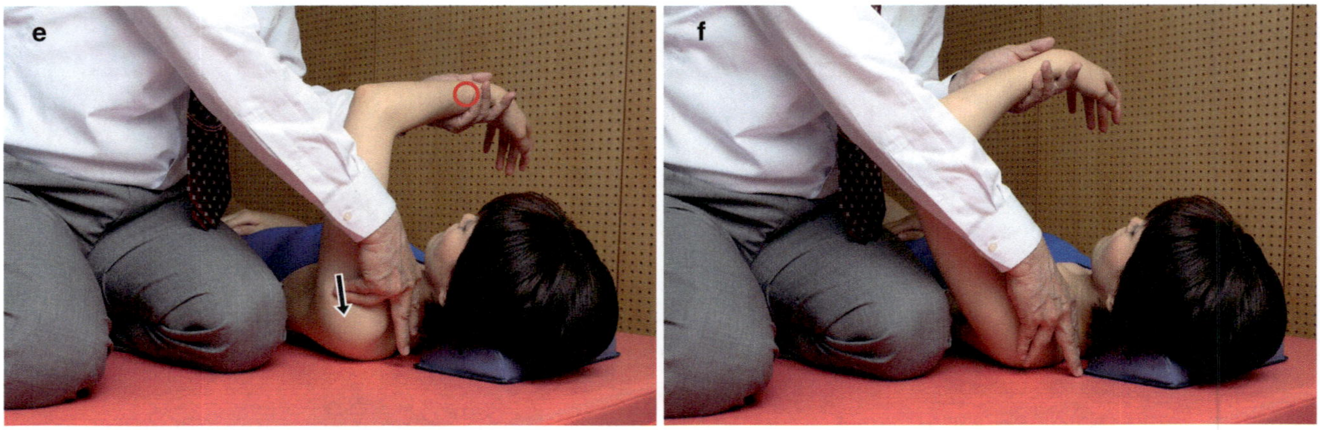

Fig. 4.17 (continued) (**e**) Continuing stimulation of the posterior deltoid muscle. (**e**) Continuing stimulation of the posterior deltoid muscle, leading to a smooth horizontal abduction. (**f**) After completing the horizontal abduction movement, change the direction of movement to horizontal adduction. (**f**) When the limit of horizontal abduction is reached, return to (**a**). Pain will occur when shifting to horizontal adduction if the horizontal abduction range of motion was too large; if that happens, reduce the range of motion by changing the direction of movement from abduction to adduction earlier in the next cycle

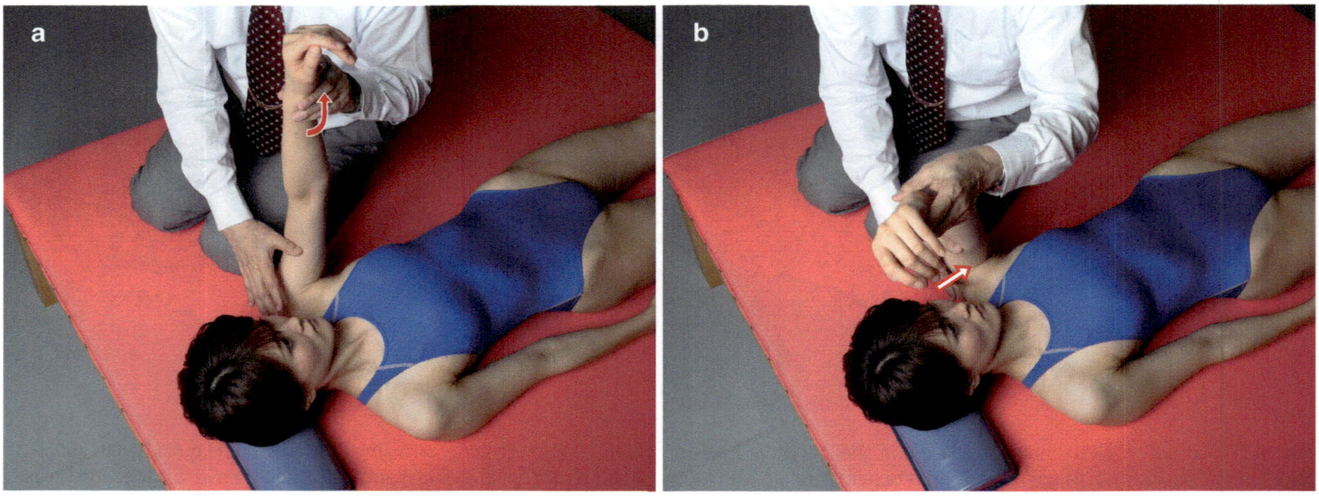

Fig. 4.18 Let's Try: Shoulder joint flexion/adduction, elbow flexion, and shoulder joint extension/abduction, elbow extension (supine position) (Video 4.4). (**a**) Support the patient's upper arm and forearm. (**a**) The therapist positions his middle finger and ring finger over the patient's humeral head with his thumb on the patient's upper arm on the affected side. Hold the forearm lightly with the fingertips of one hand and place the shoulder in a neutral position. Instruct the patient, saying, "Move your hand up above your head," and quickly pronate the forearm of the affected limb to induce supination of the forearm. (**b**) Stimulation of the internal edge of the anterior deltoid muscle. (**b**) Tap the internal edge of the anterior deltoid muscle with the middle finger and ring finger. Guide the patient's forearm to position her elbow in front of her face with her hand above her head reaching toward the opposite side.

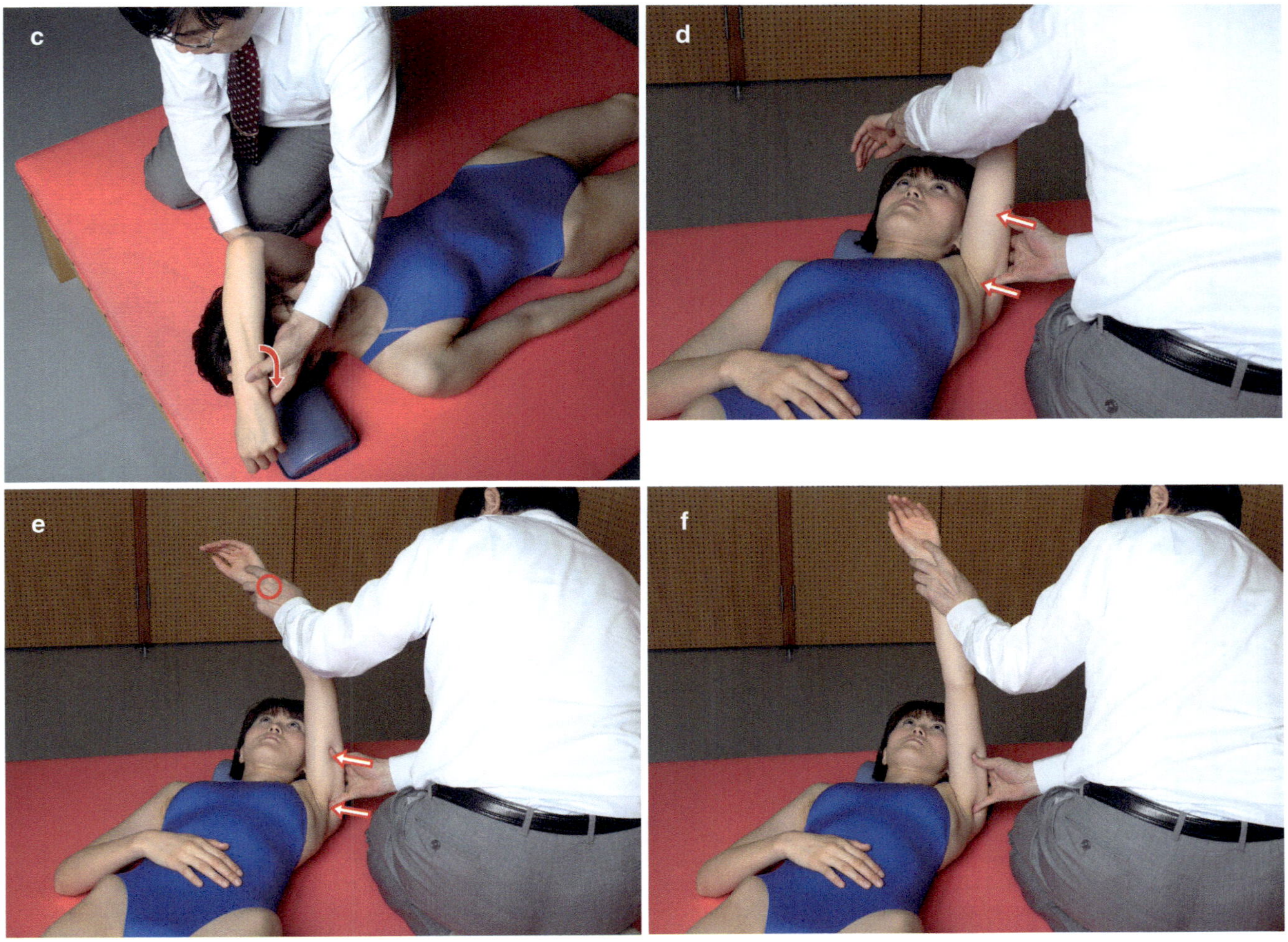

Fig. 4.18 (continued) (**c**) Switching the movement from shoulder flexion/adduction with elbow flexion/supination to shoulder extension/abduction with elbow extension/pronation. (**c**) When the shoulder flexion/adduction/forearm supination reaches its limit, the therapist places his thumb on the patient's triceps with his little finger on the triceps near the axilla; then the therapist should quickly turn the patient's forearm supination, which induces pronation. (**d**) View of the repetitive facilitation exercise for shoulder extension/abduction and elbow extension from below. (**d**) Using the thumb to push against the patient's triceps brachii muscle and resting the little finger against the tendon to induce extension and abduction of the shoulder. (**e**) Shoulder extension/abduction, elbow extension and pronation of the forearm. (**e**) While continuously stimulating the upper arm, the therapist should hold the patient's forearm without disturbing the elbow extension or pronation of the forearm. (**f**) Switching the direction of movement. (**f**) When the limits of shoulder extension/abduction, elbow extension, and forearm pronation movements are reached, change the direction of movement

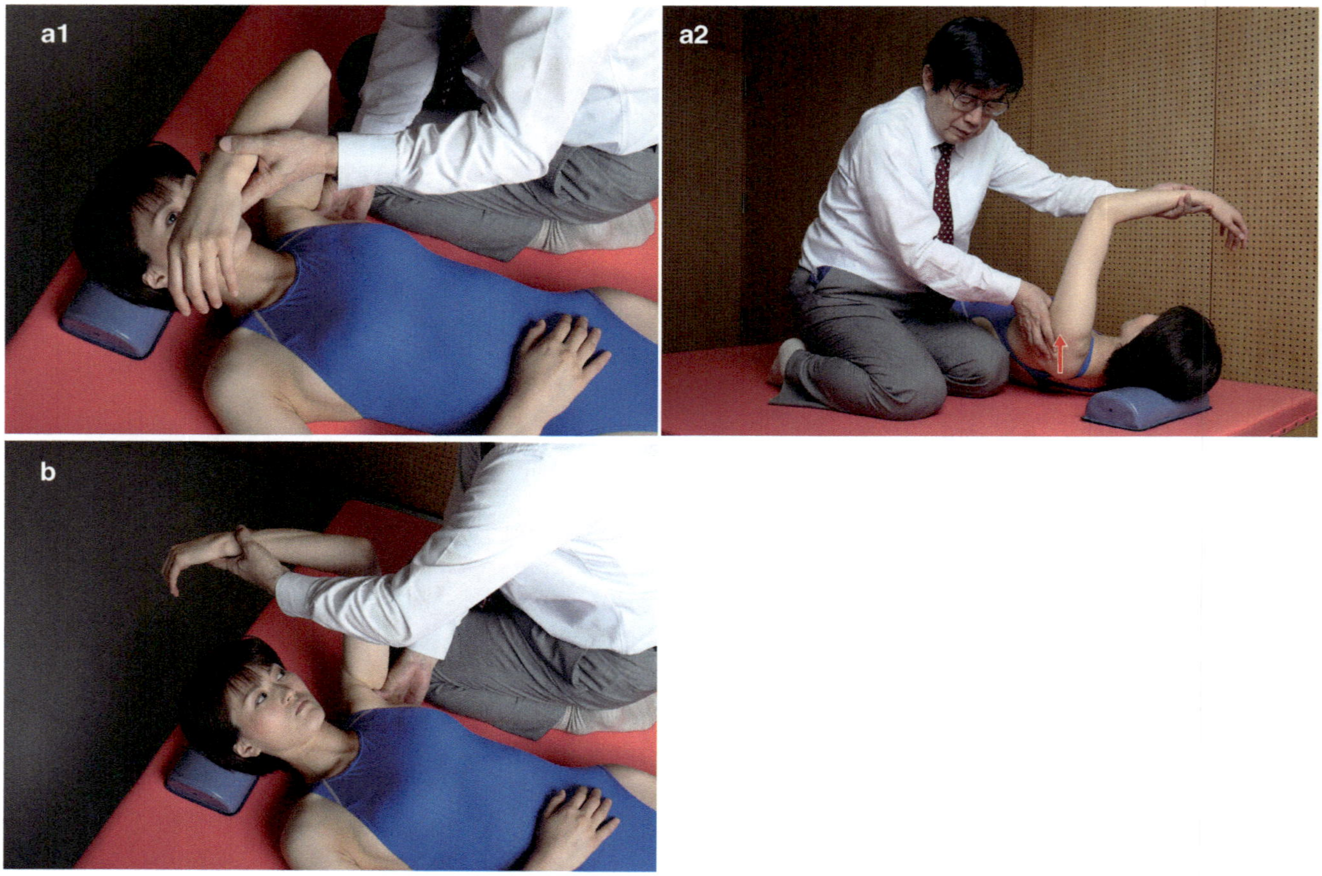

Fig. 4.19 Let's Try: External rotation of the shoulder joint (supine position) (Video 4.5). In the sitting position, it is necessary to externally rotate the shoulder to bring the patient's hand closer to her face or forehead. In the supine position, the therapist should stimulate the infraspinatus muscle to facilitate external rotation. (**a1**) Position for repetitive facilitation exercise for external rotation of the shoulder. (**a1**) The therapist sits beside the patient's shoulder. While lightly holding the patient's forearm on the affected side, the therapist should move the forearm to an intermediate position with a slight internal rotation of the shoulder. The therapist's other hand should be placed on the dorsal shoulder with the little finger, ring finger, and middle finger on the infraspinatus muscle (below the scapular spine), and the index finger on the point closing the shoulder joint and lowering the edge of the scapular spine near the axilla. (**a2**) The position of each finger during Repetitive Facilitative Exercise for shoulder external rotation is shown from an oblique direction behind the patient. (**a2**) The therapist's hand position is shown from the back. The therapist's middle finger, ring finger, and little finger are placed on the lower part of the infraspinatus and the index finger is placed close to the axilla. The thumb is placed on the axilla to permit easier manipulation of the other fingers. (**b**) Continue facilitating external rotation of the shoulder. (**b**) While stimulating the patient's infraspinatus muscle with his fingers, the therapist should instruct the patient: "Turn your hand outward." During Repetitive Facilitative Exercise, the axis of external rotation is the upper arm, which needs to remain vertical so that the patient can see the direction of movement. The therapist's hand should support the forearm without disturbing the external rotation motion during this procedure. When the movement reaches its limit, quickly return to (**a1**) and repeat the entire sequence

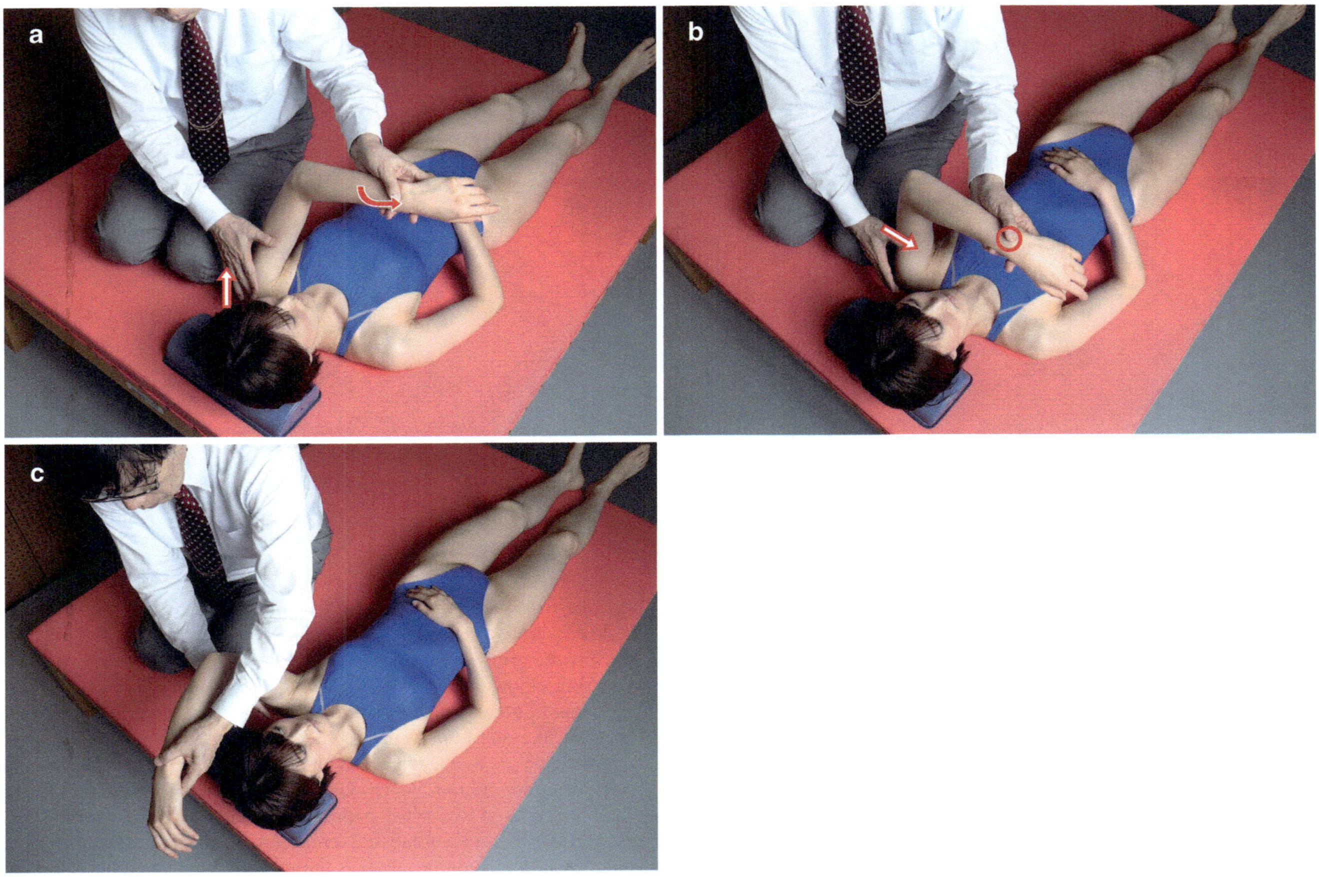

Fig. 4.20 Let's Try: External rotation and flexion of the shoulder joint (supine position) (Video 4.6). In order for the patient to become able to raise her hand above her head while in a sitting position, it is necessary to externally rotate/flex the shoulder. To achieve this, Repetitive Facilitative Exercise for supination of the forearm and flexion/external rotation of the shoulder should be performed in a supine position. (**a**) The position of the upper limb at the start of Repetitive Facilitative Exercise for external shoulder rotation/flexion. (**a**) The therapist should sit beside the patient's shoulder. Lightly hold and grip the patient's forearm in a slightly pronated position, then place the patient's shoulder in a slightly internally rotated position. In order to perform Repetitive Facilitative Exercise for shoulder flexion, the thumb on the therapist's other hand is placed on the patient's upper arm while the therapist's middle finger and ring finger are placed on the anterior deltoid muscle. While quickly pulling the forearm inward to induce internal rotation of the shoulder, perform Repetitive Facilitative Exercise for shoulder flexion by tapping the deltoid muscle with the other hand. (**b**) Using his thumb on the patient's upper arm to increase external shoulder rotation. (**b**) In order to induce shoulder flexion accompanied by sufficient external shoulder rotation, begin applying resistance when the range of the shoulder flexion reaches 90°. When external rotation of the shoulder occurs, reduce the resistance pressure and increase shoulder flexion/external rotation. (**a**) The end range position of shoulder flexion and external rotation. (**c**) When inducing shoulder flexion and external rotation, extending the elbow increases the load on the shoulder joint, which may cause pain. Therefore, the elbow must remain in a slightly flexed position. When the movement reaches its limit, return to (**a**)

4.5 Repetitive Facilitative Exercise for Upper Limb

When flexion/adduction and extension/abduction of the shoulder can be repeated rhythmically, the next, more advanced exercise becomes possible. Elbow flexion and extension are combined into the movement required to manipulate an object while in a sitting position. Repetitive Facilitative Exercise is performed for flexion/adduction of the shoulder combined with flexion of the elbow, extension/abduction of the shoulder, and extension of the elbow.

Key Points

1. Shoulder flexion, adduction, external rotation (in the supine position using the PNF pattern), and elbow flexion/extension (Fig. 4.21)

   Prior to training in practical use of the upper limb in a sitting position, begin with Repetitive Facilitative Exercise for shoulder flexion/adduction → extension/abduction in a supine position; next, perform repetitive facilitative exercise for the combination of flexion/adduction of the shoulder with flexion of the elbow → extension/abduction of shoulder with elbow extension. Emphasize that the patient should touch the opposite ear

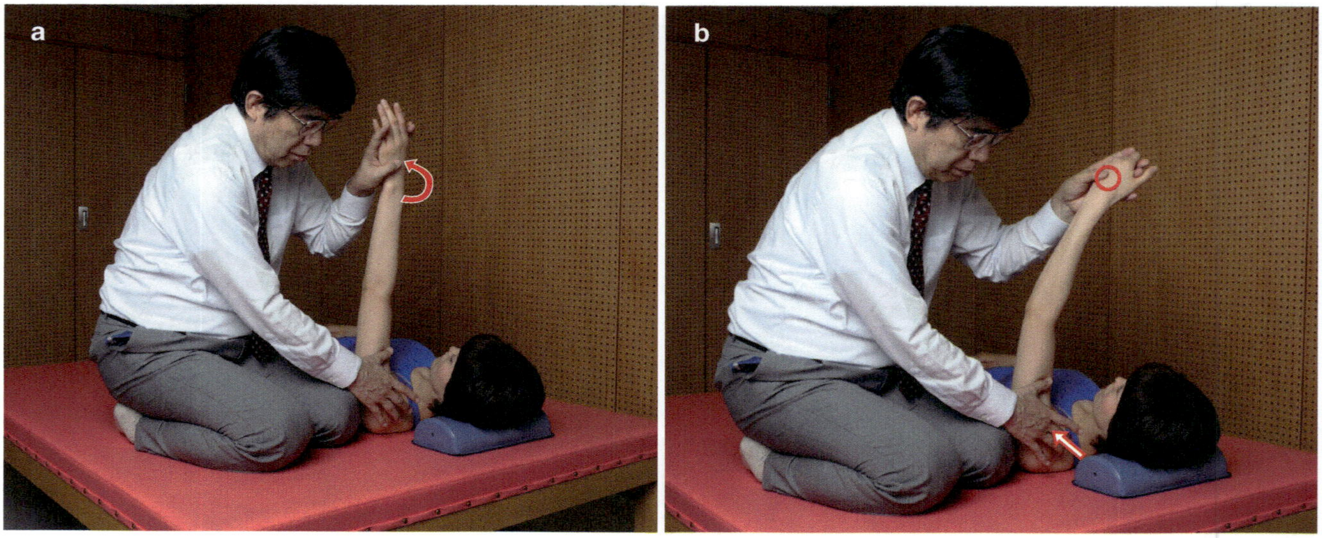

Fig. 4.21 Let's Try: Shoulder flexion/adduction/external rotation, elbow flexion, finger flexion. Shoulder extension/abduction/internal rotation, elbow extension, finger extension (supine position: PNF pattern) (Videos 4.7 and 4.8). (**a**) Shoulder and hand support. (**a**) The therapist grasps the proximal part of the palm of the affected limb using the thumb, ring finger, and little finger, while the index finger and middle finger provide support for the patient's fingers. At the beginning of the exercise, position the shoulder (flexion 90°) in abduction with internal rotation and place the forearm in a neutral position. Then the therapist places the thumb of his other hand on the patient's upper arm with the middle finger and ring finger on the medial anterior deltoid muscle (over the humeral head). Instruct the patient to "Lift upward," then turn the forearm inward and internally rotate the shoulder in order to induce external rotation of the forearm and external rotation of the shoulder. In order for the patient to thoroughly understand this complex movement pattern before beginning Repetitive Facilitative Exercise, repeatedly perform a model for the exercise by guiding the patient's upper limb while providing a clear set of instructions: "Grasp my hand lightly; now let's lift the arm above your head and tap your head on the opposite side. Your elbow comes in front of your face." The limb position at the beginning of the exercise should be a decreased shoulder angle flexing 90° to 0°. In order to prevent shoulder impingement, press the humeral head accordingly and tap the anterior deltoid muscle. (**b**) Stimulation to the anterior deltoid muscle. (**b**) Using the middle finger and ring finger of the other hand to tap the patient's medial anterior deltoid muscle, perform Repetitive Facilitative Exercise for flexion and adduction of the shoulder. If flexion of the elbow is limited, quickly induce supination of the forearm as well as external rotation of the shoulder by quickly rotating both the forearm and shoulder internally, then return to (**a**), then perform (**b**).

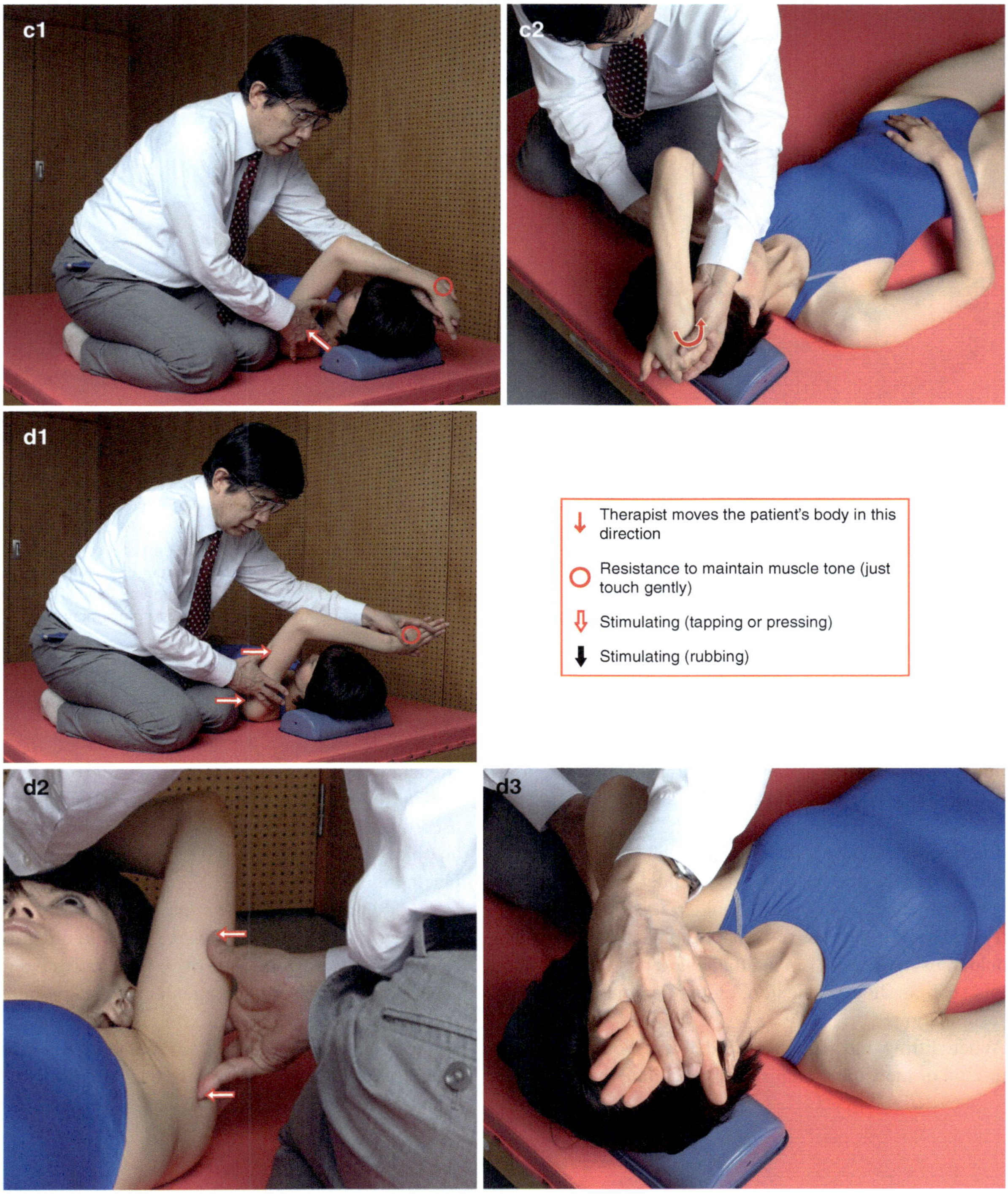

Fig. 4.21 (continued) (**c1**) Limits of flexion adduction of the shoulder, flexion of the elbow, finger flexion. (**c1**) When the patient's elbow on the affected side is in front of her face, instruct the patient to lift her hand over her head, saying, "Let's bend your elbow more in order to touch the opposite ear." (**c2**) Switch to shoulder extension/abduction/internal rotation, elbow extension, and finger extension. (**c2**) When the movement reaches its limit, instruct the patient: "Lower your hand diagonally while extending your fingers." In order to promote pronation of the forearm, dorsiflexion of the wrist joint, and extension of the fingers, the therapist should quickly press the lateral dorsal area of the patient's hand by the thumb with which he is holding the patient's fingers and palm, resulting in movement of the forearm supination and wrist flexion. (**d1**) Facilitation of shoulder extension/abduction and elbow extension from finger extension. (**d1**) When the patient's fingers start to extend, the therapist should move his own little finger to the lateral axillary region and position his thumb on the triceps brachial muscle, then gently press these spots in order to facilitate both shoulder extension/abduction and elbow extension. The direction of motion is guided using light resistance (touching) so that extension, abduction, and internal rotation of the shoulder continue smoothly. (**d2**) The stimulation points and the respective positions of the thumb and little finger are shown. (**d2**) When the therapist moves the fingers he used to promote shoulder flexion to the positions necessary for shoulder extension/abduction, the thumb on the anterior deltoid muscle is used as the axis while the little finger is rotated to the lateral axillary region; then the thumb is moved to the triceps. (**d3**) Procedure for hand and finger extension and pronation of the forearm. (**d3**) The therapist should hold the patient's affected hand between his index finger and middle finger. As the therapist's index finger touches the dorsum of the patient's hand and fingers, extension of the patient's fingers, dorsiflexion of the wrist joint, and pronation of the forearm are induced.

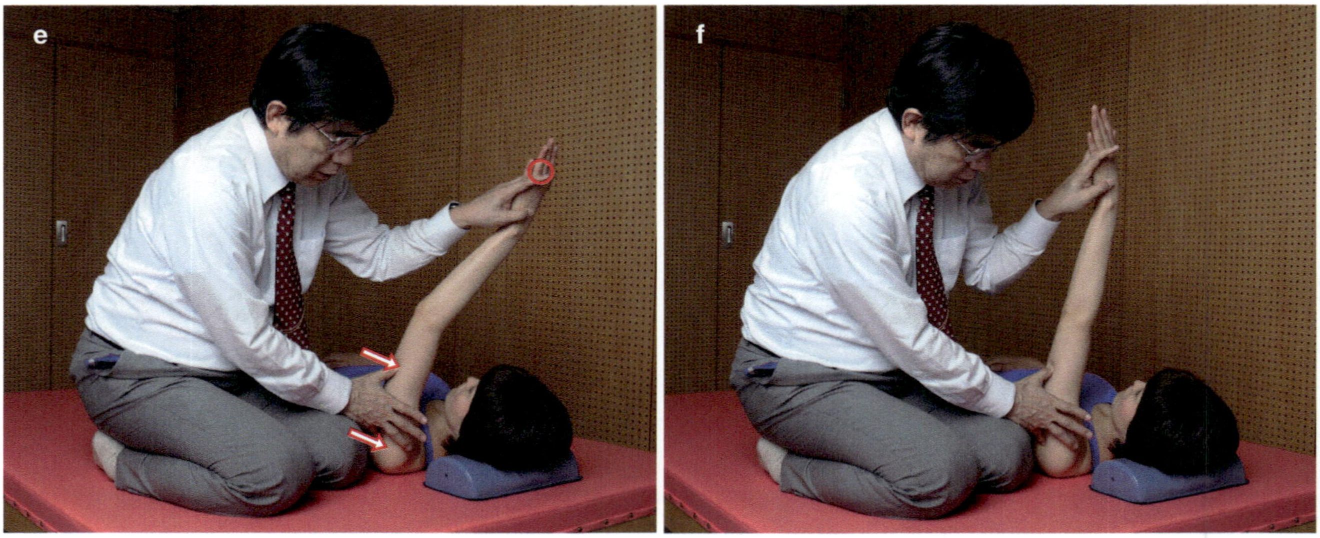

Fig. 4.21 (continued) (**e**) Continue stretching, abducting, internally rotating the shoulder, and extending the elbow. (**e**) Continue Repetitive Facilitative Exercises for extension/abduction of the shoulder and extension of the elbow, applying light resistance with the middle finger to the patient's palm (very light touch) and with the thumb and little finger of the other hand to the patient's upper arm. (**f**) Finishing extension/abduction/internal rotation of the shoulder, and elbow extension. (**f**) When the movements for extension/abduction/internal rotation of the shoulder and extension of the elbow reach their limits, return to (**a**). At the end of the exercise, try to reduce the flexion angle of the shoulder to zero, laying the arm beside the waist at the recovery level of shoulder muscle function. As the range of the shoulder extension increases, sufficient pressure should be applied to the humeral head after tapping the anterior deltoid muscle in order to prevent impingement

or top of the head with the affected hand while the shoulder is in flexion, adduction, and external rotation.

2. Shoulder flexion, elbow flexion, forearm pronation (in the sitting position lifting hands from thigh → to chin, to forehead stroke hair) (Figs. 4.22 and 4.23)

For training in the practical use of the upper limb in a sitting position, it is necessary to practice lifting the hand closer to the face and head. Since shoulder pain tends to occur during abduction of the shoulder, it is important to carry out this lifting movement without abducting the shoulder, using Repetitive Facilitative Exercise for movements such as shoulder flexion, elbow flexion, and forearm supination.

When lifting the patient's hand from their thigh to their chin or face, the therapist must facilitate the supination of the forearm by sufficiently ensuring the patient's palm is turned towards their own face.

3. Flexion of the shoulder and extension of the elbow (Reaching upwards) (Figs. 4.23 and 4.24)

This type of upper limb movement pattern is often used during activities of daily living, such as picking up an object on the table, carrying it to another location, and placing it down again.

Good shoulder joint function is necessary to support the upper limb with the elbow extended. In order to avoid causing pain during the Repetitive Facilitative Exercise, it is necessary to prevent impingement of the shoulder joint. Applying continuous electric stimulation and/or vibration stimulus in combination with Repetitive Facilitative Exercise can effectively improve the outcome.

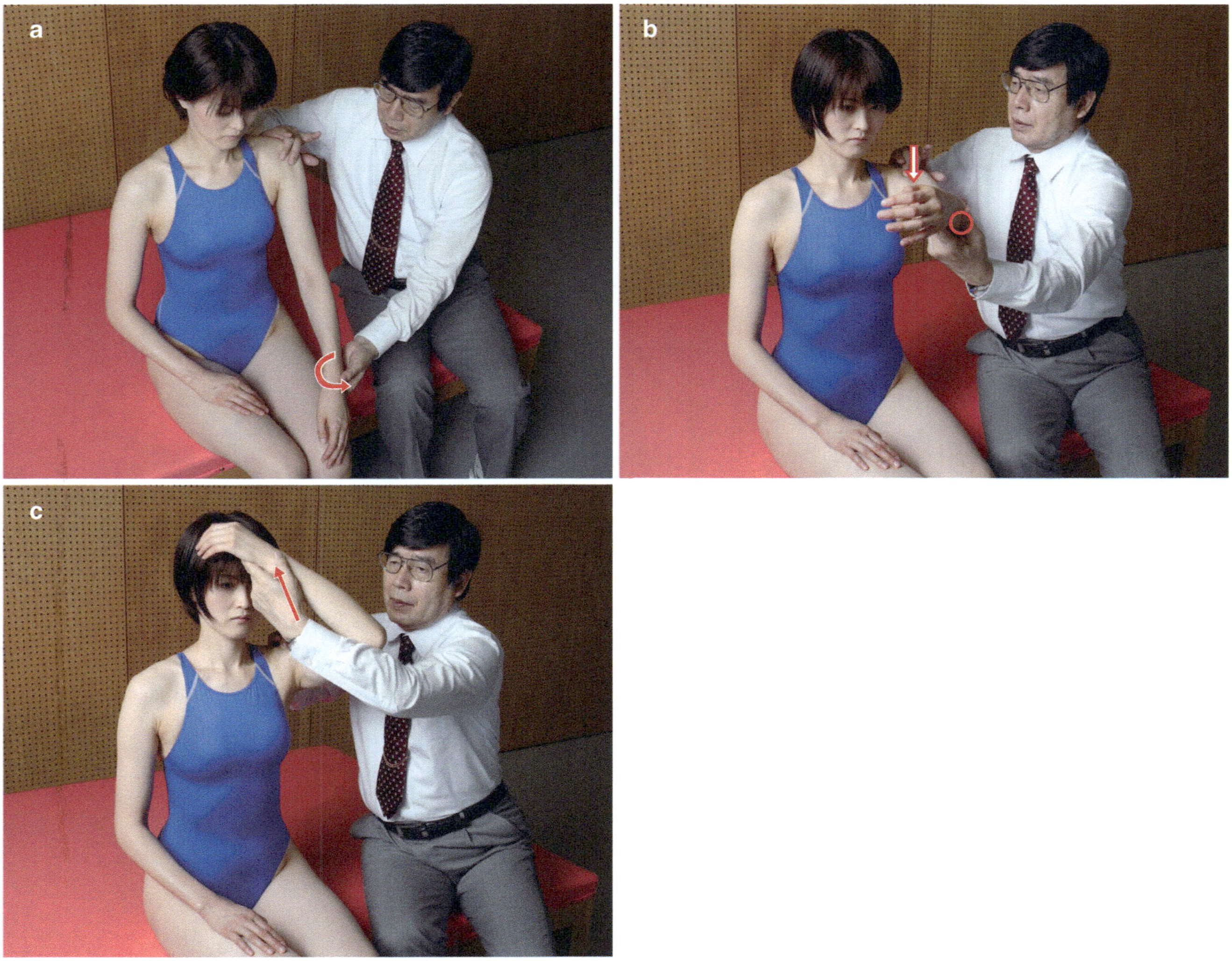

Fig. 4.22 Let's Try: Shoulder flexion/external rotation and flexion of the elbow (lifting the hand to the head, without abduction of the shoulders/in a sitting position) in a patient with relatively good shoulder function (shoulder flexion around 60°) (Video 4.9). (**a**) Holding the forearm proximal to the wrist joint and the elbow during flexion and external rotation of the shoulder. (**a**) Using his thumb and the other four fingers, the therapist grasps the patient's affected limb by the forearm (just proximal to the wrist joint). The middle finger of the therapist's other hand is placed on the patient's upper arm (above the biceps tendon). The therapist then instructs the patient, saying, "Lift your hand to your jaw," and induces supination of the forearm by rapidly inducing pronation of the forearm. (**b**) Facilitate supination of patient's forearm and elbow flexion. (**b**) The therapist grasps the patient's upper arm so that movement of the forearm on the affected limb will not be disturbed while stimulating flexion of the elbow (biceps tendon of the upper arm) with the middle finger of his other hand, which also prevents impingement. When supination of the forearm is insufficient, the therapist should instruct the patient: "Lift your hand, while turning your palm toward your face and bend your elbow." (**c**) Finish elbow flexion (lifting the hand to the chin, without abduction of the shoulder). (**c**) While the therapist holds the patient's shoulder to minimize abduction and assists with shoulder flexion/external rotation and elbow flexion, they guide the patient's hand to the top of their head. When the fingers of the affected limb touch the patient's jaw, return to (**a**). In order to prevent pain, the angle of the shoulder abduction should be kept to a minimum

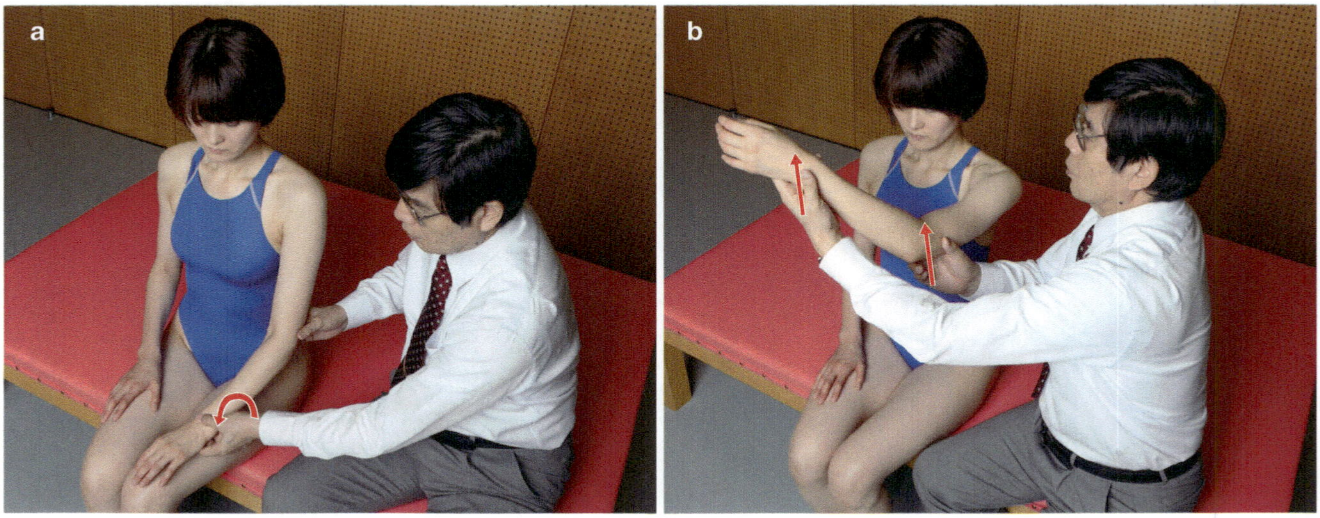

Fig. 4.23 Let's Try: Flexion and external rotation of the shoulder and supination and flexion of the elbow (lifting the hand to the head without shoulder abduction in a sitting position): In patients with limited shoulder function (shoulder flexion less than 60°) (Video 4.9). (**a**) Holding the forearm proximal to the wrist joint and the elbow at flexion as well as the external rotation of the shoulder. Using his thumb and the other four fingers, the therapist grasps the patient's affected limb by the forearm (just proximal to the wrist joint). He then places the other hand on the distal part of the upper arm (near the triceps brachii on the elbow). The therapist then instructs the patient, "Lift your hand to your jaw" and induces supination of the forearm by rapidly inducing pronation of the forearm. (**b**) The therapist says, "Facilitate supination of forearm and elbow flexion." (**b**) The therapist grasps the upper arm so that movement of the forearm and elbow flexion of the affected limb will not be disturbed. Then he assists in shoulder flexion/external rotation with his other hand on the elbow. When supination of the forearm is insufficient, the therapist should instruct the patient to "Lift your hand, while turning your palm toward your face and bend your elbow." During this time, pain might occur when the shoulder abduction angle is large.

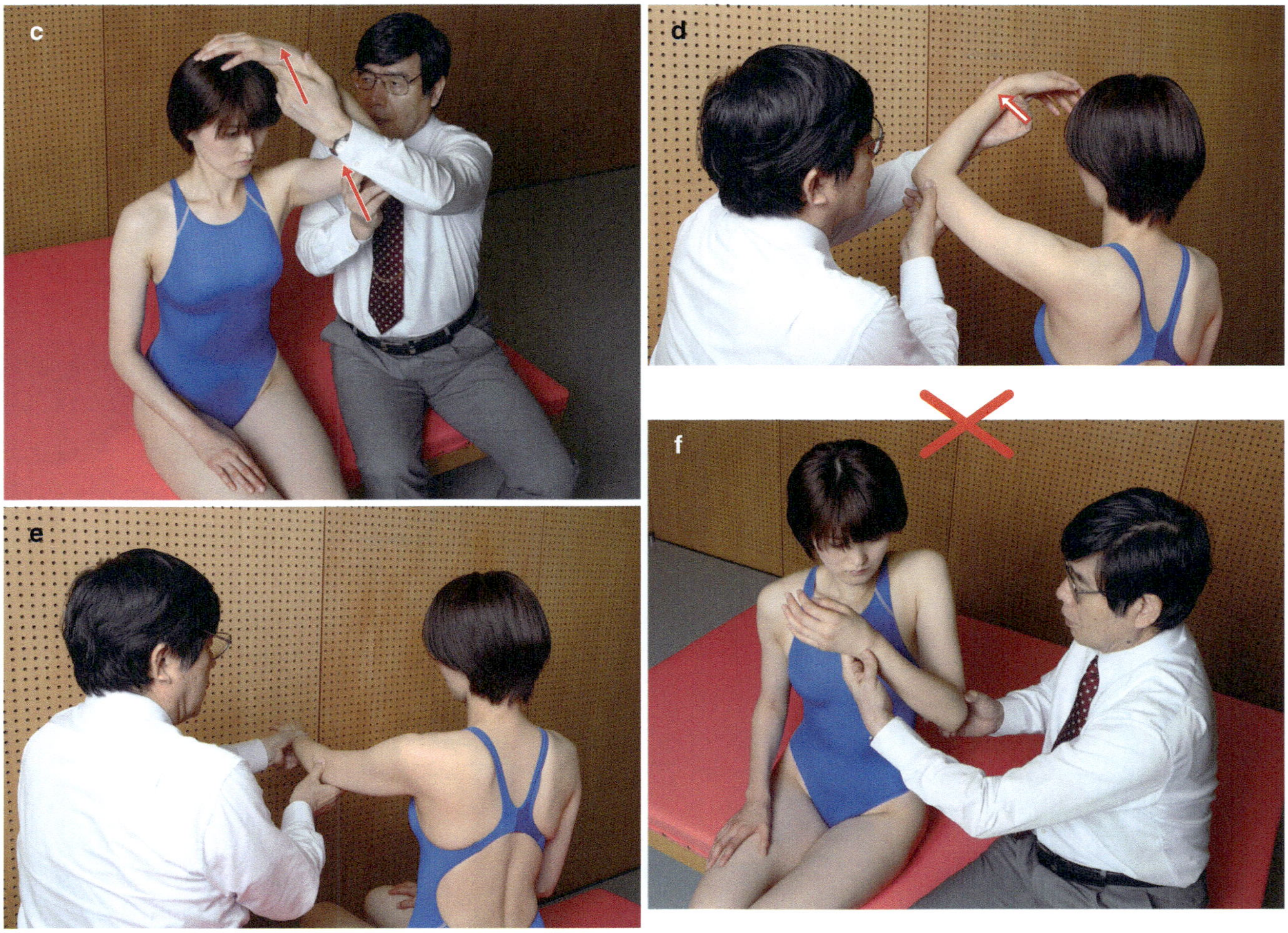

Fig. 4.23 (continued) (**c**) Finish elbow flexion (lifting the hand to the chin, without abduction of the shoulder). (**c**) The therapist minimizes the patient's shoulder abduction while assisting shoulder flexion/external rotation and elbow flexion. Then the therapist guides the patient's hand to the top of their head. (**d**) Switch to shoulder extension/internal rotation and elbow extension. (**d**) Using the middle finger of the hand holding the patient's forearm, the therapist taps the patient's radius to facilitate elbow extension. Using the other hand placed on the patient's elbow, the therapist assists the patient with shoulder extension/internal rotation. If the range of pronation for the forearm is limited, instruct the patient: "Extend your elbow until the back of your hand is in front of your face." If the therapist pulls the elbow in the direction of extension without facilitation, the movement will slow due to resistance caused by muscle tension. However, the addition of Repetitive Facilitative Exercise causes the movement to become smoother due to combined active and passive motion. (**e**) Support for shoulder extension/internal rotation and elbow extension (hand to knee). (**e**) If the range of shoulder extension on the affected side is limited, the therapist can press the tendon of the triceps brachii with his thumb in order to facilitate elbow extension. Using both hands to support the patient's hand, the therapist guides the patient to return her hand to her knee smoothly. (**f**) Poor technique. (**f**) The upper limb cannot be significantly elevated while the shoulder on the affected side is adducted and internally rotated. Instruct the patient: "First push your shoulder forward a little, then bend your elbow so that your palm is in front of your face"

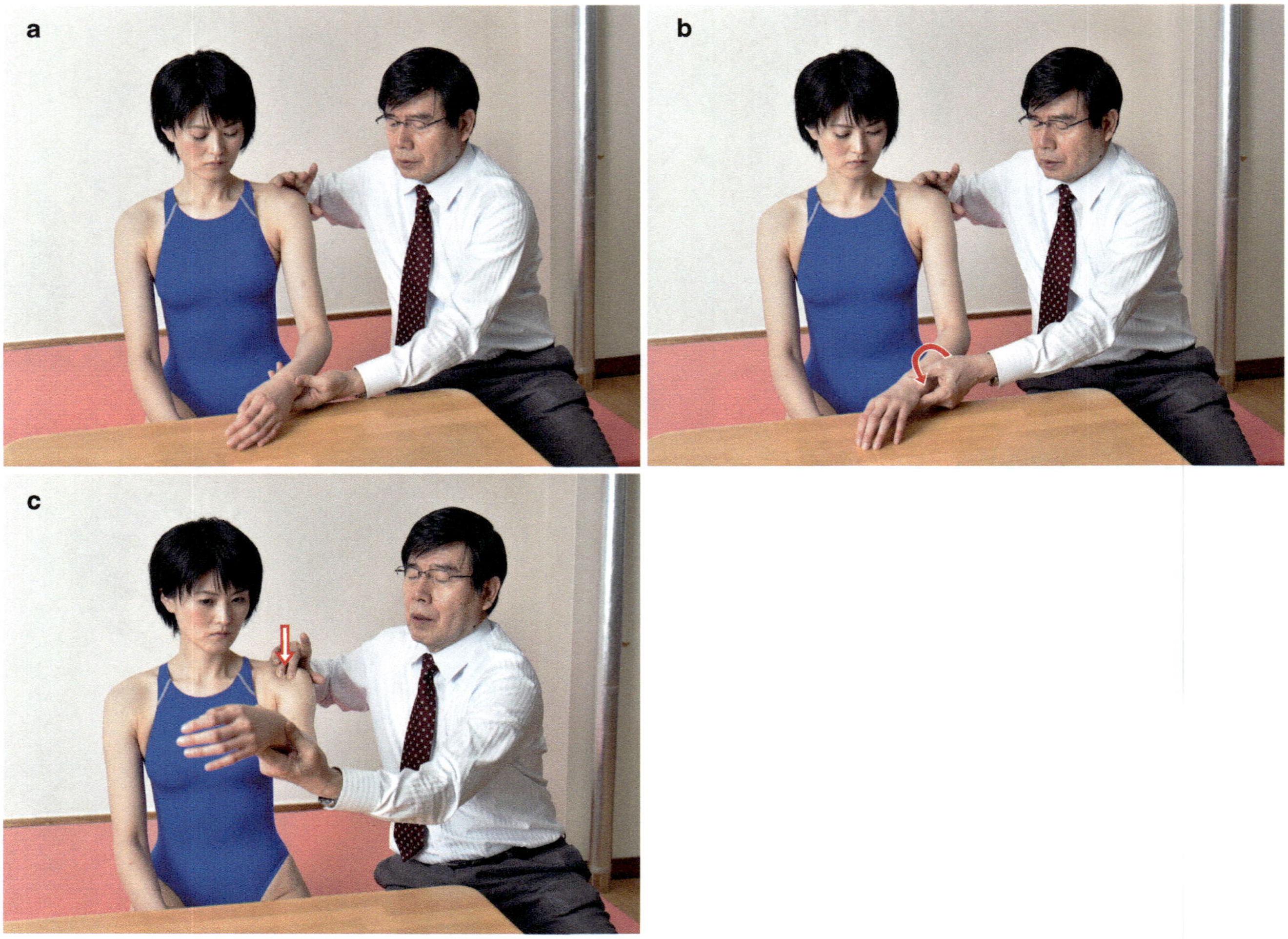

Fig. 4.24 Let's Try: Shoulder flexion, elbow extension (reaching). The therapist presses down on the humeral head using his middle and ring fingers to stimulate the front of the deltoid muscle and prevent the humeral head from being thrust upward during supination of forearm as the patient's upper limb begins to lift. (**a**) The starting position for reaching. (**a**) Flex the patient's elbow and pronate her forearm slightly. The therapist holds the patient's shoulder (humeral head) with one hand and holds the patient's wrist with his other hand. (**b**) Elbow flexion and elevation of the patient's upper limb due to inducing forearm supination. (**b**) The therapist should passively pronate the patient's forearm in order to facilitate supination of the forearm and upper limb elevation. (**c**) Shoulder flexion induced by stimulation to the front of the deltoid muscle. (**c**) By tapping the front part of the deltoid muscle with his middle and ring fingers, the therapist can facilitate the patient's upper limb elevation. After elevating the upper limbs, the patient is instructed to extend their elbow at the point of lowering it. When the range of elbow extension is limited, the therapist should supinate the patient's forearm by tapping on the radius to promote extension of the patient's elbow with pronation of the forearm. While lifting and lowering the upper limb, the therapist must press down on the humeral head to prevent impingement

4.6 Repetitive Facilitation Exercise for the Elbow

Key Point

1. Elbow flexion and extension (supine position)

   Initially, the patient may only be able to move their elbow when combined with a shoulder movement, but gradually the patient will be able to flex and extend their elbow independently.

   For Repetitive Facilitative Exercise, perform pronation and supination of the forearm and quick elbow flexion (Fig. 4.25) and extension (Fig. 4.26). To reduce the burden on the shoulder of the affected limb, the therapist should use his fingertips to induce flexion and extension of the elbow (Fig. 4.27). If the patient has shoulder pain, the weight of the upper limb can be relieved by placing the patient's elbow on the therapist's knee while performing the exercise with the patient's shoulder slightly abducted.

   If the patient cannot flex the elbow while holding the shoulder joint at 90° flexion, then elbow flexion exercises should be performed initially at a shoulder adduction/extension position ranging from 30 to 60°.

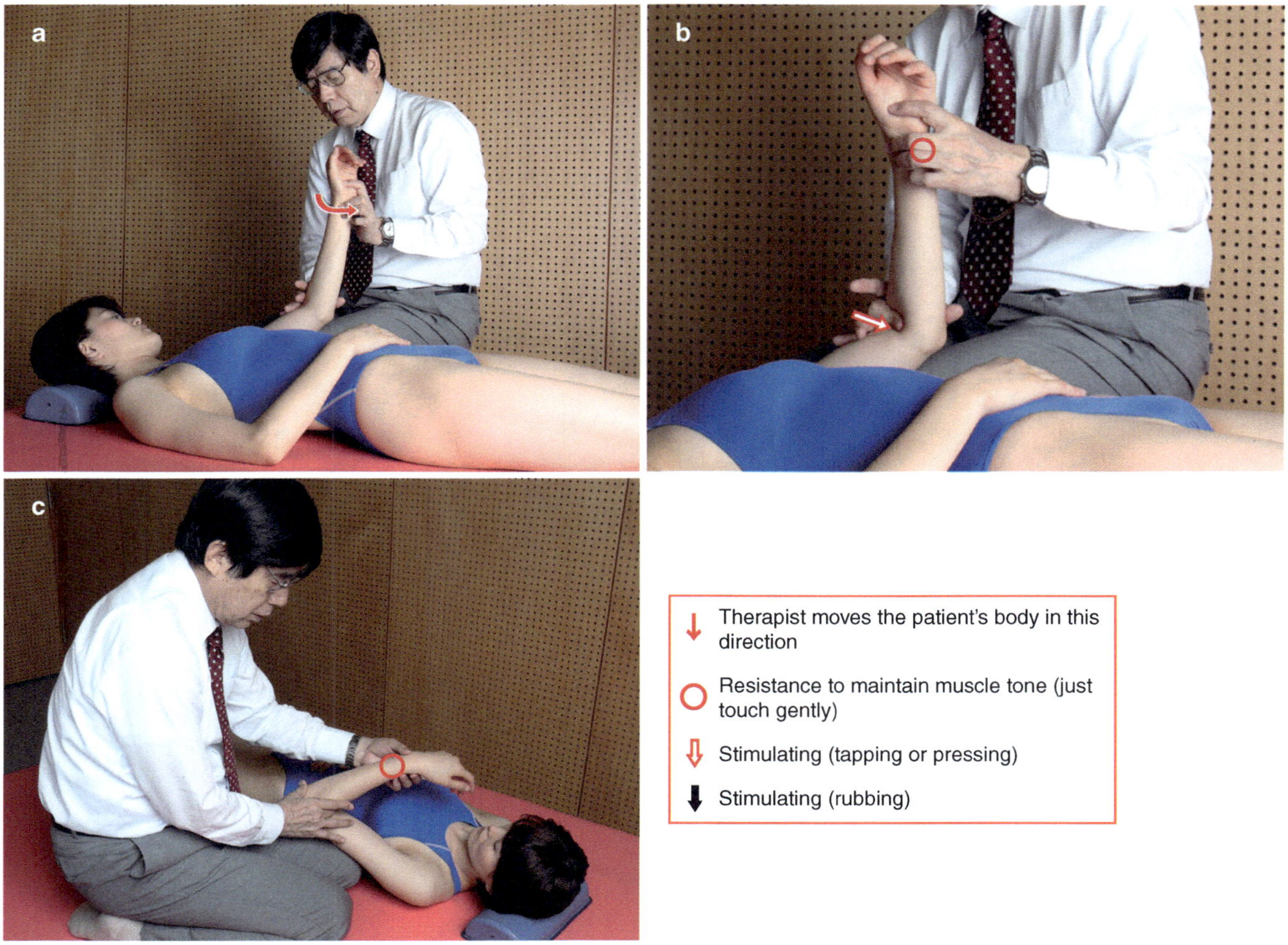

Fig. 4.25 Let's Try: Elbow flexion (hand to chin/supine position, instead of ROM-ex) (Video 4.10). (**a**) The position for holding the patient's forearm (proximal to the wrist joint) and elbow. (**a**) Using his thumb, middle finger, ring finger, and little finger, the therapist holds the forearm of the affected limb (just proximal to the wrist joint) and places his index finger on the patient's palm. Then the therapist places the middle finger of his other hand on the elbow flexor muscle (upper biceps tendon) of the affected limb. In order to reduce the mechanical stress on the shoulder joint, the patient's elbow is placed on the therapist's knee. The therapist instructs the patient by saying "Lift your hand to the chin," and induces supination of the forearm by quick pronation of the forearm. (**b**) Continuity of forearm supination and elbow flexion. (**b**) While closely following the patient's forearm supination and elbow flexion in order to avoid interfering with these movements, the therapist facilitates the movements by pressing the biceps tendon on the upper arm with the middle finger on his other hand. If the range of supination of the forearm is insufficient, the therapist should instruct the patient by saying: "Flex your elbow after turning your palm toward your face." (**c**) The completion of elbow flexion (hand to chin). (**c**) Forearm and elbow flexion of the affected upper limb positioned with the palm in front of the face. When the patient's fingertips touch her jaw, return to (**a**). If the therapist's middle finger is used to stimulate the biceps tendon of the upper arm, the therapist's finger does not prevent deep flexion of the elbow

Thereafter, change the limb position to shoulder flexion (Fig. 4.27) and extend the elbow.

2. Elbow flexion and extension (while sitting) (Fig. 4.28)

   Lift the patient's hand from the upper thigh to the chin. Rapidly pronate the forearm in order to promote elbow flexion while trying to restrain shoulder abduction.

3. Pronation and supination of the forearm (while sitting) (Fig. 4.29)

   The relationship between the forearm and elbow movement makes it easier to extend the elbow during pronation and to flex the elbow during supination.

For Repetitive Facilitative Exercise, turn the joint quickly in the direction opposite of the target muscle contraction in order to induce stretch reflex, then rub the skin over the target muscle in order to induce skin reflex.

The examples of target muscle combintions are: the round pronator muscle and pronator quadratus muscle for pronation of the forearm; the supinator brevis muscle and the brachioradial muscle for supination of the forearm (Fig. 4.30).

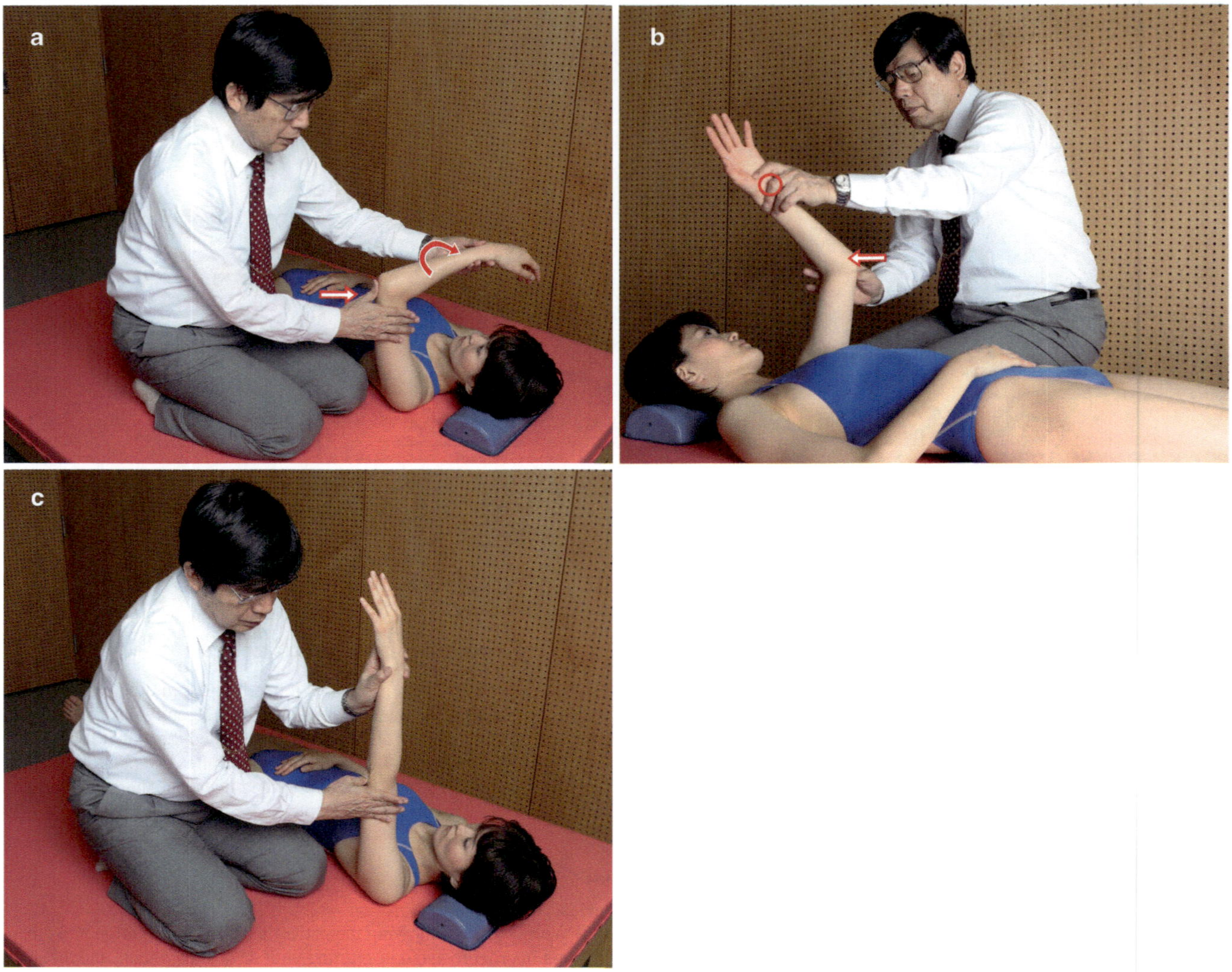

Fig. 4.26 Let's Try: Elbow extension (supine position) (Video 4.11). (**a**) Holding the patient's forearm (just proximal to the wrist joint) and elbow. (**a**) Using his thumb, middle finger, ring finger, and little finger, the therapist holds the forearm of the affected limb (just proximal to the wrist joint), then places his index finger on the patient's palm. The thumb of his other hand is placed on the extension side (triceps) of the elbow. In order to reduce mechanical stress on the shoulder joint, promote rapid supination of the forearm. Rapid flexion of the elbow should be avoided. The therapist instructs the patient by saying, "Stretch out your elbow while watching the back of the hand." Then the therapist induces pronation of the forearm and extension of the elbow by performing rapid supination of the forearm. (**b**) Continuing elbow extension and pronation of the forearm. (**b**) The therapist provides support without interfering with the patient's movement of her forearm and elbow extension while pressing the triceps tendon of the patient's upper arm with the thumb of the other hand. When using his thumb to stimulate the triceps tendon of the upper arm, the therapist should keep the other fingers extended and place the palm on the patient's upper arm without touching the flexor muscles with his fingers. When the angle of forearm rotation is insufficient, the therapist should instruct the patient, "Turn the back of your hand toward your face, then extend your elbow." (**c**) Finish extension of the elbow. (**c**) When it becomes possible for the patient to extend the elbow of the affected limb, return to (**a**)

If it is difficult to induce the intended movement with the elbow at 90° flexion, then muscle synergy can be utilized.

For supination of the patient's forearm, the therapist should pull his hand backwards and move the patient's limb into a position with the elbow flexed and the shoulder extended. For pronation of the forearm, the therapist should pull the patient's hands forward and move the patient's limb into an elbow extension and shoulder flexion position. When it is possible for the patient to perform those movements, the therapist should attempt to reduce the elbow and shoulder movements.

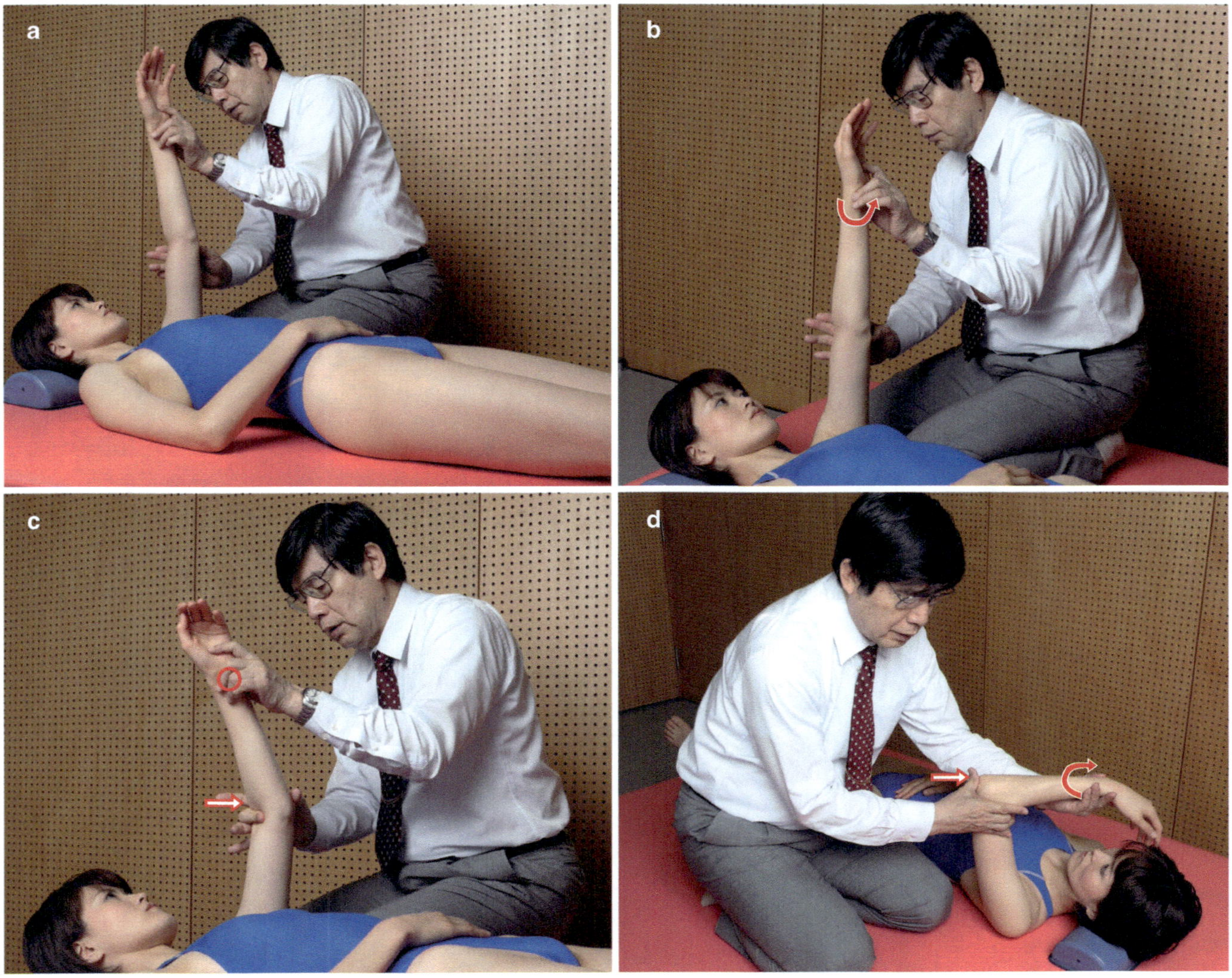

Fig. 4.27 Let's Try: Elbow flexion/extension (supine position) (Video 4.12). Repeat flexion and extension of the elbow in patients with moderate to mild paralysis. In order to prevent contracture in patients with severe paralysis, elbow flexion and extension are performed so that the affected limb will undergo the entire range of motion. For patients with severe paralysis, elbow flexion and extension are performed independently so that the affected limb will undergo the entire range of motions. (**a**) Supporting the upper arm and forearm (wrist joint). (**a**) Using his thumb, middle, ring, and little finger, the therapist holds the forearm on the affected limb (just proximal to the wrist joint) and places his index finger on the palm. The therapist then places the middle finger on his other hand just above the patient's biceps-tendon on the upper arm. The therapist instructs the patient, saying, "Put your hand to your chin" and induces forearm supination through quick pronation of the forearm. (**b**) The starting point of forearm supination and elbow flexion. (**b**) The therapist instructs the patient, saying, "Please flex your arm" and induce supination of the forearm using a quick pronation. The therapist should support the patient's movement without interfering with the patient's forearm supination and elbow flexion, while pressing the biceps tendon of the patient's upper arm with the middle finger on his other hand. When the angle of rotation of the forearm is insufficient, the therapist should instruct the patient, saying, "Turn your palm toward your face, then bend your elbow." (**c**) Continuing elbow flexion. (**c**) After placing his middle finger on the biceps tendon in the patient's forearm and holding his thumb against the olecranon, the therapist induces elbow flexion by stimulating the biceps tendon in the forearm of the affected limb. The therapist provides support for the patient's limb and adjusts the speed at which the elbow flexes so that the weight of the affected limb does not activate the stretch reflex in the triceps brachialis muscle. (**d**) Switching from elbow flexion to elbow extension. (**d**) When elbow flexion reaches its limit, the therapist quickly moves to spinate the forearm while simultaneously stimulating the triceps tendon with the thumb in order to induce elbow extension of the affected limb. During this procedure, the therapist should instruct the patient, "Please extend your elbow and turn your arm so that you can see the back of your hand." **Additional tips**: Move the thumb from the olecranon to the tendon at the distal part of the triceps brachii (near the olecranon) and stimulate that point.

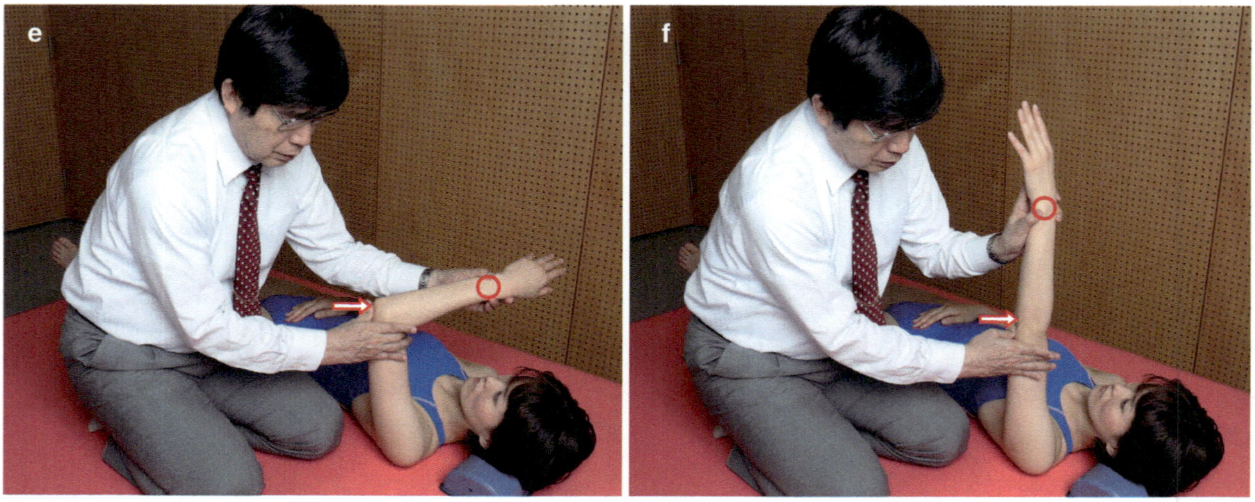

Fig. 4.27 (continued) (**e**) Continue elbow extension and pronation of the forearm. (**e**) The therapist should support the patient's effort to move the affected limb without interfering with the pronation of the forearm or elbow extension. During the movement, the therapist should press the triceps tendon of the patient's upper arm with the thumb on his other hand. When the angle of pronation for the forearm is insufficient, the therapist should instruct the patient "Turn the back of your hand toward your face and extend your elbow." (**f**) The extension limit of the elbow. (**f**) When it becomes possible for the patient to pronate the forearm and extend the elbow, return to (**a**)

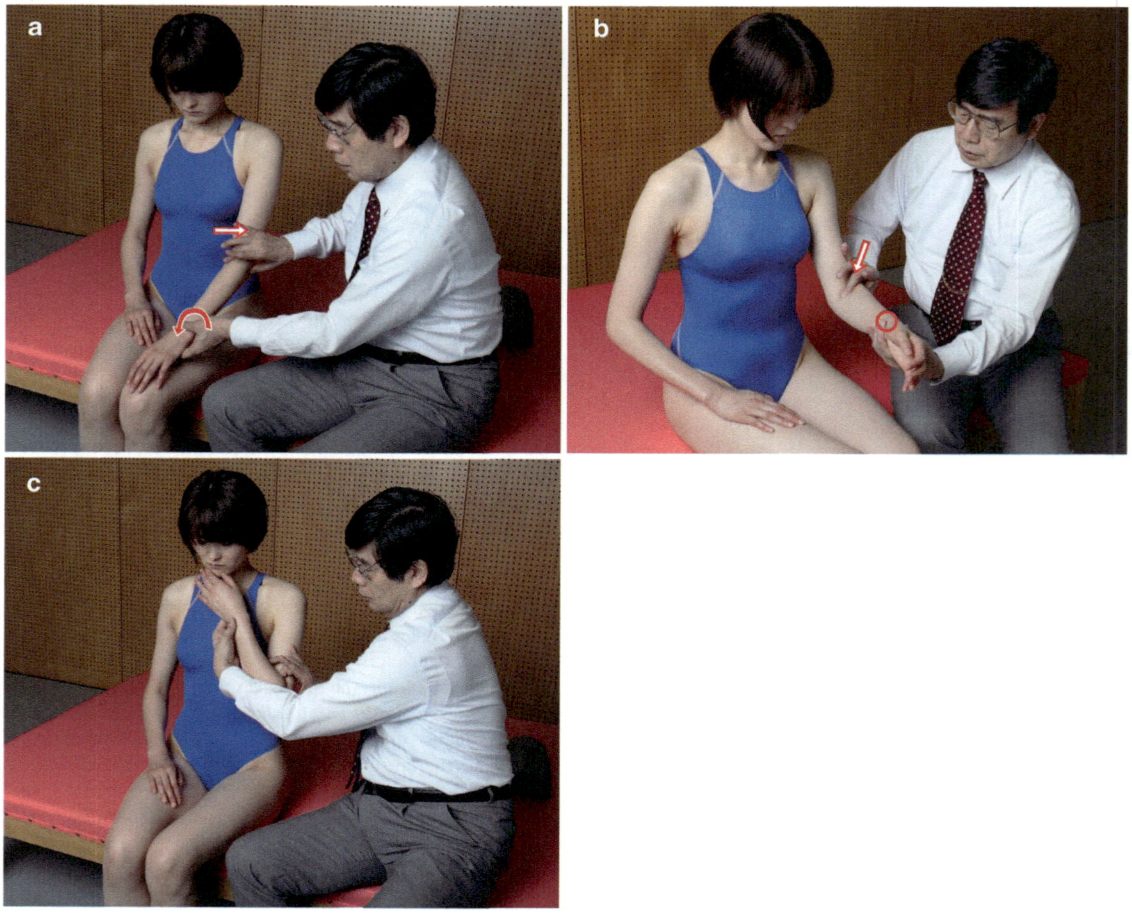

Fig. 4.28 Practical elbow flexion (lifting the hand to the chin with gentle support while in a sitting position). (**a**) How to hold the forearm (proximal to the wrist joint) and the elbow. (**a**) The therapist uses his thumb, middle, ring, and little finger, to hold the forearm of the affected limb (just proximal to the wrist joint). He then places the middle finger in front of the patient's elbow (the biceps tendon). (**b**) Facilitation of forearm supination and elbow flexion. (**b**) Holding the patient's arm to avoid disturbing forearm supination formed in the affected upper limb, the therapist stimulates the biceps tendon of the upper arm with the middle finger of his other hand. The therapist gently supports the upper limb and follows the movement after it lifts from the therapist's hand. (**c**) Finishing elbow flexion with gentle support. (lifting the hand to the chin). (**c**) When the fingers of the affected limb touch the patient's jaw, return to step (**a**)

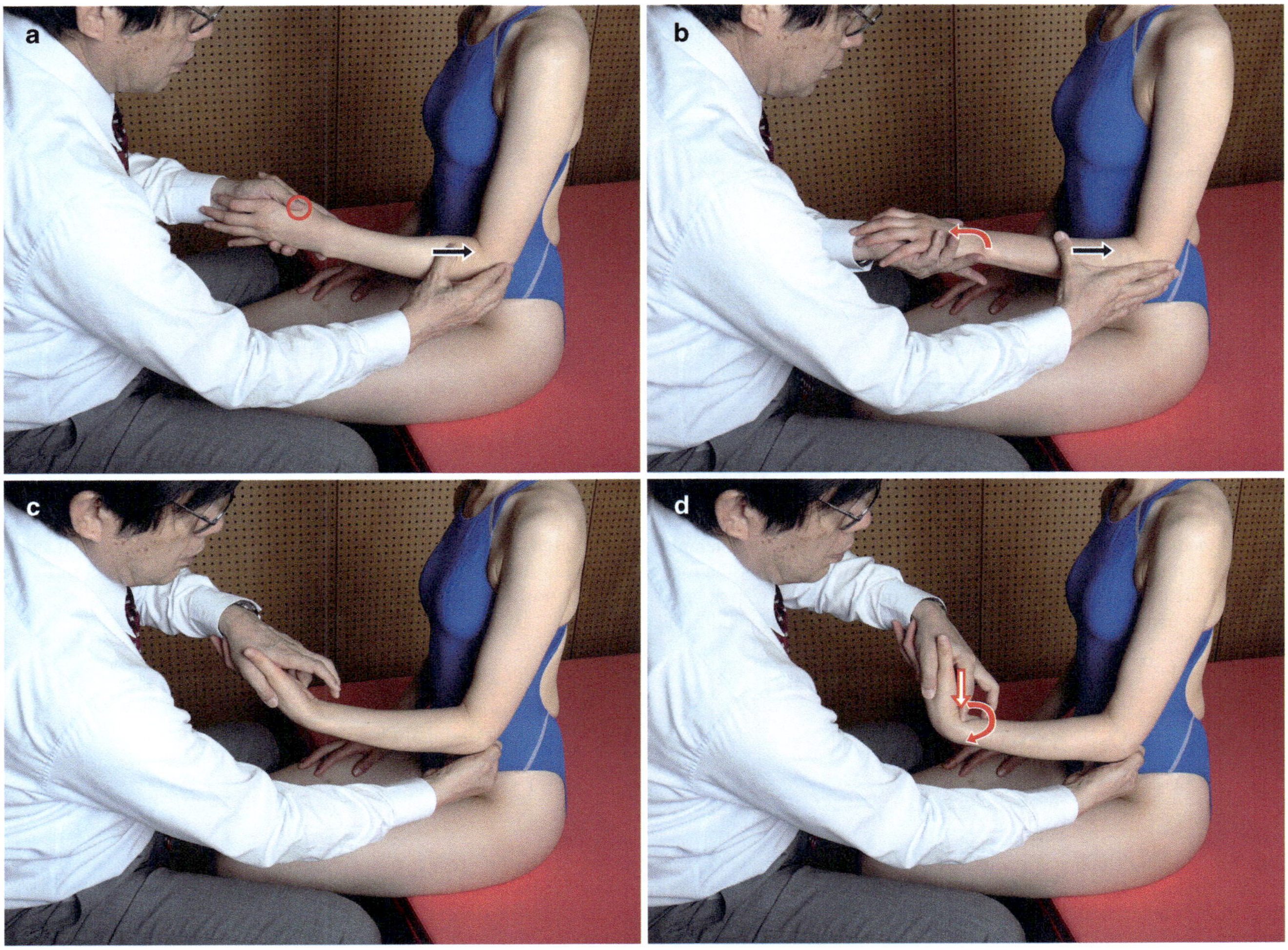

Fig. 4.29 Let's Try: Supination-pronation of the forearm (Sitting position) (Videos 4.13, 4.14, and 4.15). (**a**) Supporting the forearm and the palm. The therapist should sit facing the patient and grasp the patient's affected hand, gripping the patient's thumb and other four fingers as if shaking hands. The patient's thumb should be abducted and the other four fingers extended, while the elbow should be flexed and the forearm in a neutral position (0°). Then the therapist should place the index finger and middle finger of his other hand on the patient's elbow (olecranon) and place his thumb on the outside of the patient's forearm. The therapist should tell the patient, "Keep your fingers extended and try to relax your hand as much as possible. Try not to grip my hand." To begin the exercise, the therapist should quickly pronate the forearm of the affected limb and instruct the patient, "Please turn your forearm outwards." **Additional tip**: For the patient with high muscle tonus in the pronator teres muscle, it may be difficult to induce supination of the forearm. However, pronation of the forearm can be more easily induced by maximal supination of the forearm to suppress spasms in the pronator teres muscle. During this maneuver, the therapist should gently support the patient's forearm near the wrist in order to avoid injury. (**b**) Stretching the supinator muscle. (**b**) The therapist should quickly perform passive pronation of the forearm and simultaneously rub the lateral forearm using his thumb. When supinating the forearm of the affected limb, the therapist should use his thumb to lightly apply resistance (only touching). **Additional tips**: When rotation of the forearm is insufficient, the therapist should use his thumb to scrape against the skin overlying the brachioradialis muscle. (**c**) From the end range of forearm supination towards forearm pronation. (**c**) When supination of the forearm on the affected limb is fully extended, the therapist should place his thumb on the olecranon with his middle finger and ring finger on the front side of the forearm (pronator teres) while guiding supination of the forearm to the end of its range. The therapist should move his index finger to the radial side of the wrist joint. The therapist taps the radial side of the wrist joint with his index finger to supinate the forearm quickly, while instructing the patient to "Turn your forearm inwards." (**d**) Stretching stimulation of the pronator teres muscle in the forearm. The therapist taps the radial side of the wrist with his index finger and quickly supinates the forearm and gives the instruction: "Turn your forearm inwards." When pronation of the forearm stops, guide the forearm through the full range of movements.

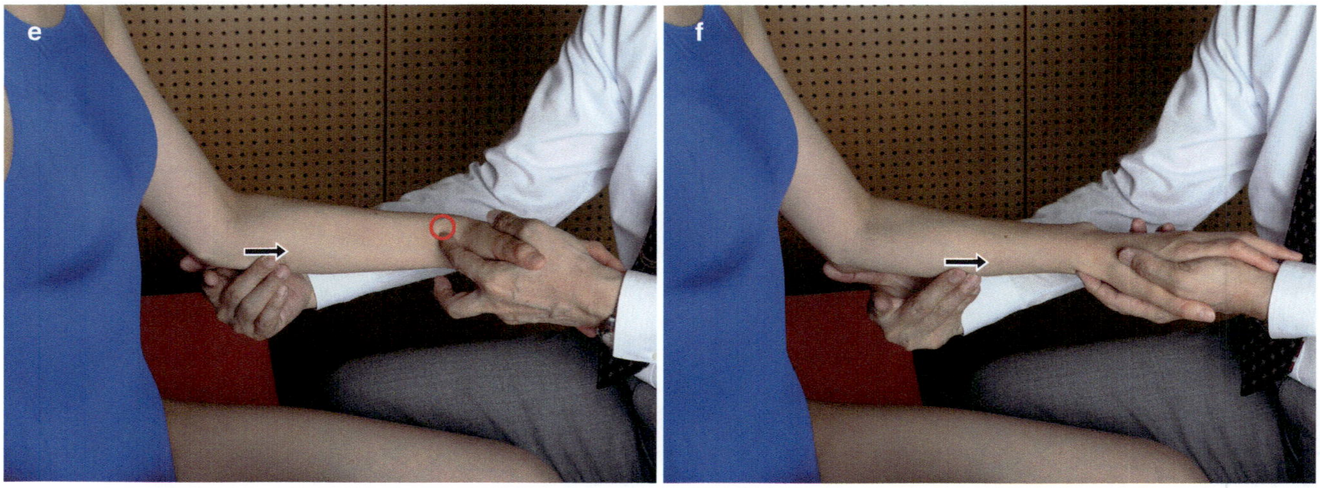

Fig. 4.29 (continued) (**e**) Stimulation of the pronator teres muscle viewed medially. (**e**) To facilitate pronation of the forearm, the therapist should use his middle and ring fingers to rub the ulnar side of the forearm. (**f**) Treatment method and hand positions (medial view). From forearm pronation to the start of the exercise. (**f**) Using his middle and ring fingers, the therapist should rub the medial forearm and use a stroking motion toward the wrist joint. When the pronation movement stops, guide the forearm through the full range of movement. Return to the starting position shown in (**b**)

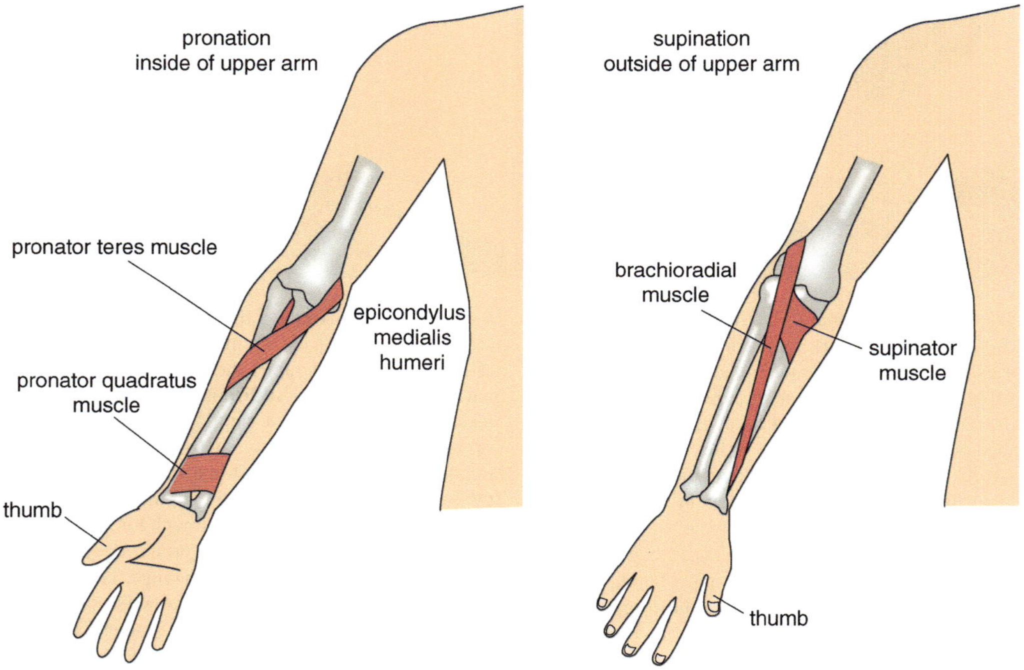

Fig. 4.30 During Repetitive Facilitative Exercise for supination of the forearm, the therapist should rub the skin overlying the brachial radial muscle. For pronation, the therapist should rub the ulnar side of the forearm lengthwise in the direction of the long axis to its full extent

4.7 Repetitive Facilitative Exercise for the Hand and Wrist Joint

Repetitive Facilitative Exercise for the fingers, wrist joint, and forearm become easier when PNF patterns are utilized.

Modified combinations of PNF patterns are finger flexion, palmar flexion of the wrist joint, supination of the forearm, and fingers extension dorsiflexion of the wrist joint pronation of the forearm.

Key Points

The finger extensor muscle and the dorsiflexion muscles of the wrist joint are stretched when the wrist is pushed with the fingers flexed. On the other hand, the finger flexor muscle and the flexor muscle group of the wrist joint are stretched when the wrist is pushed and dorsiflexed with the fingers extended (Fig. 4.31).

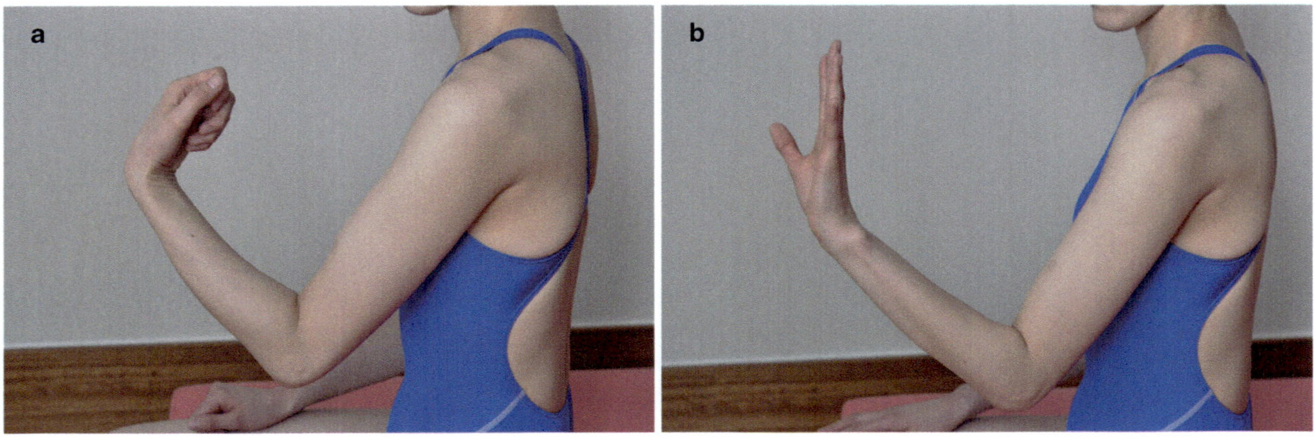

Fig. 4.31 Let's Try: Functional anatomy of fingers and wrist joint. (**a**) Position in which the extensors of the hand, fingers and wrist joint are extended. The extensor muscles of the fingers are stretched when the forearm is in a supinated, the wrist in a palmar flexed position and the fingers are also flexed. (**b**) Position in which the flexors of the hand, fingers and wrist joint are extended. The finger flexor muscles are stretched when the forearm is pronated, the wrist is in a dorsal flexed position and the fingers are extended

Most of the muscles involved in moving the fingers and wrist joint are located in the forearm. The group of muscles that extend the fingers as well as those for dorsal flexion of the wrist are on the dorsal/radial side of the forearm; the group of muscles that flex the fingers and those for palmar flexion of the wrist are located on the ventral/ulnar side of the forearm.

When using stretching reflexes and skin muscle reflexes for Repetitive Facilitative Exercise of the finger, it is necessary for the therapist to adjust the position of the patient's fingers and wrist joint while maintaining access to the stimulation site. The PNF pattern used for Repetitive Facilitative Exercise for the forearm and fingers is shown in Fig. 4.32.

1. Dorsiflexion of the wrist joint (Fig. 4.33)
   When performing dorsiflexion of the wrist joint, the therapist should gently flex the fingers of the patient's affected hand and wrap his own hand around the patient's fingers to hold them in that position. Then he quickly performs a palmar flexion movement of the wrist joint, while simultaneously engaging the stretch reflex by quick supination of the forearm and stimulating the skin on the forearm. This procedure can also suppress flexor spasms, but if the patient has severe flexor spasms, perform this method in combination with vibration stimulation or electrical stimulation to suppress spasticity of the flexor muscle before attempting to increase the patient's ability to perform voluntary movement.
2. Pronation of the forearm, dorsiflexion of the wrist joint extension of the fingers, supination of the forearm palmarflexion of the wrist joint, and finger flexion (in a supine position: active assistive exercise finger movement using PNF patterns involving facilitation) (Fig. 4.34).

The hand and fingers contain a large number of joints and contracture tends to develop if the treatment plan does not include a sufficient range of motion exercises.

This exercise is primarily aimed at improving the range of joint motion in patients with severe paralysis. However, this exercise is also intended to improve paralysis of the fingers and forearm, contributing to the practical functionality of the upper limb in patients with moderate or less paralysis who can flex and extend their fingers easily.

For patients with flaccid or mild paralysis, it is more effective to be treated with quick movements. For patients with severe paralysis and spasticity, slow movements are better.

In patients with spasticity of the flexor muscles in the upper limbs, this spasticity will cause finger flexion when the patient tries to extend the fingers and dorsiflex the wrist joint at the same time. However, if this combination of movements is performed while the forearm is pronated, it becomes easier to suppress the finger flexion caused by spasticity.

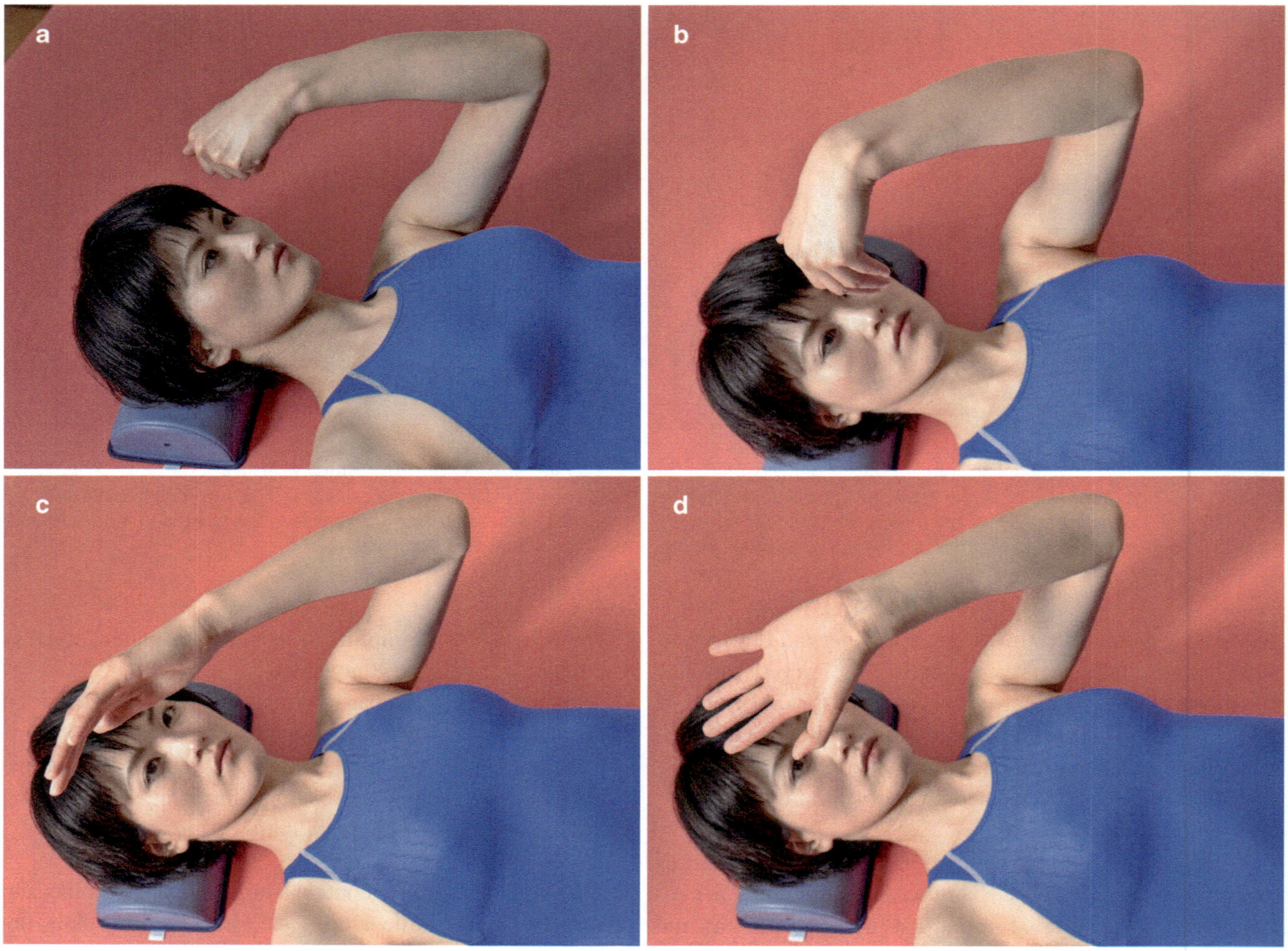

Fig. 4.32 Let's Try: PNF pattern for pronation of forearm, dorsiflexion of the wrist and extension of the fingers. (**a**) Starting position of exercise. (**a**) Finger flexion, wrist flexion, forearm supination. **Additional tip**: According to the original PNF method, exercises involving pronation of the forearm dorsiflexion of the wrist joint extension of the fingers start from the periphery. However, in patients with persistent paralysis of the upper limbs, it is easier to begin those exercises from the more proximal part where voluntary control of movement is still present. (**b**) From supination of the forearm to dorsiflexion of the wrist joint. (**b**) Dorsiflex the wrist joint and pronate the forearm. The fingers can be in a flexed position. (**c**) The exercise combines pronation of the forearm, dorsiflexion of the wrist and extension of the fingers. (**c**) Instruct the patient to extend her fingers if the extension movement of the wrist joint is insufficient. (**d**) The position at the end of the exercise. (**d**) Encourage the patient to extend the fingers while the forearm is pronated, and the wrist is in a dorsal flexed position

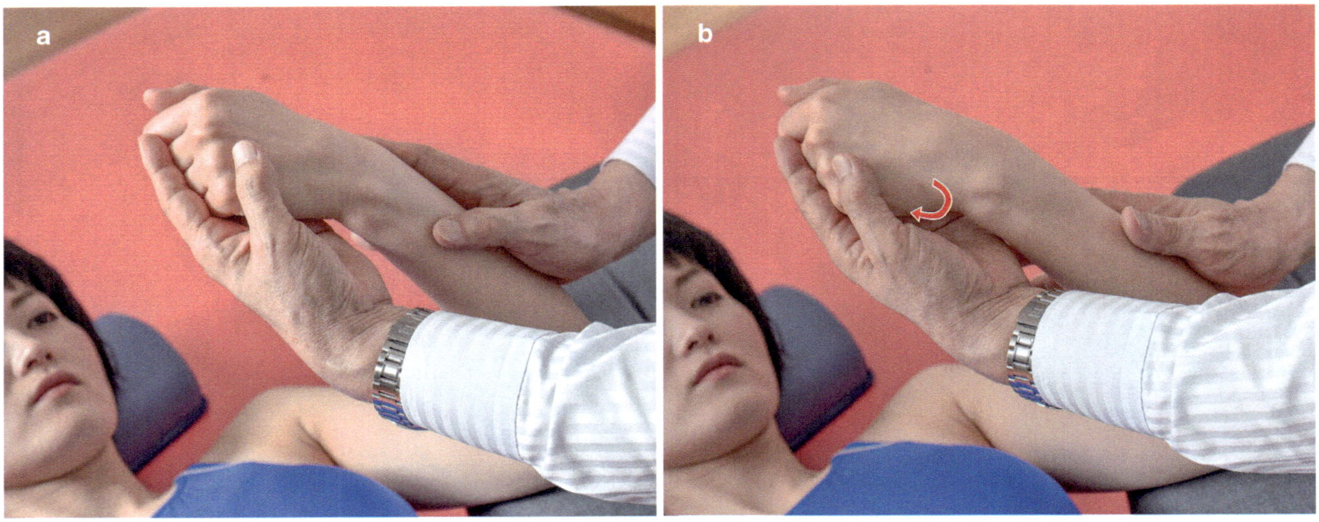

↓ Therapist moves the patient's body in this direction

○ Resistance to maintain muscle tone (just touch gently)

⇩ Stimulating (tapping or pressing)

⬇ Stimulating (rubbing)

Fig. 4.33 Let's Try: Dorsiflexion of the wrist joint (supine position) (Video 4.16). (**a**) Supporting the fingers and forearm. (**a**) The therapist should hold the elbow of the patient's affected upper limb against his knee in order to stably position the patient's forearm in front of her face. The therapist should then place his thumb on the back of the patient's affected hand and place his index finger, middle finger and ring finger over the patient's finger joints. Then the therapist should flex the patient's fingers forward together, as if wrapping his hand around the patient's hand. Both the wrist joint and the forearm are kept in neutral positions at 0°. Additional tips: Figure (**e**→**d**, **e**) shows how to hold the patient's affected fingers. (**b**) Stretching the muscle group involved in finger extension and wrist dorsiflexion. (**b**) Palmar flexion of the wrist and supination of the forearm should be performed quickly by pushing on the lateral dorsum of the patient's hand. During this procedure, the therapist should instruct the patient: "Please bend your wrist upward, while you turn your forearm away from your body." The finger extensor muscles are maximally stretched at this position, so be sure to maintain a gentle touch. If the therapist applies strong force, the patient will experience pain.

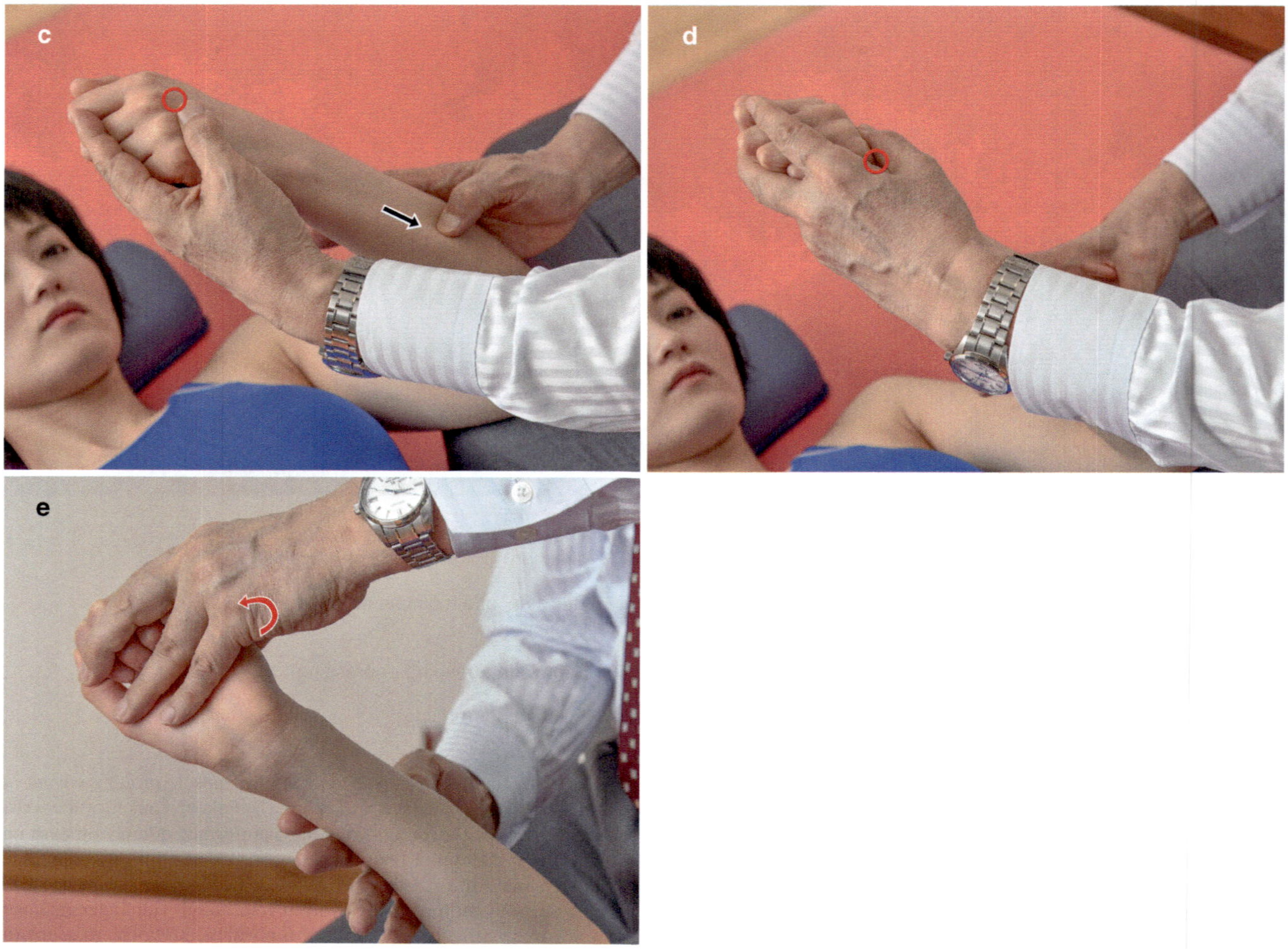

Fig. 4.33 (continued) (**c**) Stimulation for the muscle group involved in finger extension and wrist dorsiflexion. (**c**) Using the thumb of his other hand, the therapist should rub the posterior side of the forearm as if scratching a longer area. **Additional tips**: If this approach is performed without sufficiently holding the affected limb against the therapist's knee, the stimulation given to the forearm will be weak. (**d**) Continuing pronation of the forearm and dorsiflexion of the wrist joint. (**d**) Apply light resistance to the movements involved in dorsiflexion and forearm pronation of the affected limb. (**e**) The limit of pronation of the forearm and maximal dorsiflexion of the wrist. (**e**) When dorsiflexion of the wrist and forearm pronation are stopped, then slowly perform dorsiflexion of the wrist joint and pronation of the forearm until these movements reach their respective limits. Then return to step (**b**). The patient's affected hand is placed between the therapist's which is positioned as follows: the thumb is on the back of the patient's with his index finger on the proximal phalanges, the middle finger on the intermediate phalanges, and the ring finger on the distal phalanges

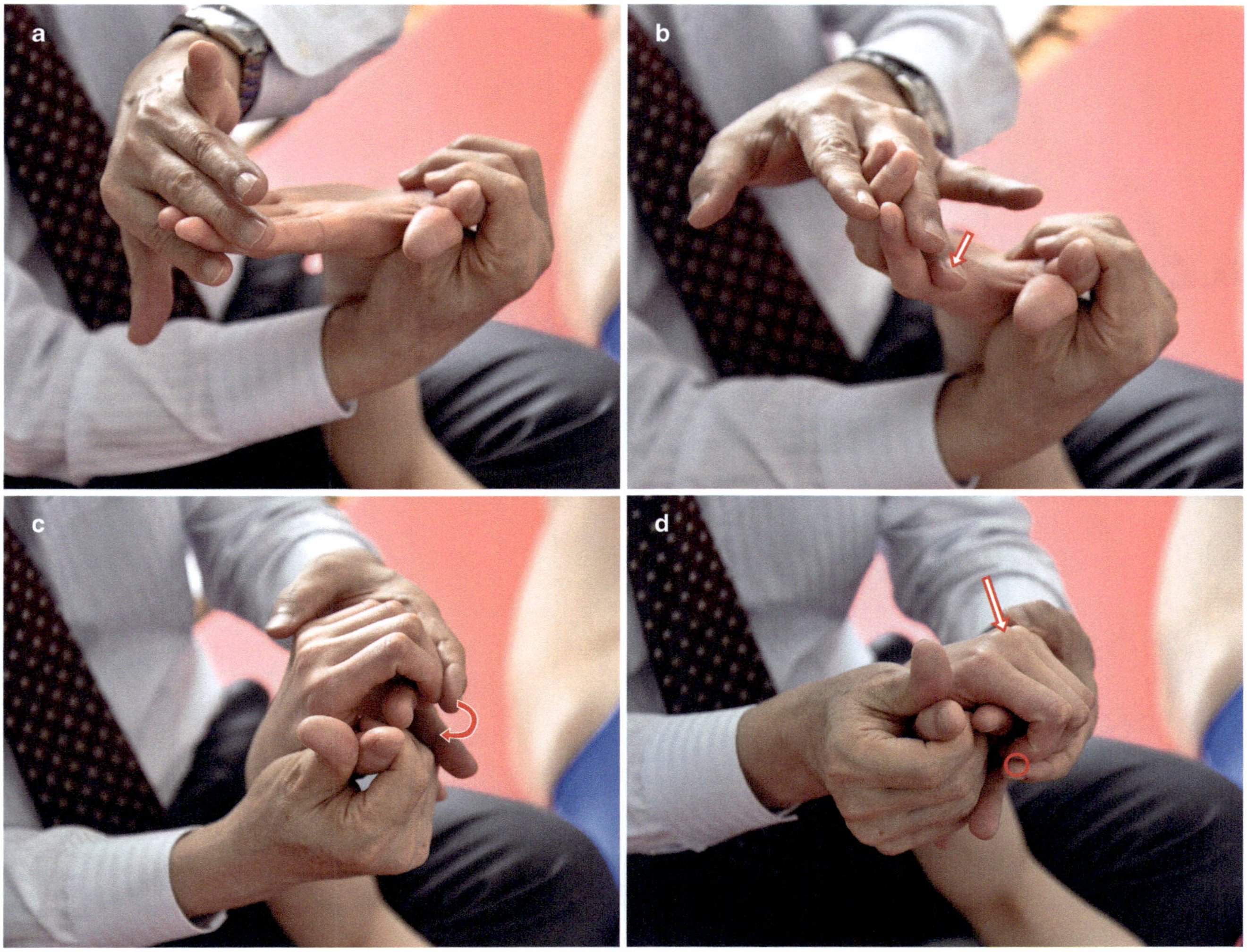

Fig. 4.34 Let's Try: Pronation of the forearm/dorsiflexion of the wrist joint extension of the fingers followed by supination of the forearm palmar flexion of the wrist joint flexion of the fingers (Videos 4.17, 4.18, 4.19, 4.20, 4.21, 4.22, and 4.23). (**a**) Supporting the thumb and other four fingers. (**a**) The therapist should hold the metacarpal bones and thenar eminence of the thumb on the patient's affected hand with one hand, while holding the other four fingers of the patient's hand with the index finger and middle finger of his other hand. Hold the patient's affected hand in this position with pronation of the forearm, dorsi flexion of the wrist joint and finger extension. Place the elbow of the patient's affected limb on the therapist's knee in order to reduce the burden on the shoulder and make it easier to facilitate the manipulation of the fingers. If there is spasticity or shortening in the finger flexor muscles, the therapist should support the affected hand in this position for as long as possible, maintaining supination of the forearm, dorsiflexion of the wrist and flexion in the fingers. (**b**) To promote flexion, the therapist uses his ring finger to tap the patient's finger. (**b**) The therapist should instruct the patient "Please grasp my finger", then hold the patient's fingers between his index and middle fingers and tap the patient's fingers on the affected hand. The therapist should then quickly perform pronation of the patient's forearm and tap the patient's fingers on the affected hand. (**c**) Following flexion of the fingers and supination of the forearm. (**c**) The therapist lifts his ring finger, which was tapping the affected hand, from the patient's palm and away from the patient's flexed fingers, which are lightly held between the therapist's index and middle fingers. Using his other hand, the therapist gently holds the patient's thumb and palm in order to support the patient's hand while the finger flexion movement causes supination of the forearm. When the fingertips of the affected hand approach the palm and the tips (nails) are aligned, flex the fingers as much as possible as if the patient's fingers are wrapping around the therapist's fingers. During these manipulations, the therapist's thumb is placed on the metacarpal of the affected hand (near the middle metacarpal joint (MP joint)). (**d**) Using his thumb, the therapist facilitates extension of the patient's fingers and pronation of the forearm. (**d**) The therapist gently loosens his index finger and middle finger from the patient's grasp and gently pushes the wrist joint in the direction of palmar flexion using his thumb on the metacarpal bone (close to the MP joint) to induce dorsiflexion of the wrist with finger extension and pronation of the forearm.

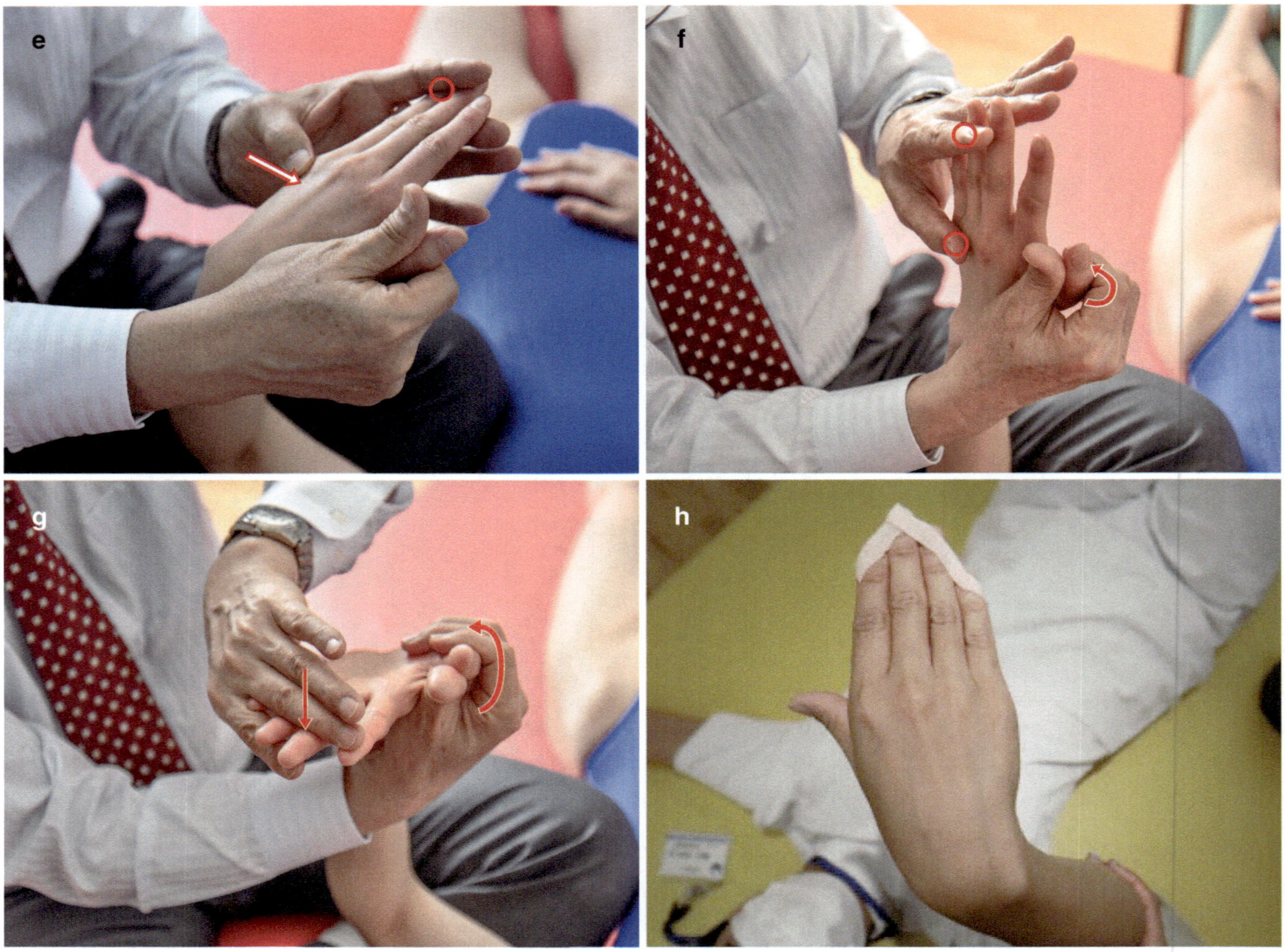

Fig. 4.34 (continued) (**e**) Continuing extension of the fingers, dorsal flexion of the wrist joint and pronation of the forearm. (**e**) Finger extension (including tenodesis), dorsal flexion of the wrist and pronation of the forearm occur. The therapist should assist the pronation of the patient's forearm using his other hand, which is holding the patient's thumb and thenar eminence. (**f**) Continuing extension of the fingers, dorsiflexion of the wrist joint and pronation of the forearm. (**f**) The therapist's thumb gently touches the metacarpals of the affected hand (close to the MP joint) and the dorsal side of the fingers. Be careful to avoid disturbing dorsiflexion of the wrist joint and extension of the fingers. When the wrist joint of the affected limb approaches the neutral position, the therapist pronates the forearm using the hand that is holding the patient's thumb and thenar eminence. (**g**) The limits of finger extension, dorsiflexion of the wrist and pronation of the forearm. (**g**) When dorsal flexion of the wrist joint and extension of the fingers and pronation of forearm reach their respective limits during active movement of the affected hand, the therapist should use the hand that is holding the patient's thumb and thenar eminence to passively perform pronation of the patient's forearm. Then using his middle finger, which is on the palmar side of the hand, the therapist should continue with passive dorsiflexion of the wrist, extension of fingers and pronation of the forearm. When the exercise reaches the limits of passive movement, return to (**b**). (**h**) Taping makes facilitation easier. (**h**) Repeating the procedures from (**a**)–(**g**) by active and passive movement is more efficient than conventional passive finger exercises. It becomes easier to prevent contractures and to treat edema, paralysis and sensory impairments. Tape the patient's fingertips from the little finger to the index finger in order to facilitate the movements until a therapist get used to the exercise. From the index to the little finger, the fingertips line up as an inverted V shape when the fingers are extended. When they are bent, however, the fingertips are aligned. Therefore, use a two-step process to tape the nails together. First, paste the tape on the nails of the index and middle fingers in an extended position; then use a separate piece of tape to attach the middle, ring and little finger. Taping the fingertips using this two-step process will avoid disturbing the natural movement of the fingertips from an inverted V to a straight line

4.8 Repetitive Facilitative Exercise for the Hand

The treatment plan should be based on the level of finger paralysis and changed as the patient's voluntary control of movement improves.

Key Points

1. Manipulation method for the hand and fingers

   The tendons that flex the fingers run along the volar side of the wrist joint while the extensor tendons run along the dorsal side of the wrist joint. Therefore, it is easier to extend the fingers when the wrist is in palmar flexion, while it is easier to flex the fingers when the wrist is dorsiflexed.

   In order to flex and extend the fingers extensively begin the exercise from the distal finger and proceed to the proximal finger in order to induce the intended movement.

2. Flexing the fingers individually (in the sitting and supine positions, respectively) (Figs. 4.35 and 4.36)

   For patients with severe paralysis who cannot flex or can only slightly flex the fingers, try to encourage intentional movement by flexing and extending the fingers even if all the fingers move together.

   For patients with flaccid paralysis, strong stimulation (vibration, rubbing, quick operation) will be necessary.

   When flexing individual fingers, the other fingers may also start flexing strongly due to increased spasticity. If this happens, stop the exercise and gently extend the non-

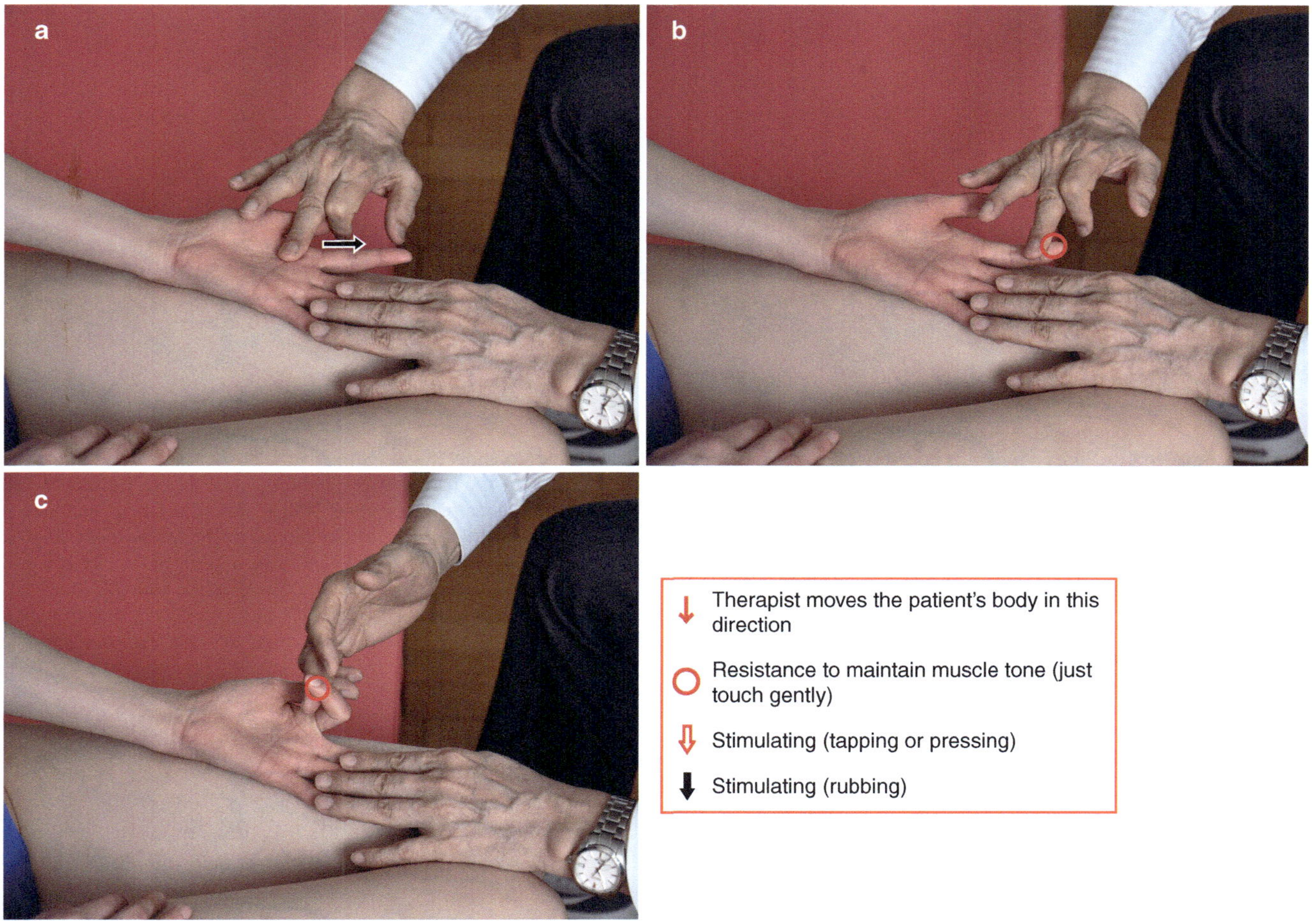

Fig. 4.35 Basic Technique: Flexing of individual fingers (in a sitting position) (Video 4.24). In order to improve the fine finger movements necessary to manipulate objects, the patient must be able to flex each individual finger. A patient with severe flaccid paralysis can start to improve flexion in the fingers by making a fist. In Repetitive Facilitative Exercise, finger flexion exercises for each of the four fingers can be effectively facilitated by having the patient grasp the therapist's middle finger. This technique for the index finger is shown as a typical example. (**a**) The index finger flexion and finger support (sitting position). (**a**) The therapist should place the patient's affected hand on the patient's thigh with the palm up, and press the patient's middle, ring and little finger lightly with the fingers on his other hand. (**b**) Flexion of the index finger. (**b**) The therapist should instruct the patient: "Please bend your index finger", then induce bending of the proximal interphalangeal (PIP) joint by rubbing the patient's affected hand from the palm to the tip of index finger. (**c**) Continuing flexion of the index finger. After the therapist's finger reaches the patient's fingertip, the therapist follows flexion of the patient's index finger. The therapist's fingers maintain physical contact with the patient's fingertips during flexion of the index finger. When the finger adjacent to the targeted finger begins to bend, the patient should be instructed, "Please relax", and then the therapist returns to the starting position (**a**)

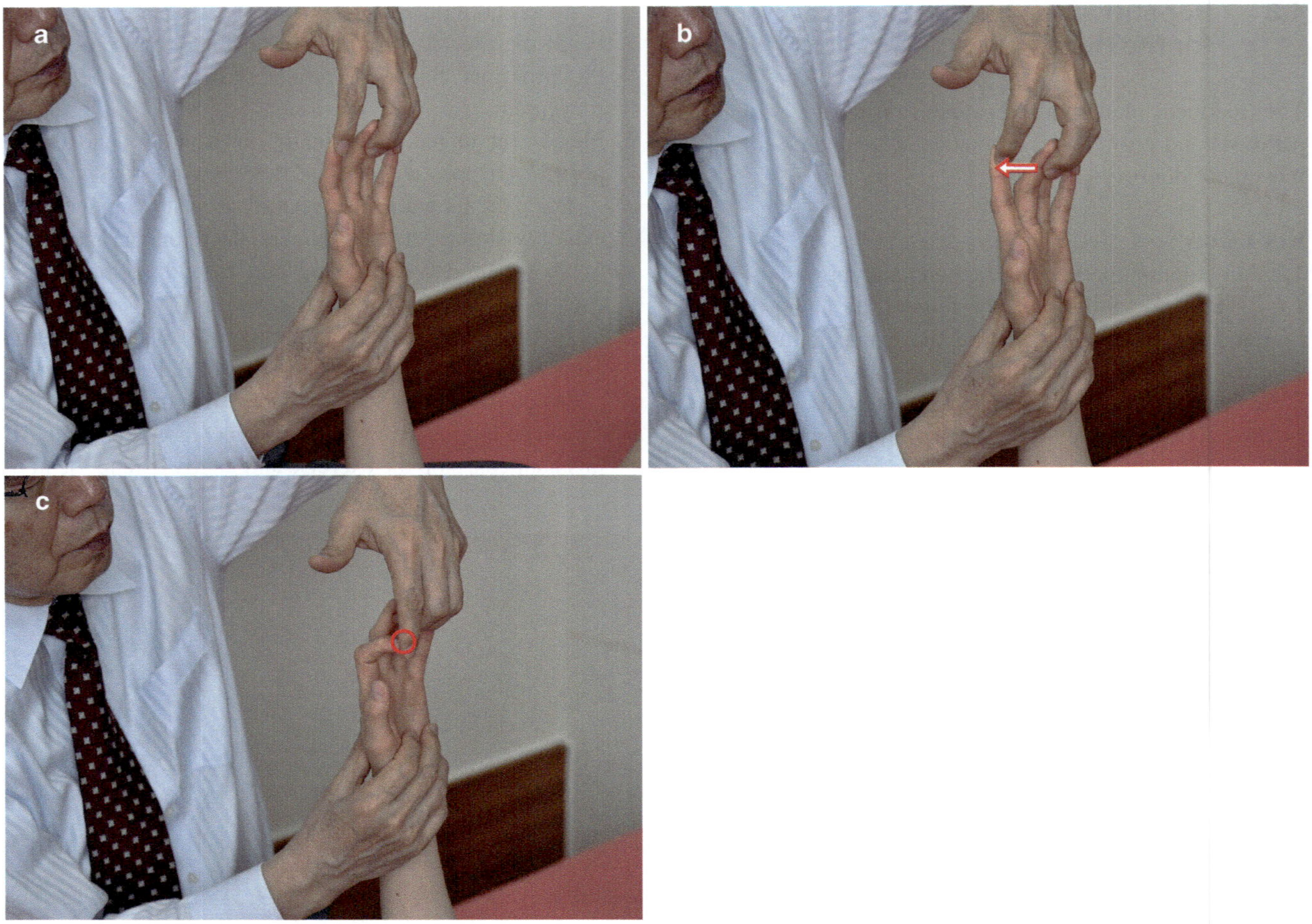

Fig. 4.36 Basic Technique: Flex the individual fingers (supine position). (a) Flexion of the index finger and how to support the fingers. (a) The therapist should place his index and middle fingers on the palm of the patient's affected hand and his thumb on the back of the patient's hand, then keeping the wrist joint held at dorsiflexion. The therapist places each of his fingertips on the individual fingertips of the patient's affected hand. The degree of dorsiflexion to induce the affected wrist joint depends on the state of the patient's flexor muscles. If the flexor muscles are flaccid, the dorsiflexion angle should be large, but in a patient with spasticity, the angle should be small. (b) Inducing a flexing movement by extending the index finger quickly. (b) Using his own index finger to apply pressure, the therapist should quickly extend the index finger on the patient's affected hand. Before this procedure, the therapist should instruct the patient "Please flex your finger" and induce a flexion movement. (c) Continuing flexion of the index finger. (c) The therapist should closely follow the movement of the patient's fingers so that his own finger, which is touching the tip of the patient's index finger, will not separate from the patient's fingertip during flexion movement. When the finger adjacent to the targeted finger begins to flex, the therapist should instruct the patient: "Please relax" and return to (a)

target fingers. Then start the individual finger flexing exercise again. Repeat the exercise as much as possible without inducing severe spasticity.

The effectiveness of repeatedly flexing the finger with low voluntary movement has not yet been proven. However, given that there is strong coupling between the primary motor area and primary sensory area and that it has previously been shown that the primary sensory cortex and premotor field will act on finger movement when the primary motor cortex is damaged, it seems reasonable to assume that repeatedly combining finger flexion with stretch reflex and the patient's intentional movement will contribute to recovery from paralysis.

3. Extension of individual fingers (supine position)

In patients who can grasp objects with all fingers but have difficulty extending the fingers and in those who can flex and extend all fingers together but cannot flex or extend the fingers individually, it is useful to perform Repetitive Facilitative Exercises in a supine position.

Considering the types of upper limb/hand movements performed during activities of daily living, it is important to improve voluntary control of the thumb, index finger, and middle finger in order to become able to manipulate various types of knobs.

(a) Extension/abduction of the thumb (Figs. 4.37 and 4.38)

Thumb abduction of the affected hand becomes easier with the wrist in palmar flexion.

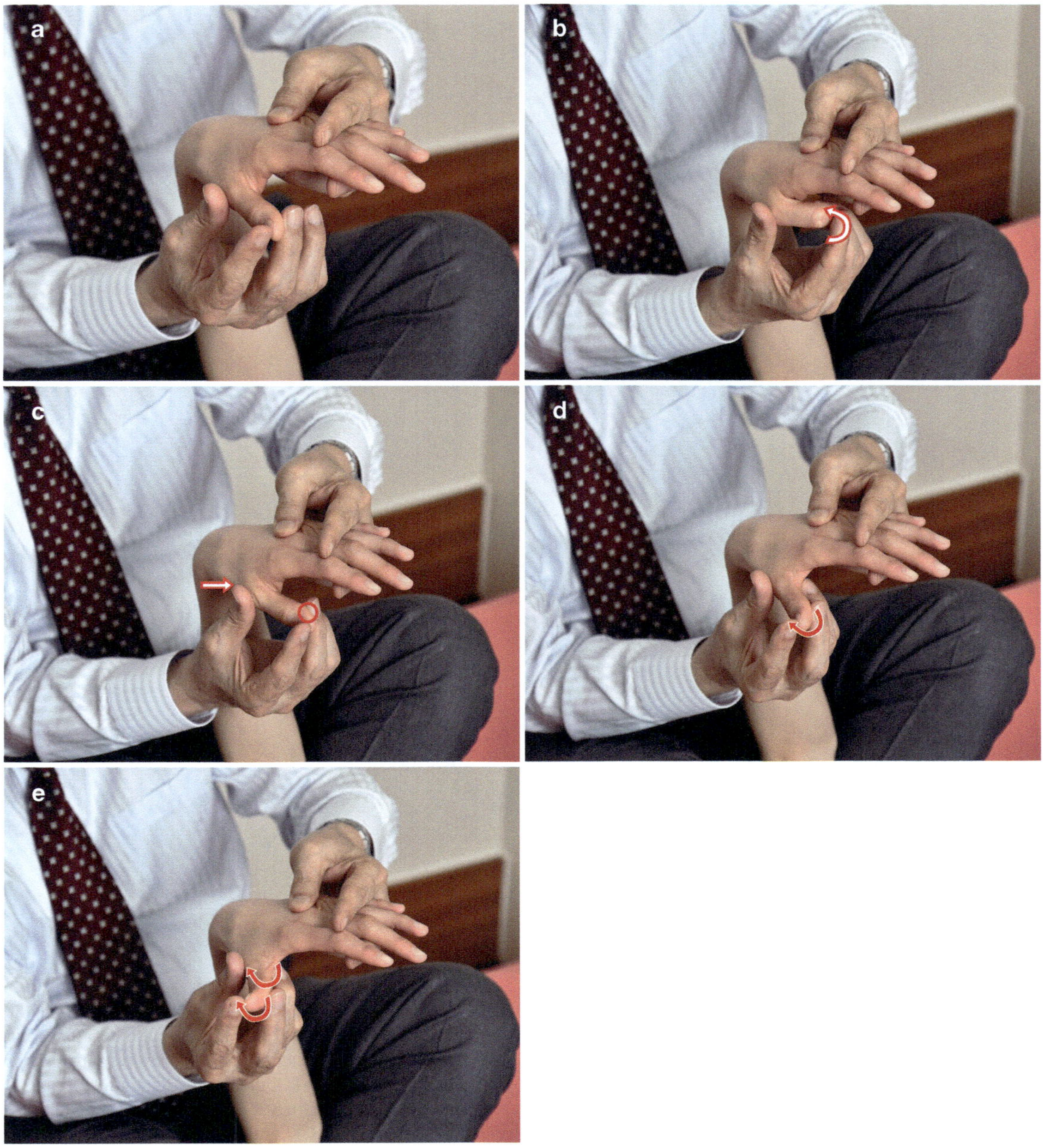

Fig. 4.37 Basic Technique: Extension and abduction of thumb instead of ROM-ex (supine position). Extension/abduction of the thumb is necessary for pinching things using the thumb and index finger. However, shortening of the soft tissue between the thumb and index finger may cause many restrictions to the range of motion needed for this maneuver. Even if the movement can only be achieved by a passive exercise due to severe paralysis, the therapist should still instruct the patient, "Please bend your thumb and finger" or "Please extend your thumb and finger." Ask the patient to make an effort to move their own fingers and perform the Repetitive Facilitative Exercise together. (**a**) Supporting the thumb and the other four fingers. (**a**) The therapist places his thumb on the metacarpal bone of the patient's thumb and places his index finger and middle finger on the tip of the patient's thumb. Using his other hand, the therapist holds the patient's palm between his thumb and ring and little fingers, while supporting the patient's fingers between his index and middle fingers. The patient's wrist should be held in a palmar flexed position. (**b**) Facilitating abduction of the thumb by quick adduction. (**b**) The therapist quickly adducts and flexes the thumb of the affected hand, which is held between the therapist's thumb and the index finger. With the thumb pushing diagonally forward, the therapist then instructs the patient "Please move your thumb outward," while the therapist pushes against the MP joint. (**c**) Facilitation of abduction of the thumb by pushing the MP joint. (**c**) The therapist pushes obliquely forward on the MP joint of the thumb following the thumb movement to avoid disturbing the extension/abduction movement of the thumb (including the tenodesis response). (**d**) Assisting extension and abduction of thumb. (**d**) When the limits of the extension/abduction movement of the thumb of the affected hand have been reached, the therapist uses his index and middle finger to support further extension/abduction of the patient's finger if the range of extension/abduction has been insufficient. (**e**) Assisting up to the maximal extension and abduction of the thumb. (**e**) The therapist uses his index and middle fingers to abduct the patient's thumb. When the movement reaches its limit, the therapist uses his ring finger, which has been repositioned on the lateral front (palm side) of the MP joint, to gently abduct the thumb further

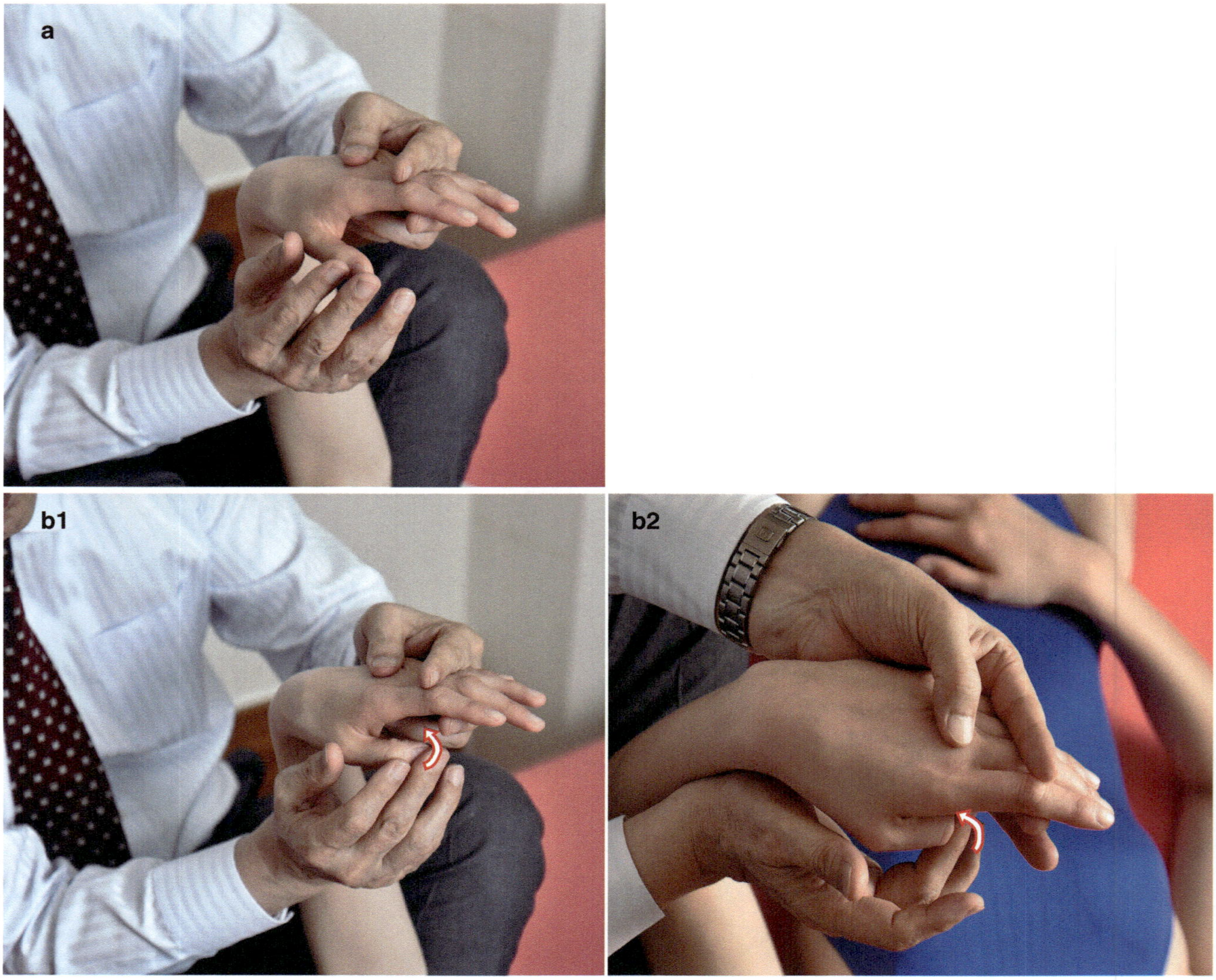

Fig. 4.38 Let's Try: Extension and abduction of the thumb (example in a patient with well-preserved voluntary movement) (supine position) (Video 4.25). (**a**) Supporting the finger extension with the wrist in palmar flexion. (**a**) Supporting the other four fingers. The therapist should fix the patient's elbow on his knee in order to stabilize the affected upper limb. Then the therapist should place his thumb on the dorsum of the patient's hand, while supporting the patient's fingers between his ring and little fingers. The wrist joint should be in palmar flexion. All of the patient's fingers from the index to the little finger should be held together in an extended position between the therapist's index and middle fingers. The therapist should place the thumb of his other hand on the patient's metacarpal bone near the MP joint and place his index finger on the proximal phalanx of the thumb with the tip of his middle finger on the thumb nail. Positioning the patient's affected hand in front of her face makes it easier to facilitate extension of the thumb. (**b1**) Flexion of the thumb. (**b1**) Quickly flex the thumb on the affected hand and simultaneously instruct the patient, "Please extend your thumb." Flex the thumb MP joint slightly. The maneuver becomes difficult if the thumb is fully flexed. (**b2**) Viewed from another direction. (**b2**) Supporting the four fingers and quickly bending the thumb. Note: In order to see the therapist's hand position in this photograph more easily, the image shows the location of the therapist's thumb and index finger is slightly apart from the patient's thumb.

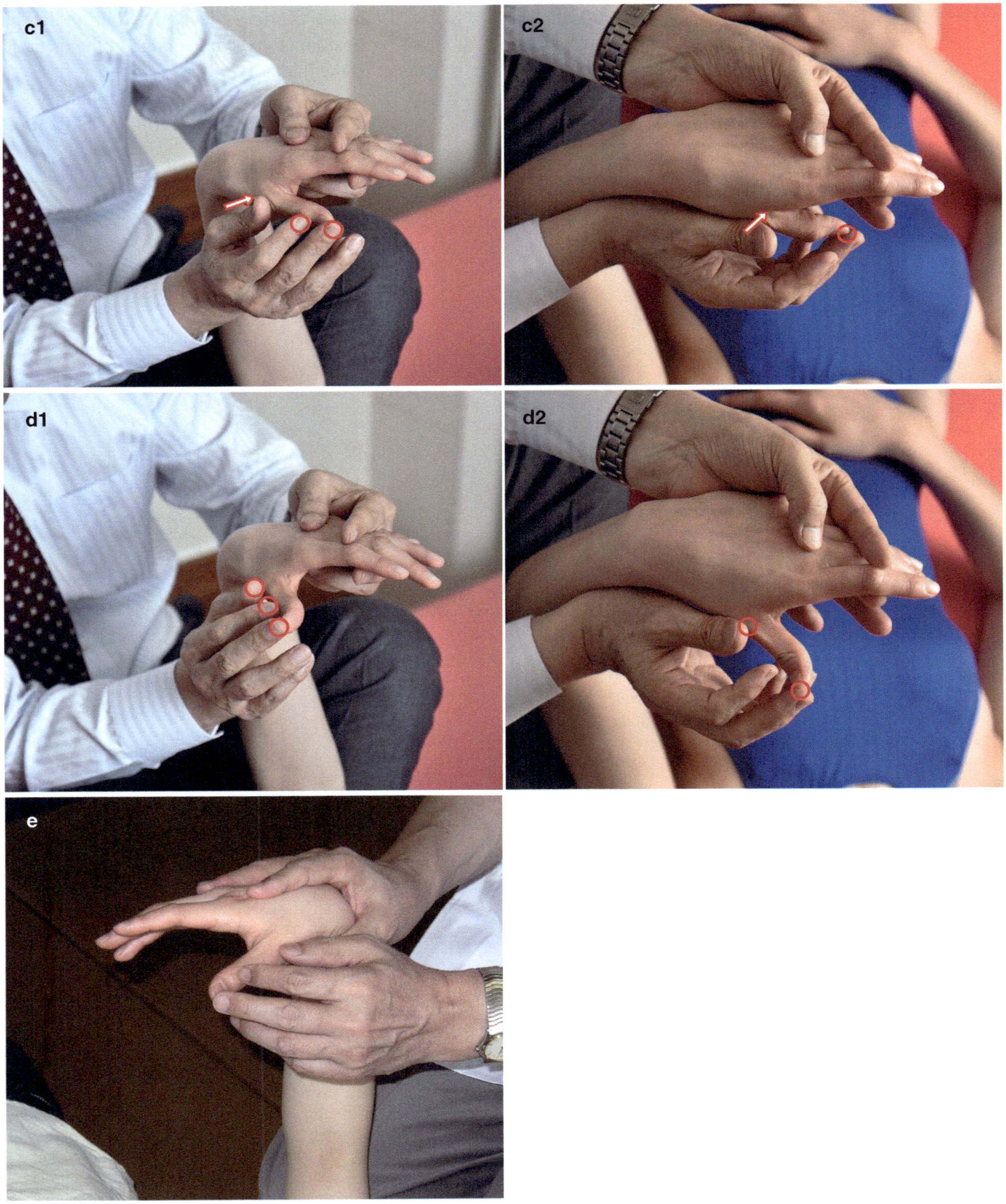

Fig. 4.38 (continued) (**c1**) Extension and abduction of thumb. (**c1**) The therapist applies pressure close to the MP joint so that the metacarpal bone is pushed in the direction of adduction, which induces extension of the interphalangeal (IP) joint. (**c2**) Viewed from another direction. (**c2**) The therapist gradually reduces pressure on the metacarpals as the thumb extends. (**d1**) Extension and abduction of the thumb. (**d1**) The therapist uses his thumb to hold the carpometacarpal joint of the patient's thumb in an adducted position, while allowing the thumb to extend from the IP joint. By allowing abduction of the MP joint at the end of extension, it is possible for the patient to fully extend the thumb. The goal is for the whole thumb to be able to extend and abduct. When extension/abduction reaches the limit, return to (**b**). (**d2**) Viewed from another angle. (**d2**) By simply touching the appropriate points on the patient's thumb, the therapist can apply weak resistance for extension and abduction. (**e**) Supporting the wrist joint only. (**e**) How to fix the wrist joint at palmar flexion. The therapist should grasp the wrist joint of the affected hand with his thumb, ring and little finger at the back of the patient's wrist. Then hold the back of the patient's hand with his index and middle fingers so as to keep the palm flexion. The support position is good for patients who can extend their index and middle fingers, which do not interfere with flexing and extending the thumb

Since the interphalangeal joint (IP) often flexes when the thumb is abducted, it is important to keep the IP joint extended during abduction of the thumb.

(b) Thumb/finger opposition (palmar abduction) (Figs. 4.39 and 4.40)

To induce palmar abduction and bring the thumb into opposition, tap the outer edge of the thenar eminence of the thumb.

In order to restore sufficient functionality for the affected hand to become useful in daily life, the

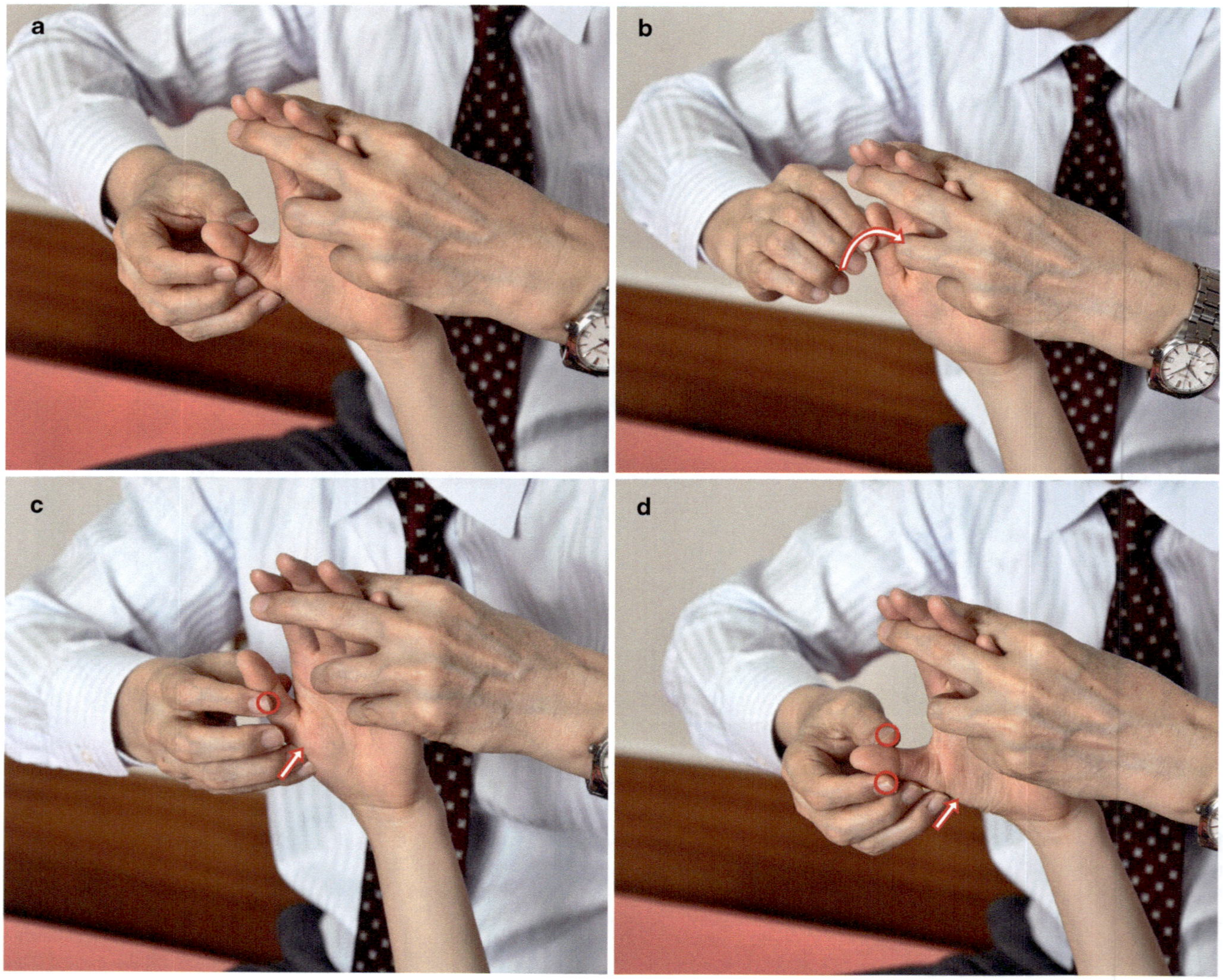

Fig. 4.39 Let's Try: Palmar abduction of the thumb (Spine position) (Video 4.26). (**a**) Supporting the thumb and the other fingers. Palmar abduction of the thumb is a necessary movement for manipulating objects. However, it is unlikely that the thumb will spontaneously recover the ability to confront the index finger (opposition of the thumb). Usually, the Repetitive Facilitative Exercise is done with the wrist joint at 0°. However, in patients with severe spasticity of the flexor muscle, the thumb will be flexed when the wrist joint is at 0°. Therefore, the wrist should be maintained in palmar flexion in order to keep the thumb extended. (**a**) The therapist should lift his elbow and position his forearm perpendicular to the axis of the patient's thumb in palmar abduction. The therapist should then use his thumb and index finger to hold the thumb of the patient's affected hand, while placing his ring and little finger on the radial side of the patient's thumb. Using the thumb and fingers on his other hand, the therapist should try to keep the patient's fingers extended. (**b**) Adduct the thumb quickly. (**b**) The therapist should use his thumb and index finger to quickly adduct the thumb on the patient's affected hand. Then the therapist internally pronates his forearm and releases his ring and little fingers from the patient's thenar muscles, while instructing the patient, "Please extend your thumb outward." (**c**) Tapping at the base of the thenar muscles and inducing palmar abduction. (**c**) The therapist supinates the forearm of the hand holding the patient's thumb and taps the patient's thenar muscles on the radial side with his ring and little finger. If the patient's palmar abduction movement is insufficient, the therapist should apply rubbing to the patient's thenar muscles in the direction of the wrist joint with his ring and little finger. In order to tap the same part of the thenar muscles quickly and repeatedly, it is important to tap while holding the ring and little finger in a flexed position and to rotate the forearm supination and pronation as needed. (**d**) The end of palmar abduction. (**d**) When palmar abduction of the thumb reaches its limit, return to (**b**)

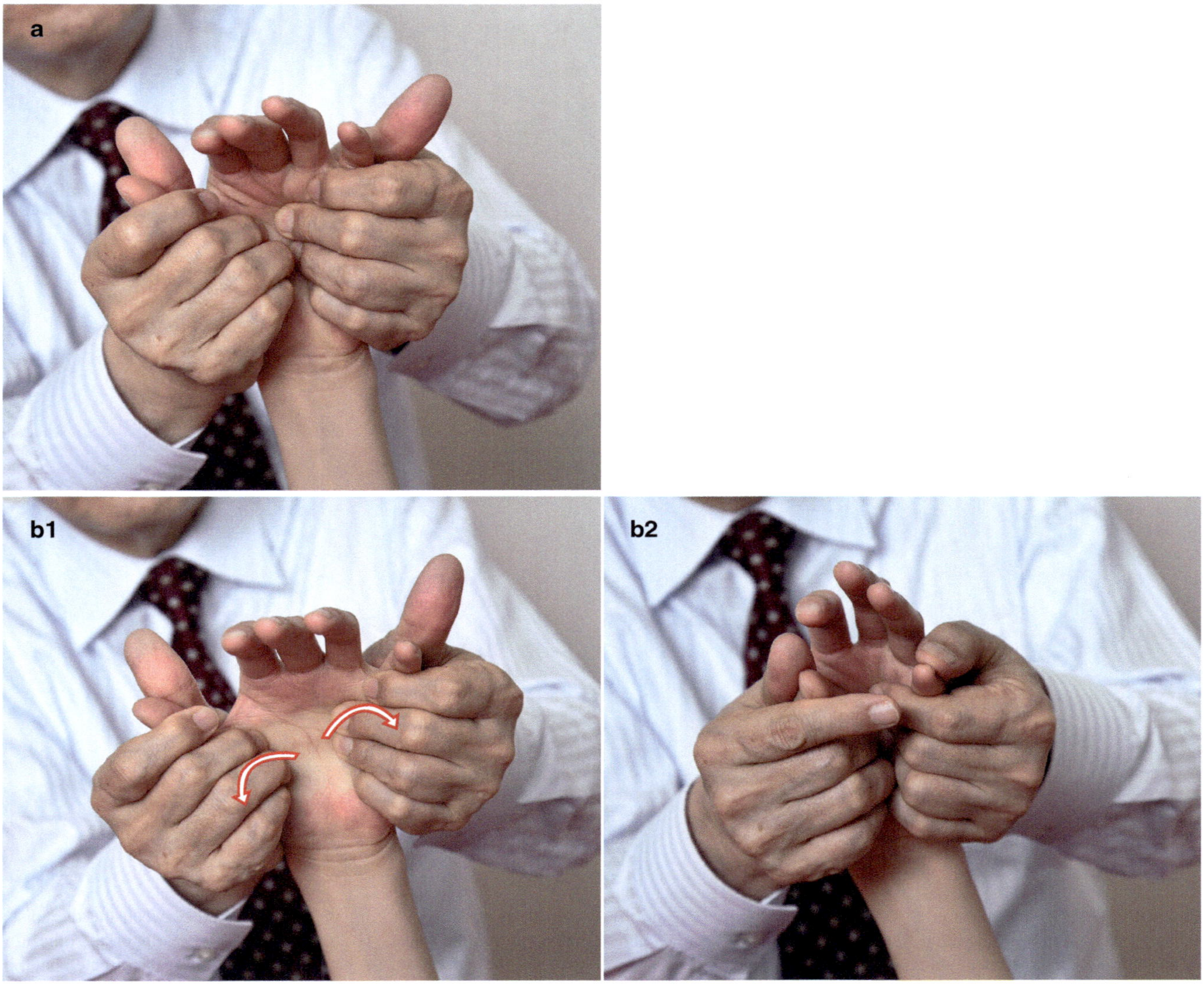

Fig. 4.40 Let's Try: Abduction of the fingers and stretching the arch (supine position) (Video 4.27). (**a**) Maintaining a range of motion sufficient to allow the thumb to pinch the thumb with each finger. (**a**) In order to sufficiently enable opposition of the thumb, it is necessary to prevent shortening or atrophy of the skin between the thumb and the index finger by performing gentle radial abduction of the thumb beginning in the early stages after onset. In order to maintain the thumb's ability to oppose each of the other fingers include the little finger, the therapist should use the ring and little fingers of both hands to grasp the thenar muscles and hypothenar eminence muscle of the patient's hand and stretch the tissue outward. While stretching the patient's fingers outward, the therapist should instruct the patient "Please spread your fingers outwards." During this maneuver, the patient's palm will form a large arch and the space between the thumb and the index finger will be stretched. (**b1**) Stretching the thumb and other fingers outward. (**b2**) Place the thumb and little finger in opposition. (**b**) When the therapist is stretching the patient's affected hand outward using both of his hands, he should slide the fingers of both of his hands outward so as to avoid applying excessive force on the patient's hand. When the distance between the thumb and little finger widens, the therapist should perform the maneuver with one hand placed on the patient's thumb and the other hand on the patient's little finger so that the patient's fingers assume opposing positions. During this procedure, the therapist should guide the patient's movement without holding the tips of the patient's thumb and little finger since that could injure the patient's finger joints

thumb needs to move in opposition to the remaining fingers (palmar abduction).

It is important for the tip of the thumb to be able to make a circle with each of the fingertips of the other four fingers, with emphasis on the index finger, the middle finger, and the ring finger.

When the patient cannot pinch the thumb with each finger, it may be because abduction of the thumb is insufficient due to atrophy of the skin between the thumb and index finger or because voluntary control of the fingers is insufficient.

Even after thumb abduction of the affected hand has become possible, it is often quite difficult for the patient to bring the tip of the thumb into opposition with each of the other four fingertips.

(c) Extension of the index finger (Fig. 4.41)

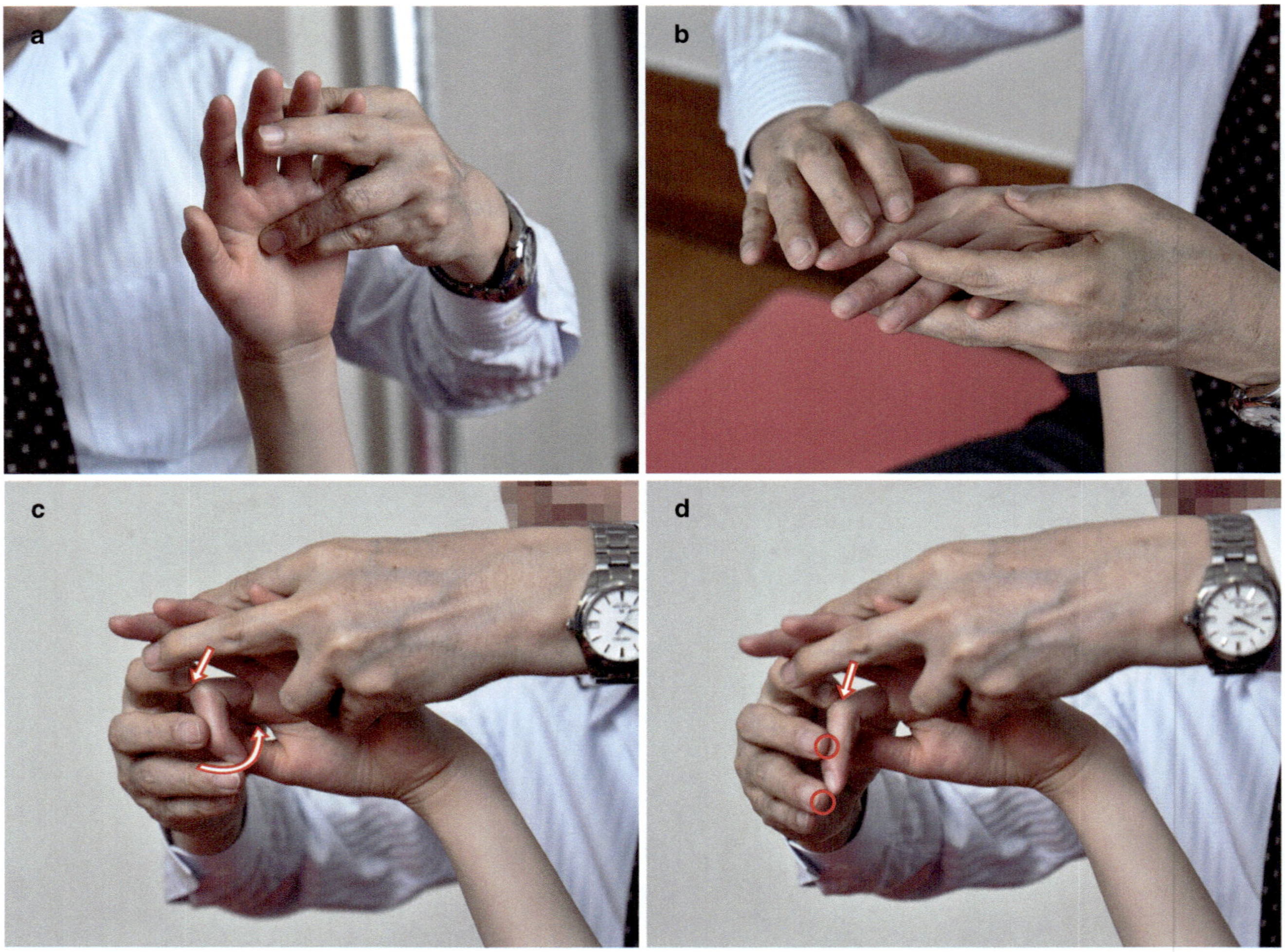

Fig. 4.41 Let's Try: Extension of the index finger (supine position) (Video 4.28). (**a**) Supporting the patient's palm and fingers (Shown here in an intermediate position so that the position of the therapist's fingers can be seen.) The therapist's basic finger position and the method used to support the patient's finger extension during Repetitive Facilitative Exercise is shown in (**a**), and the sequence of procedures for Repetitive Facilitative Exercise are shown from (**b**) to (**f**). Stimulating the patient's fingers by rapid finger bending (**b**)–(**c**) and MP joint bending (**d**) can induce extension of the patient's fingers. The therapist should lift his elbow and position the forearm perpendicularly to the movement of the patient's fingers and perform operations (**b**)–(**c**) while performing supination or pronation of his forearm. The therapist can evaluate the results of this procedure by observing the patient's range of finger extension and adjust the speed and strength of the procedure as needed. For patients with limited voluntary movement (finger extension is slight or slow), MP joint flexion should be performed gently at a slow speed. In patients with good voluntary movement (ability to move quickly as intended) tapping on the MP joint is effective. However, tapping should be quite gentle and the therapist's fingers should not actually break contact with the patient's finger (fingertip or nail) while the patient is extending her finger. (**a**) Using therapist's thumb, ring finger and little finger, the therapist holds the palm of the patient's affected hand. The patient's other fingers, except the index finger, are supported between the therapist's index and middle fingers. (**b**) Supporting the index finger, the other fingers and the wrist joint. (**b**) Keep the wrist joint of the affected hand in palmar flexion. The therapist places the thumb of his other hand on the back of the patient's affected hand. Then, without his middle finger making contact, he places his index finger on the base of the patient's PIP joint and put his ring finger on the patient's distal phalanx (nail). The MP joints of the other three fingers should be maintained in an extension position so that the other fingers are extended and form a straight plane with the back of the hand. When the MP joint of the other fingers are flexed, the index finger is prevented from extending. (**c**) Method of promoting extension of the index finger. (**c**) Using the therapist's ring finger, the therapist quickly supinates his forearm then flexes the patient's index finger. (The MP joint should initially be at 0° to 10° since hyperextension makes this maneuver difficult.) The therapist should instruct the patient, "Please extend your finger," while using his own index finger to tap or push the PIP joint of the patient's index finger in order to flex the MP joint of the patient's finger. (**d**) Continuing extension of the index finger. (**d**) The therapist pronates his forearm to follow the extension of the PIP and DIP joints in response to MP joint flexion of the index finger. Using his ring finger, the therapist should continue to lightly touch the patient's index finger.

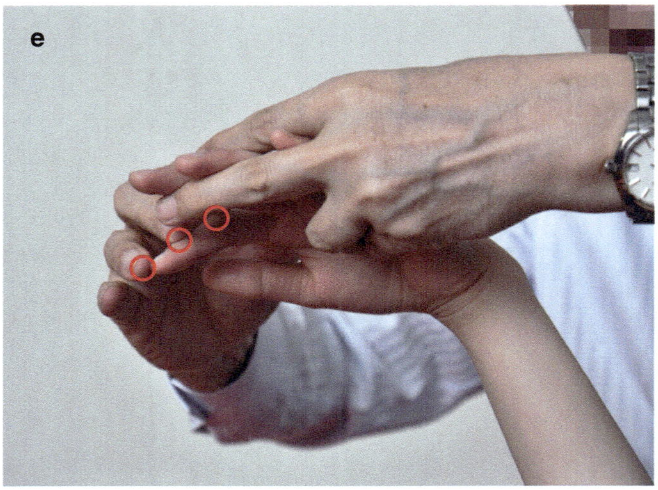

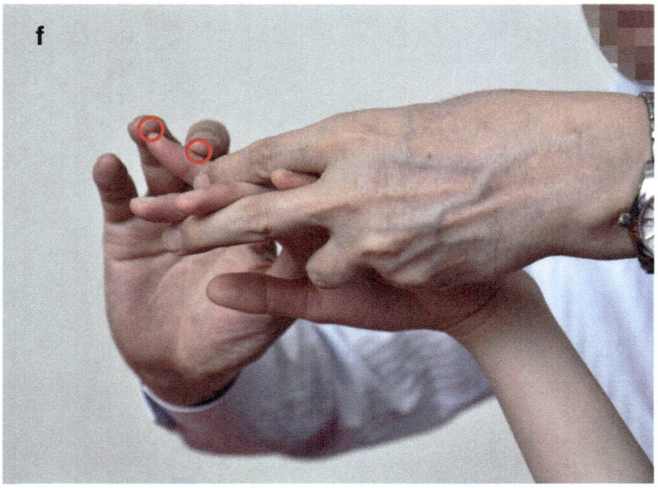

Fig. 4.41 (continued) (**e**) Continuing extension of the index finger. (**e**) The therapist should pronate his forearm so that his index and ring finger remain in contact with the patient's PIP and DIP joint while the index finger of the affected hand is extending. (**f**) From the end of index finger extension to the starting position. (**f**) The therapist should continue touching the patient's index finger lightly and when extension of the index finger reaches its limit, return to (**c**)

When starting extension exercises for the index finger, rest the triceps of the patient's upper arm on the therapist's knee and position the affected hand in front of the patient's face. After quickly flexing the index finger, allow its extension from the distal finger joint to the proximal finger joint.

(d) Extension of the middle finger (Fig. 4.42)

Extension of the middle finger tends to become easier after quickly flexing it, then allowing its extension from the distal finger joint to the proximal finger joint.

(e) Extension of the ring finger (Fig. 4.43)

Again, extension of the ring finger tends to become easier after quickly flexing it, then allowing its extension from the distal finger joint to the proximal finger joint.

(f) Extension of little finger (Fig. 4.44)

As described above, extension of the little finger tends to become easier after quickly flexing it, then allowing its extension from the distal finger joint to the proximal finger joint.

4. Flexion and extension of individual fingers (while sitting)

When both flexion and extension of the fingers become possible, training in flexing and extending individual fingers can be started.

The therapist sits in front of the patient and places the patient's affected hand on the patient's thigh with the palm facing up and the wrist slightly flexed (0–10° palmar flexion).

If it is difficult to suppress movement of the fingers that are not being treated, ask the patient to hold those other fingers with the unaffected hand in order to suppress movement of those fingers.

First instruct the patient, "Do not try to extend this finger so much that you also move the other fingers." Emphasize that the patient should only make an effort to move the treated finger without inducing movement of the adjacent fingers.

During the initial finger flexing and extending exercise, it is necessary to avoid moving the adjacent fingers from an excessive effort to fully extend or flex the fingers. Ask the patient to adjust their effort appropriately.

In patients who lack a range of movement in flexion and extension, the therapist needs to not only provide support for the flexion and extension movement but also induce quick flexion or extension so as to lead to the next exercise. Manipulate the affected finger so that it easily flexes and extends rhythmically.

When there is limited flexion/extension of the fingers, the therapist can hold the palm of the patient's affected hand between the therapist's index and middle fingers. The wrist joint of the affected hand should be dorsiflexed while performing the finger flexion facilitation exercise. The movement should then change to palmar flexion during the finger extension facilitation exercise.

(a) Flexion/adduction and extension/abduction of the thumb (Fig. 4.45)

When the wrist joint of the affected limb is in palmar flexion, the thumb can easily abduct and extend, and when the wrist joint is dorsiflexed, the thumb tends to flex.

When the range of flexion/extension of the thumb is limited, combining palmar flexion/dorsiflexion of the wrist joint with flexion/extension of the thumb makes the Repetitive Facilitative Exercise easier.

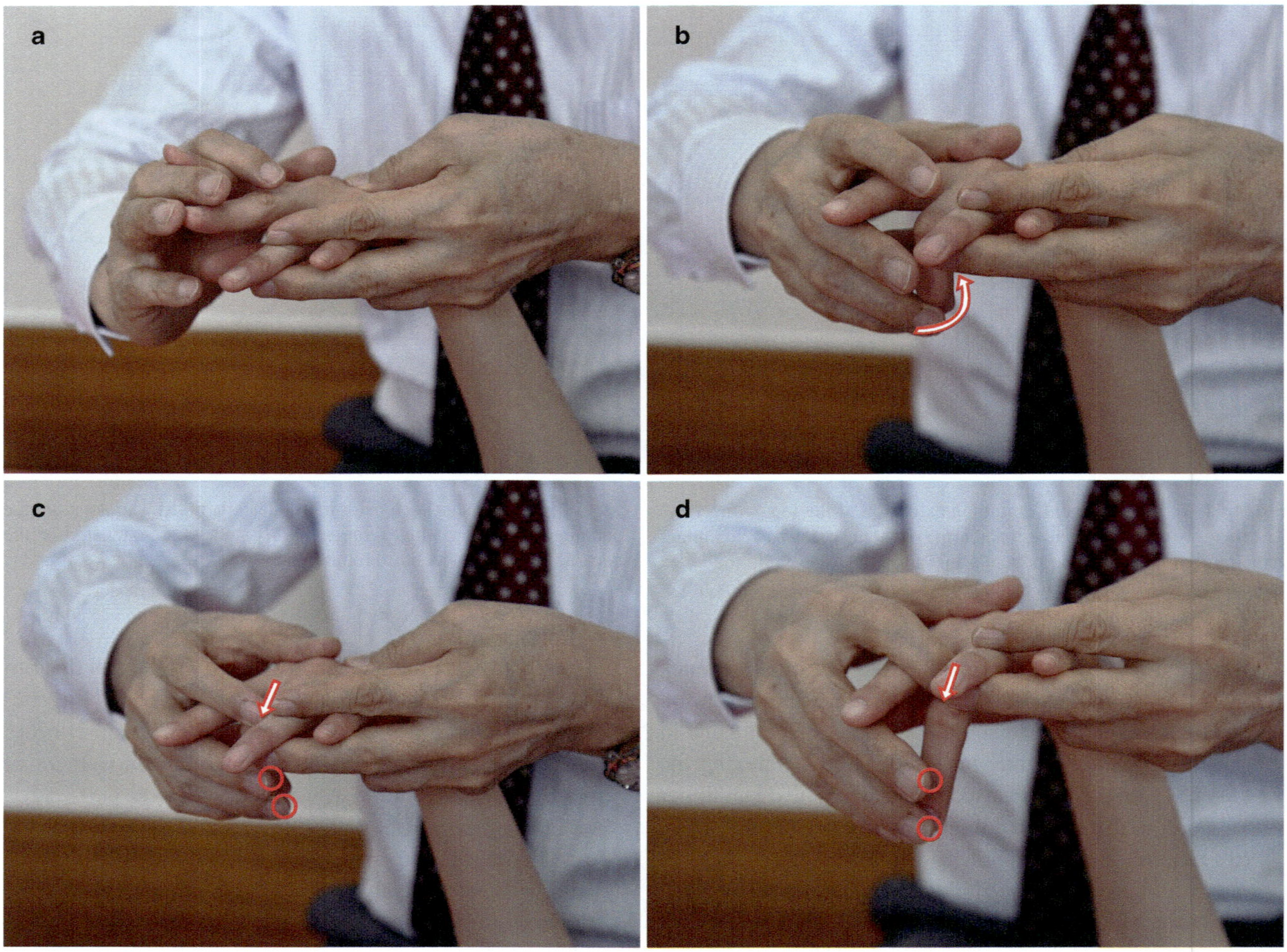

Fig. 4.42 Let's Try: Extension of middle finger (supine position) (Video 4.29). (**a**) Support for the middle finger as well as the other fingers. (**a**) Using therapist's thumb, ring and little finger, the therapist should grasp the palm of the patient's affected hand and place the patient's wrist in palmar flexion. The therapist should hold the patient's ring and little fingers between his index and middle fingers. Then the therapist should place the thumb of his other hand on the back of the patient's affected hand. The therapist's index finger should be placed on the PIP joint of the patient's middle finger's proximal phalanx. While his middle finger should be placed under the patient's index finger to keep its extension, and his ring finger should be placed on the tip of the patient's distal phalanx (nail). The patient's finger is then placed in an extended position so that the MP joint forms a straight plane with the back of the hand. When the MP joint for the other fingers are flexed, the middle finger is prevented from extending. (**b**) Quick flexion of the middle finger. (**b**) Using his ring finger, the therapist quickly flexes the middle finger on the patient's affected hand. (**c**) Repetitive Facilitative Exercise for extension of the middle finger. (**c**) When the middle finger on the patient's affected hand is strongly flexed, the therapist should instruct the patient, "Please extend your middle finger," while simultaneously using his index finger to tap the patient's middle finger at the proximal PIP joint. (**d**) Continuing extension of the middle finger while the finger is flexed at the MP joint. (**d**) The therapist should hold the patient's wrist in palmar flexion. Placing his index finger at the PIP joint of the patient's middle finger on the affected hand to flex the MP joint, the therapist can perform Repetitive Facilitative Exercise for extension movement of the PIP and DIP joint. The therapist's ring finger placed on the DIP joints and nails maintain light contact with the patient's middle finger during finger extension movement.

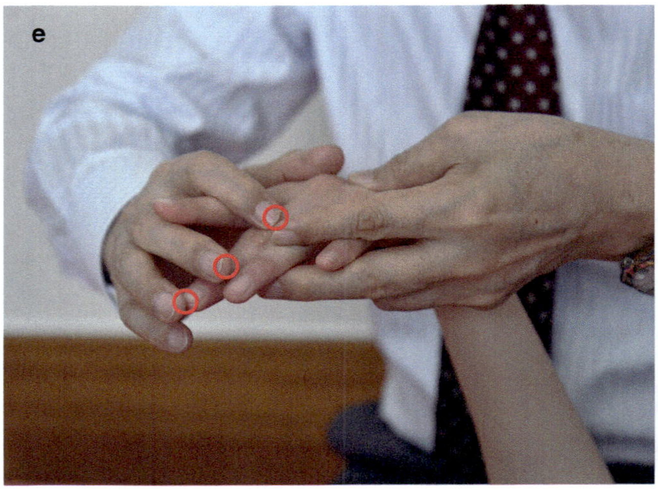

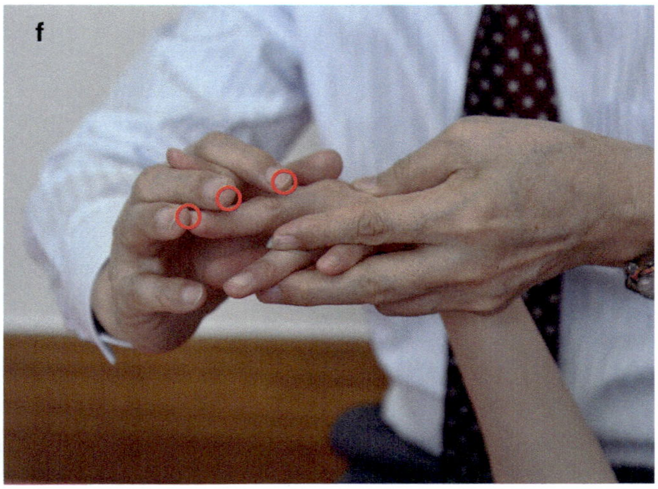

Fig. 4.42 (continued) (**e**) Continuing extension of the middle finger. (**e**) The therapist pronates his forearm and keeps each of his fingers in contact with the patient's middle finger while inducing a smooth maximal extension of the middle finger. (**f**) The end of middle finger extension. (**f**) When extension of the patient's middle finger reaches the limit, return to (**b**)

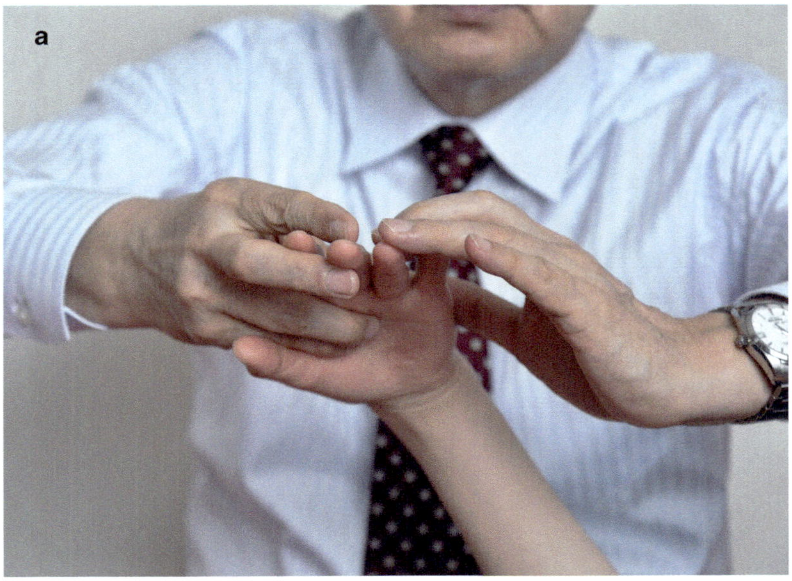

Fig. 4.43 Let's Try: Extension of ring finger (supine position) (Video 4.30). (**a**) Holding the patient's three fingers on the affected hand. (**a**) Holding the patient's affected hand between his thumb, ring and little finger on one hand, the therapist should place the patient's wrist joint in palmar flexion. Using his other hand, the therapist holds the patient's middle and index fingers between his own index and middle fingers, while placing his thumb on the back of the patient's affected hand. With the therapist's hands in this position, both of his thumbs are on the back of the patient's hand. Then the therapist's index finger is placed on the PIP joint of the proximal phalanx of the patient's ring finger, the therapist's middle finger is placed under the patient's middle phalanx, and his ring finger is placed on the tip of the patient's distal phalanx (nail). The patient's finger should be extended so that the MP joint is extended and forms a horizontal plane from the finger to the back of the affected hand, providing support for the ring finger as well as the other fingers.

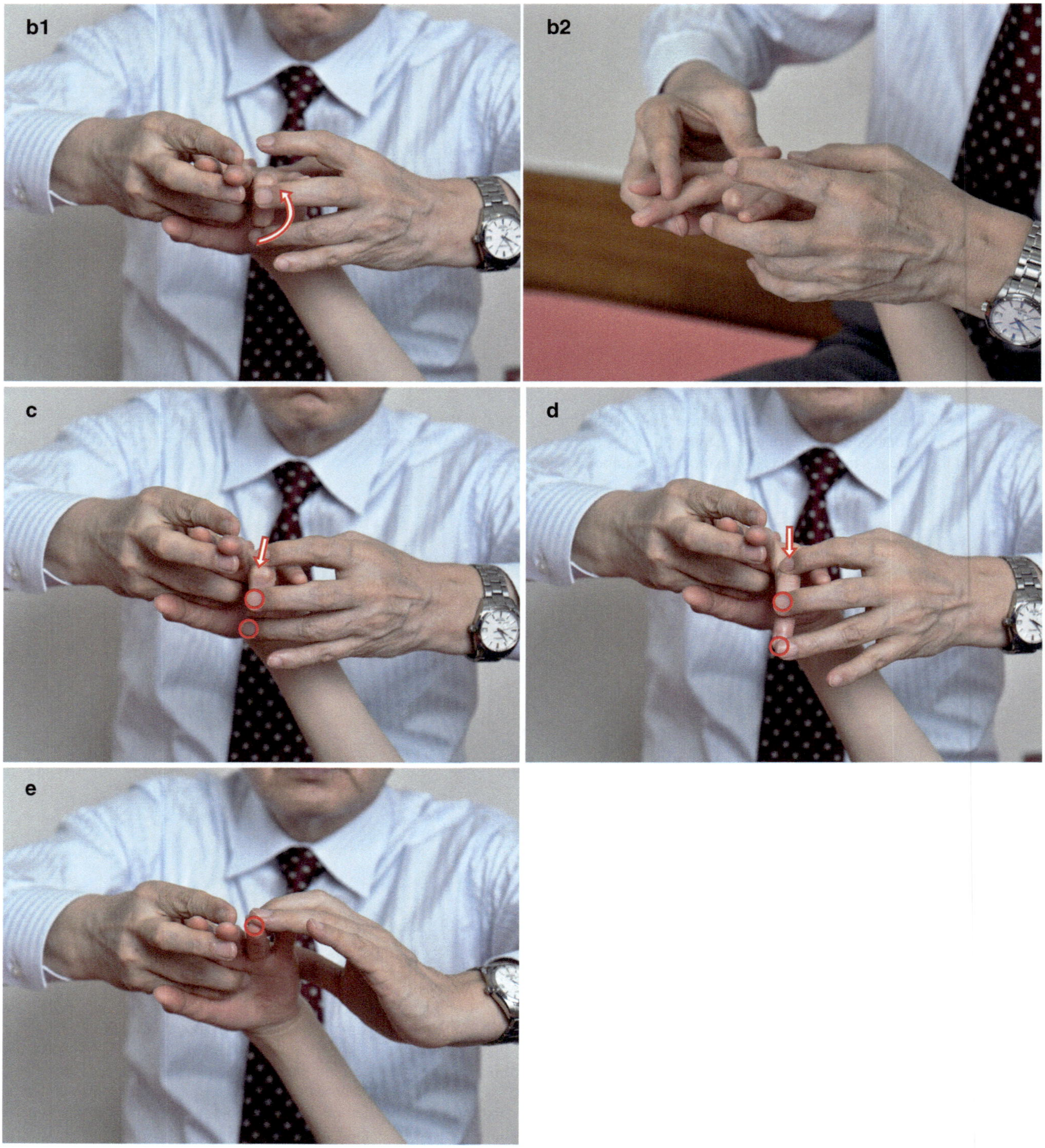

Fig. 4.43 (continued) (**b1**) Rapid flexion of the ring finger. (**b1**) Using therapist's own ring finger, the therapist should quickly flex the patient's ring finger. (**b2**) Facilitation for finger extension is performed by bending the MP joint of the ring finger (A different view). (**b2**) The therapist's index and middle fingers are used to hold the index and middle fingers on the patient's affected hand. The therapist's thumb, ring and little finger are used to keep the wrist joint in palmar flexion. (The position of these fingers is not visible in this photograph because the therapist's fingers are under the patient's palm.) (**c**) Facilitation for finger extension using MP joint flexion of the ring finger. (**c**) To facilitate extension of the ring finger, the therapist instructs the patient, "Please extend your ring finger," while flexing the MP joint and tapping or pressing the patient's ring finger with his index finger. (**d**) Continuing extension of the ring finger. (**d**) The therapist pronates his forearm in order to follow the extension of the PIP joint and DIP joint caused by flexion of the MP joint of the patient's index finger. Continuous light contact from the therapist's ring finger is maintained with the patient's ring finger. (**e**) The end of extension of the ring finger. (**e**) When extension of the ring finger on the affected hand reaches its limit, return to (**b**)

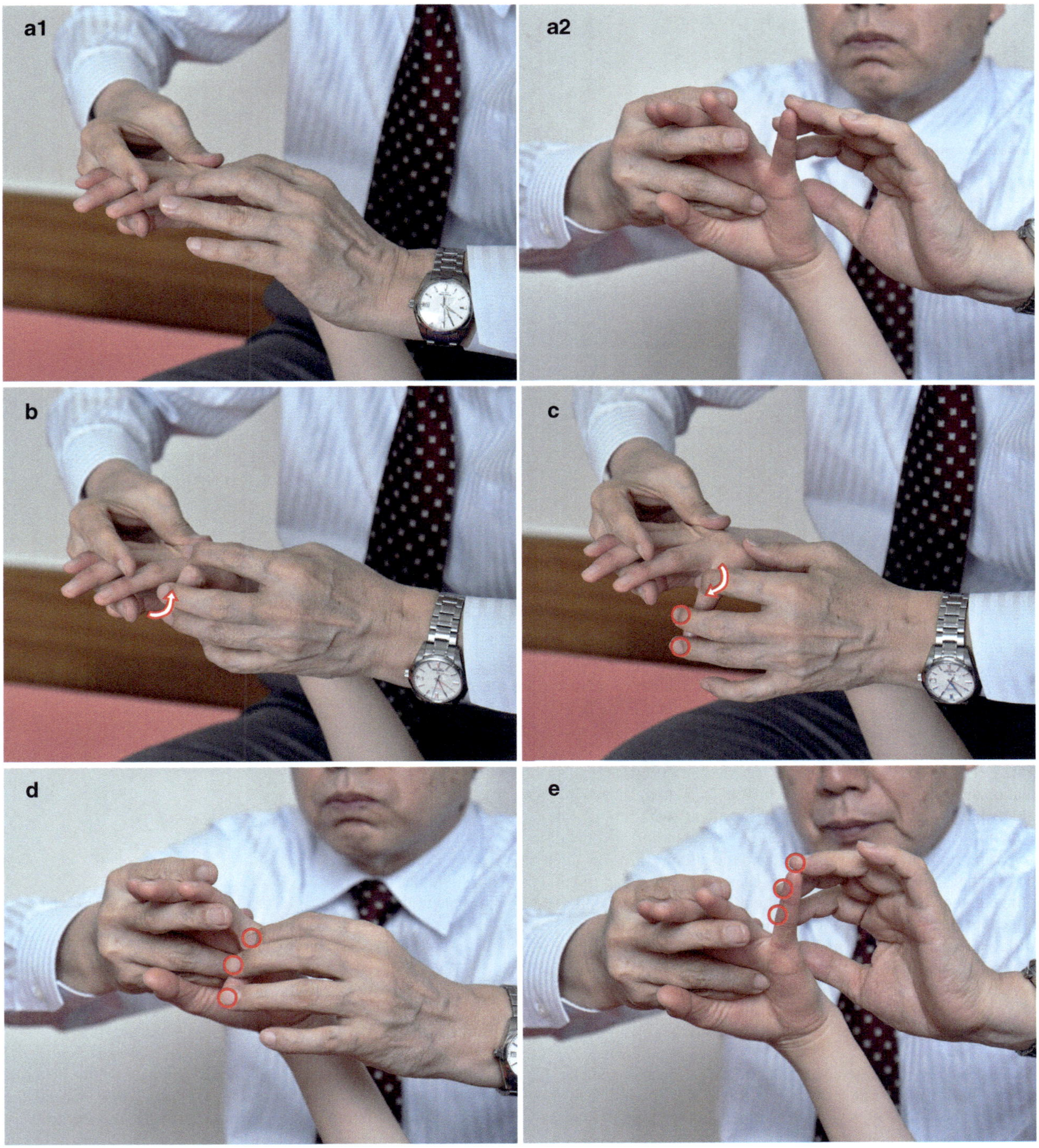

Fig. 4.44 Let's Try: Extension of the little finger (supine position) (Video 4.31). (**a1**) Support for the little finger as well as the other fingers. (**a1**) Using the thumb, ring and little finger on one hand, the therapist holds the patient's affected hand with the wrist joint in palmar flexion. The therapist then uses his index and middle fingers to hold the patient's index, middle and ring fingers. The therapist places the thumb of his other hand on the back of the patient's affected hand and places his index finger on the PIP joint of the proximal phalanx of the patient's little finger, and his ring finger on the tip of the distal phalanx (nail). The patient's MP joint should be extended to form a horizontal plane from the fingers to the back of the affected hand. (**a2**) Support for the little finger as well as the other fingers (shown from a different angle). (**a2**) The angle of this photograph was changed in order to show the positional relationships between the therapist's fingers and the patient's hand. (**b**) Quickly bending the patient's little finger. (**b**) Using the therapist's ring finger, he quickly flexes the patient's little finger on the affected hand. (**c**) Facilitation for finger extension by flexion of the MP joint of the little finger. (**c**) The therapist should instruct the patient, "Please extend your little finger," while tapping or pushing with his index finger to flex the MP joint in order to induce extension of the patient's little finger. (**d**) Continuing finger extension by flexion of the MP joint of the little finger. (**d**) The therapist continues to use his index and ring fingers to lightly touch the patient's little finger while rotating his forearm so as to follow the extension of the patient's little finger. (**e**) The end of extension of the little finger. (**e**) When extension of the little finger on the affected hand reaches its limit, return to (**b**)

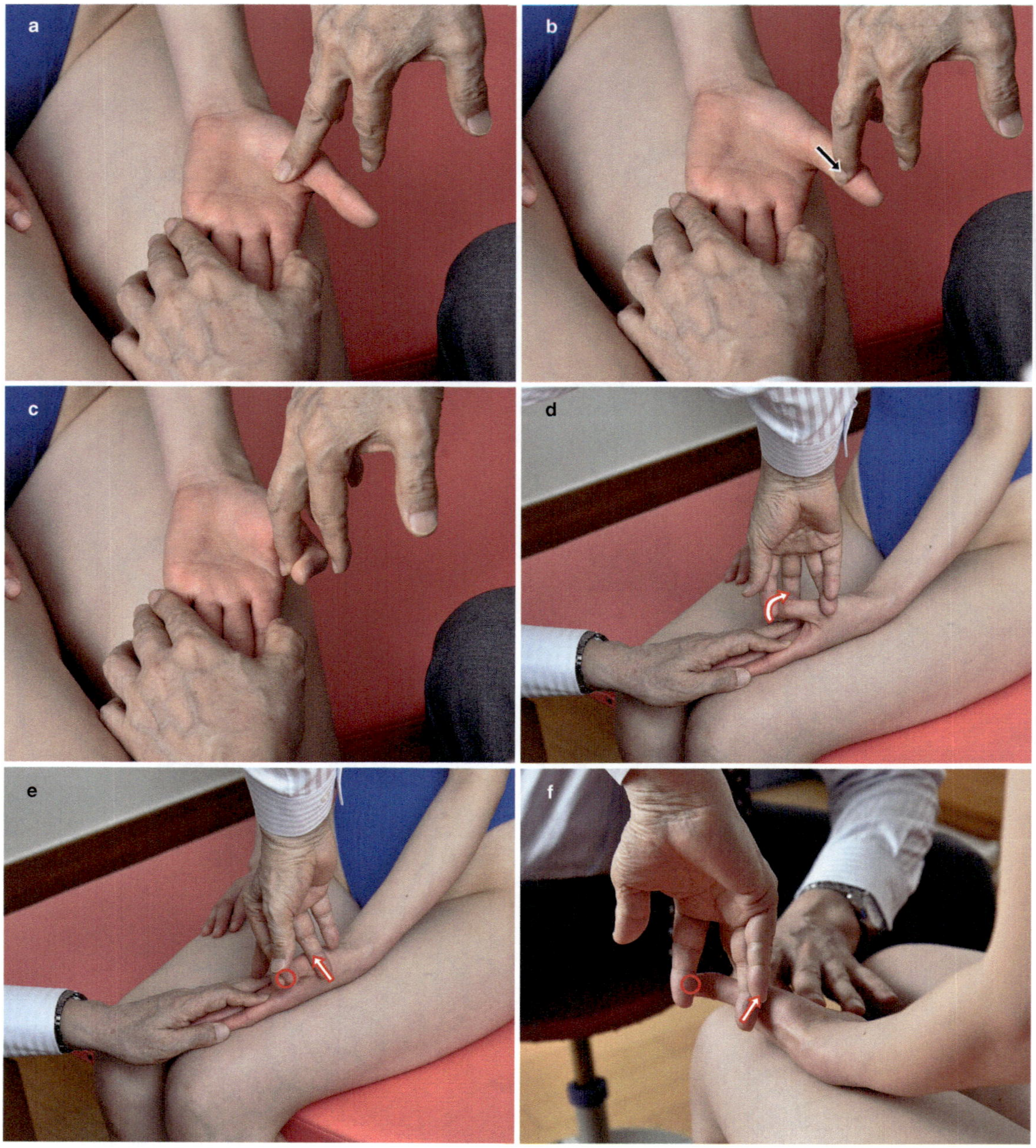

Fig. 4.45 Let's Try: Flexion/adduction and extension/abduction of the thumb (sitting position, mild paresis) (Video 4.32). Flexion/adduction and extension/abduction of the thumb is important for cooperative movements such as writing and pinching small objects. Set the thumb's MP joint on the affected hand to a flexion position with the carpometacarpal joint in adduction and then start the extension and abduction movement from the IP joint. Allow abduction of the carpal metacarpal joint at the end. Keep the thumb extended and abduct. (**a**) Support for the thumb as well as the other fingers. (**a**) The therapist should place the affected hand lightly on the patient's thigh with the palm side up. Using four fingers on one hand, the therapist should hold the patient's fingers gently against her thigh, while the therapist's other hand should be positioned vertically above the patient's palm. The therapist should then place his middle finger in the first interdigital space between the patient's thumb and index finger while his index and ring fingers should be on the outside of the thumb. (**b**) Stretch stimulation to the flexor muscle of the thumb. (**b**) Stroke the thumb of the affected hand rapidly from the MP joint towards the distal phalanx and apply extensional stimulus to the flexor while instructing the patient, "Bend your thumb, please." (**c**) Follow the flexion movement of the thumb. Using his middle finger, the therapist lightly touches the MP joint of the patient's thumb and slides forward to the IP joint to induce flexion of the thumb and increase the depth of flexion movement. (**d**) Quick flexion of the thumb after finger flexion stops. (**d**) In order to facilitate extension of the thumb, the therapist should use his index finger to quickly flex both the IP joint and MP joint of the thumb when thumb flexion reaches its limit. (**e**) Facilitation of thumb extension by adduction of the thumb's metacarpal bone. (**e**) Following (**d**), the therapist uses his ring finger to push against the patient's metacarpal bone near the MP joint in order to adduct the metacarpal joint, then facilitates extension and abduction of the thumb. (**f**) Continuation and ending thumb extension and abduction. (**f**) In response to extension/abduction of the patient's thumb, the therapist should pronate his forearm and keep his index finger and ring finger in contact with the affected thumb. When thumb extension/abduction reaches the limit, return to (**a**)

(b) Flexion and extension of the index finger (Fig. 4.46)
When the range of flexion/extension of the index finger is limited, combining palmar flexion/dorsiflexion of the wrist joint with extension/flexion of the index finger makes the Repetitive Facilitative Exercise easier.

(c) Flexion and extension of the middle finger (Fig. 4.47)
When the range of flexion/extension of the middle finger is limited, combining those movements with dorsiflexion/palmar flexion makes the Repetitive Facilitative Exercise easier.

(d) Flexion and extension of the ring finger (Fig. 4.48)
When the range of flexion/extension of the ring finger is limited, combining those movements with dorsiflexion/palmar flexion makes the Repetitive Facilitative Exercise easier.

(e) Flexion/extension of the little finger (Fig. 4.49)
When the range of flexion/extension of the little finger movement is limited, combining those movements with dorsiflexion/dorsiflexion makes the Repetitive Facilitative Exercise easier.

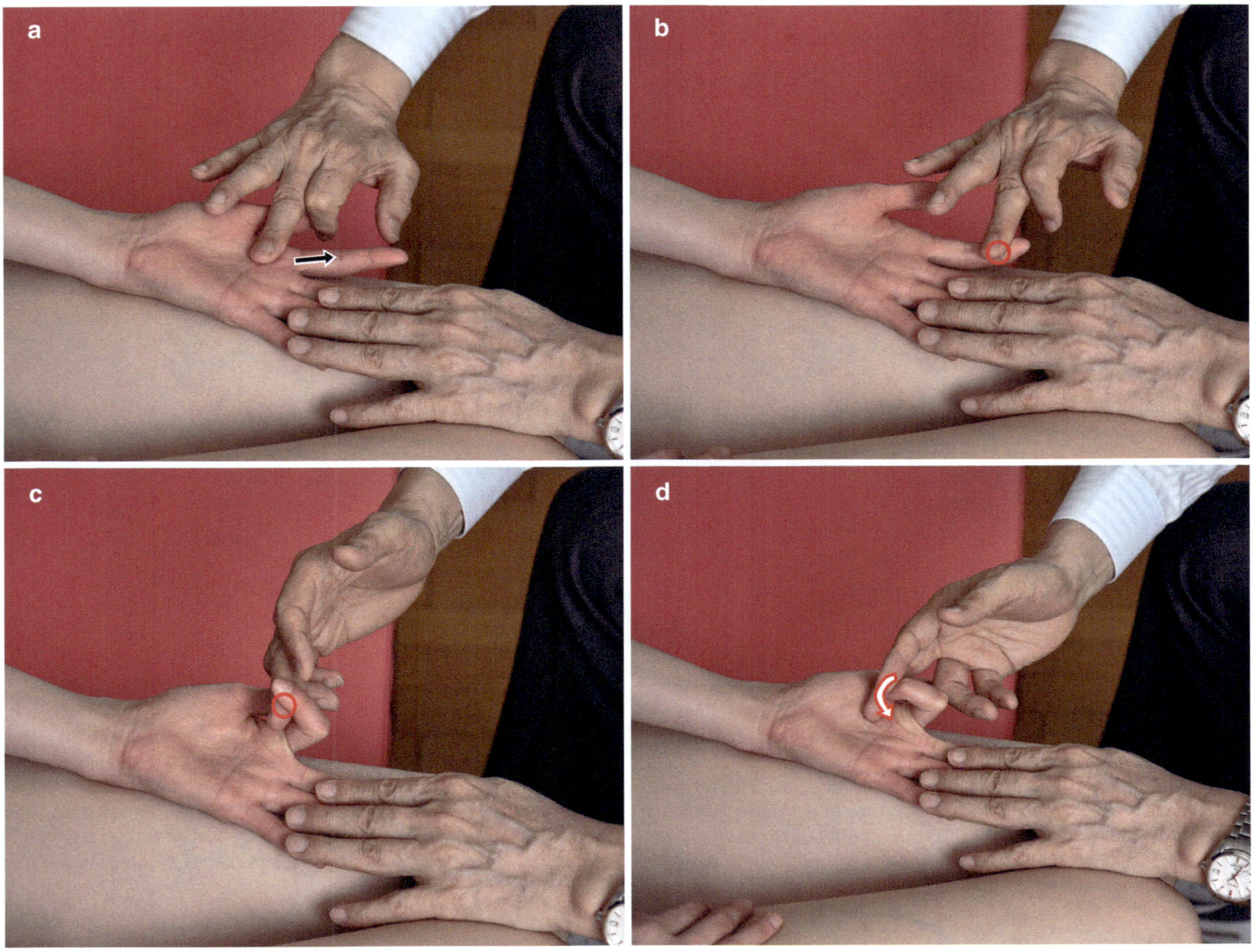

Fig. 4.46 Let's Try: Flexion and extension of the index finger (sitting position, mild paresis) (Videos 4.33 and 4.34). (**a**) Supporting the index finger as well as the other fingers. (**a**) The therapist places the middle, ring and little fingers of the affected hand lightly on the patient's thigh. Using the fingers on one hand, the therapist holds the patient's fingers lightly against her thigh, while the middle finger of his other hand is placed on the patient's palm proximal to the MP joint of the index finger. If the thumb of the affected hand bends and adducts and therefore becomes an obstacle to Repetitive Facilitative Exercise, ask the patient to assist in holding down the middle, ring and little fingers. Then, the therapist grips the patient's thumb between his index and middle finger. **Additional tips**: The therapist should bend his elbow and position his forearm perpendicular to the motion of the index finger, then perform procedures (**b**) to (**f**) while moving to supinate or pronate the forearm as needed. (**b**) Stretching stimulation to the flexor muscles of the index finger and the start of flexion. (**b**) Rapidly stroke the patient's index finger from the MP joint towards to the distal phalanx, and instruct the patient, "Please bend your index finger." When the index finger begins to bend, the therapist should maintain contact between his middle finger and the patient's index finger. (**c**) Continuing flexion of the index finger. (**c**) While the index finger of the affected hand flexes, the therapist's middle finger maintains contact with the tip of the patient's index finger to promote tension during flexion of the index finger. (**d**) Quickly and passively flex the patient's index finger after flexion of the index finger has reached its limit. (**d**) When flexion of the index finger reaches its limit, the therapist should release contact between his middle finger and the patient's index finger and then use the tip of his index finger (nail) to push the tip of the patient's index finger (nail) so that the patient's index finger quickly flexes. As the patient's index finger flexes, the therapist should instruct the patient, "Please extend your finger."

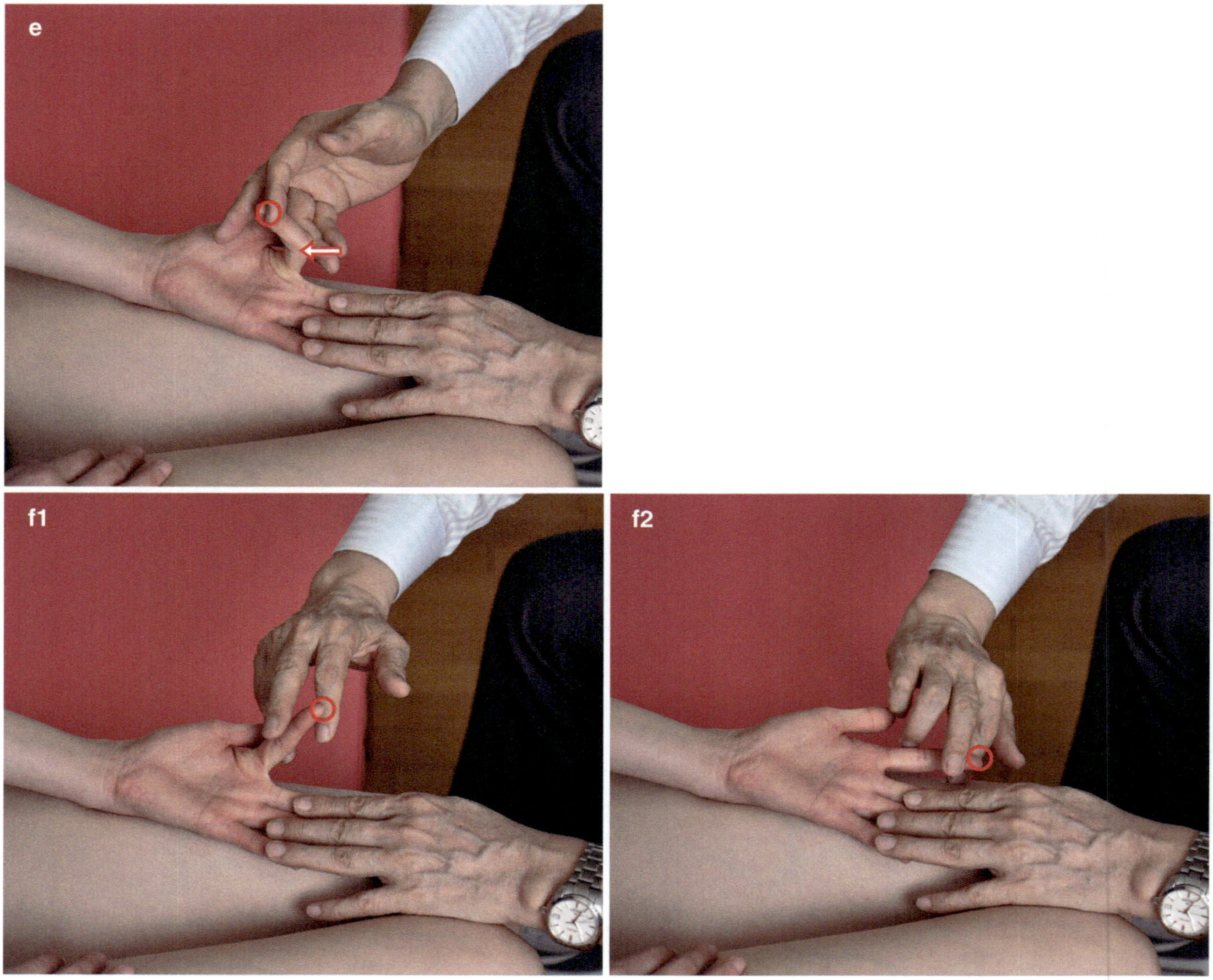

Fig. 4.46 (continued) (**e**) Use of the therapist's ring finger to facilitate extension of the patient's index finger by flexion of the MP joint. (**e**) The therapist should use his ring finger to push the patient's index finger upward to promote flexing at the MP joint to facilitate finger extension. At the same time, the therapist should relax the pressure on the tip of the patient's index finger (nail) and allow extension of the patient's affected hand. During this movement, the therapist should gently place his index finger on the patient's index finger in order to prevent loss of tension in the extension muscle, while simultaneously releasing contact between his ring finger and the patient's index finger. **Additional tips**: The therapist should use his ring finger to assist in pushing the patient's index finger upward from its flexed position. The therapist's ring finger will remain slightly flexed, while he pronates his forearm then allows his ring finger to slip through to the other side of the patient's index finger. (**f1**) Continuing extension of the index finger and contact with the therapist's index finger. Following the extension of the index finger, the therapist's index finger continues to touch the patient's index finger. (**f2**) End of extension of the index finger. When extension of the index finger reaches its limit, return to (**a**)

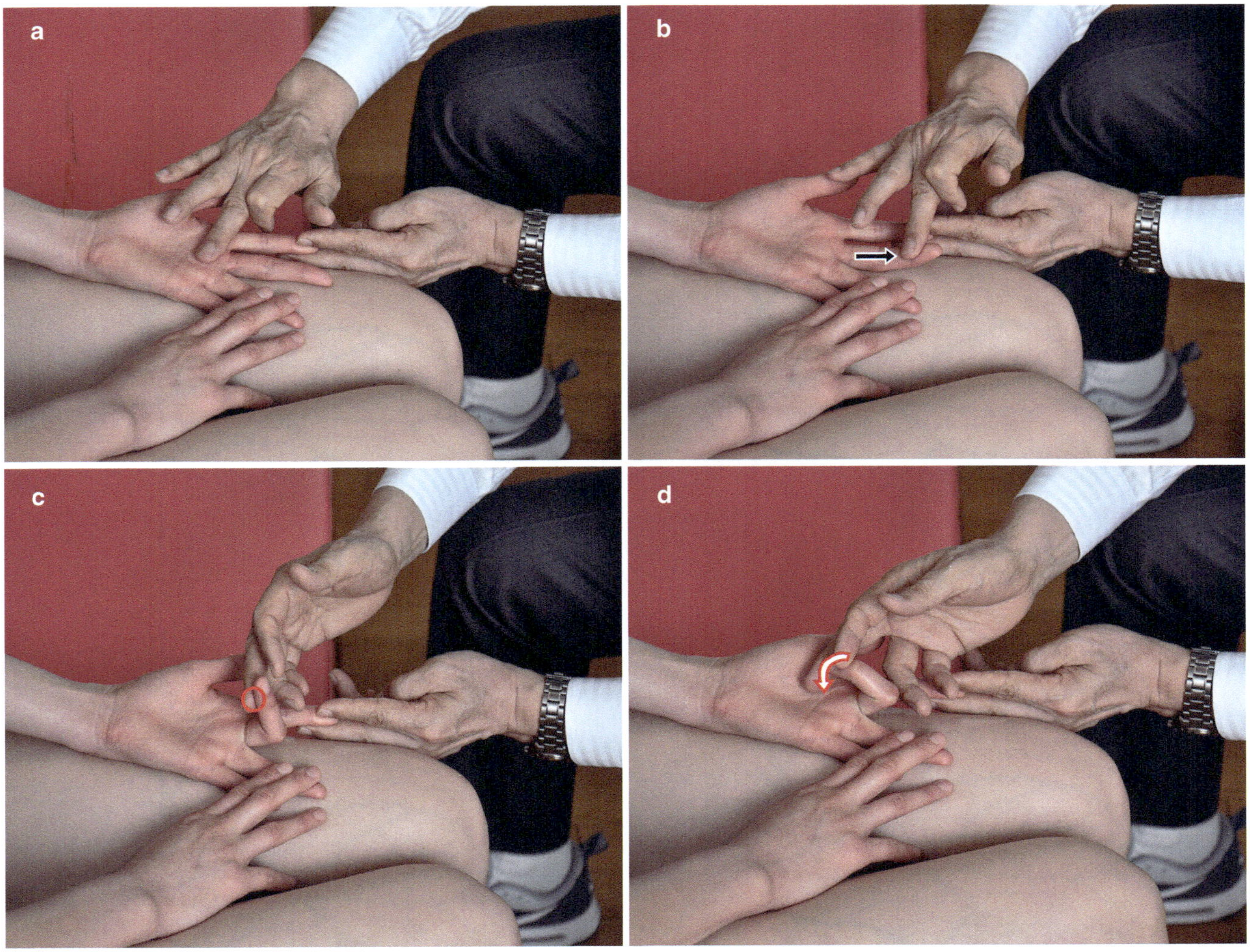

Fig. 4.47 Let's Try: Bending and extension of the middle finger (sitting position, mild paresis) (Video 4.35). (**a**) Support for the middle finger as well as the other fingers. (**a**) The therapist uses his index and middle fingers to hold the patient's index finger, while the patient uses her unaffected hand to hold the ring and little finger on her affected hand. The therapist then positions his middle finger above the patient's middle finger. (**a**) Facilitation to flex the patient's middle finger using stretching stimulation. (**b**) The therapist rapidly strokes the patient's middle finger from the MP joint toward to the distal phalanx. The therapist instructs the patient, "Bend your index finger, please." (**c**) Continuing flexion of the middle finger. (**c**) Using his middle finger, the therapist gently touches the tips of the patient's middle finger and follows by flexing it while instructing the patient to: "Continue bending your index finger, please." (**d**) When bending the middle finger voluntarily reaches its limit, the therapist quickly and gently flexes the finger. (**d**) When the patient's middle finger has reached the extent it can bend to, the therapist should use his index finger to flex the patient's middle finger to facilitate extension of her middle finger. The therapist instructs the patient to: "Extend the finger, please."

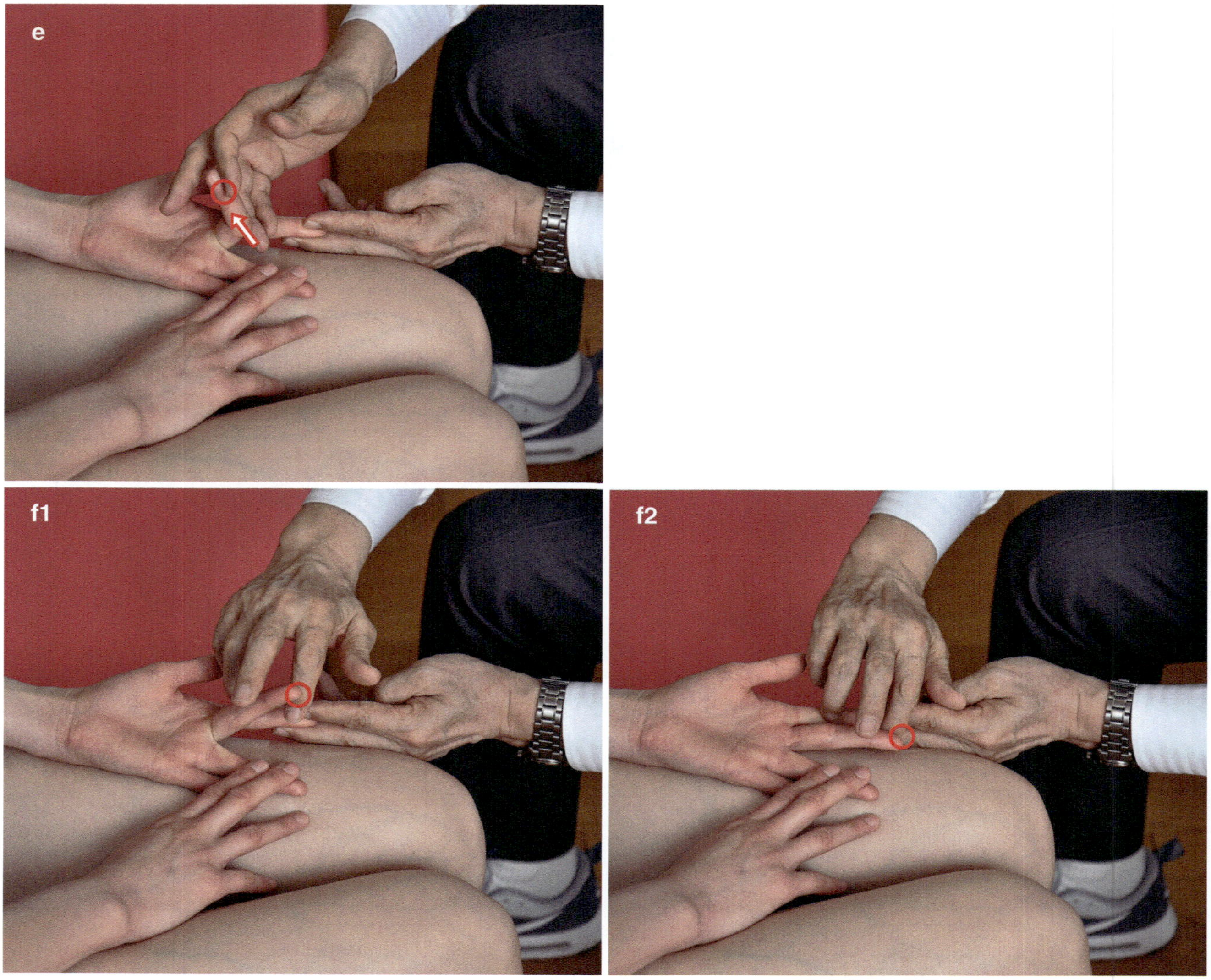

Fig. 4.47 (continued) (**e**) Middle finger extension by assistive flexing of the MP joint. (**e**) The therapist should use his ring finger to quickly flex the MP joint of the patient's middle finger to facilitate extension of the middle finger. As the patient's middle finger extends, the therapist should use his index finger to maintain contact with the patient's middle finger. The therapist instructs "Please continue extending the finger." Then the therapist pronates his forearm, which releases his ring finger from the patient's middle finger. (**f1**) Continuing extension of the middle finger. The therapist should carefully follow the extension movement of the patient's middle finger, maintaining contact between his index finger and the tip of the patient's middle finger. At the same time, the therapist should allow his ring finger to pass through to the other side of the patient's middle finger. (**f2**) Ending extension of the middle finger. When extension of the middle finger reaches its limit, return to (**a**)

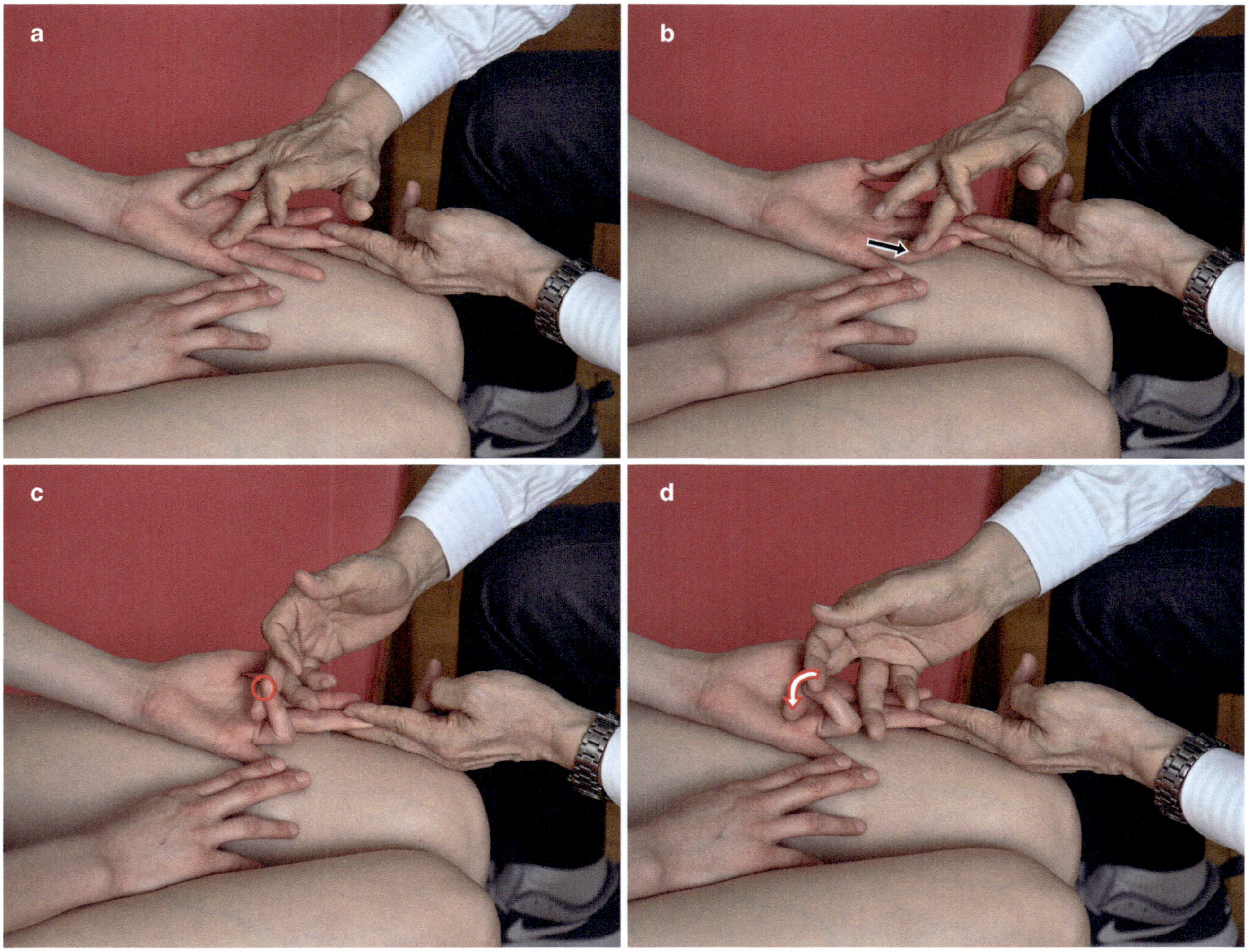

Fig. 4.48 Let's Try: Flexion and extension of the ring finger (sitting position, mild paresis) (Video 4.36). (**a**) Support for the ring finger as well as the other fingers. (**a**) The therapist uses his index and middle fingers to hold the patient's index and middle fingers, while the patient holds her little finger using her unaffected hand. The therapist places his middle finger close to the metacarpal MP joint of the patient's ring finger. (**b**) Stretching stimulation for the flexor muscle of the ring finger. (**b**) The therapist rapidly strokes the patient's ring finger from the MP joint toward the distal phalanx. The therapist instructs the patient to "Bend your ring finger, please." (**c**) Flexion of the ring finger. (**c**) During flexion of the patient's ring finger, the therapist uses his middle finger to continue making light contact with the patient's ring finger. (**d**) Facilitation to extend the ring finger by flexing it quickly. (**d**) When the ring finger reaches its flex limit, the therapist should flex the patient's finger quickly and gently using his index finger. The therapist instructs the patient "Please extend your ring finger."

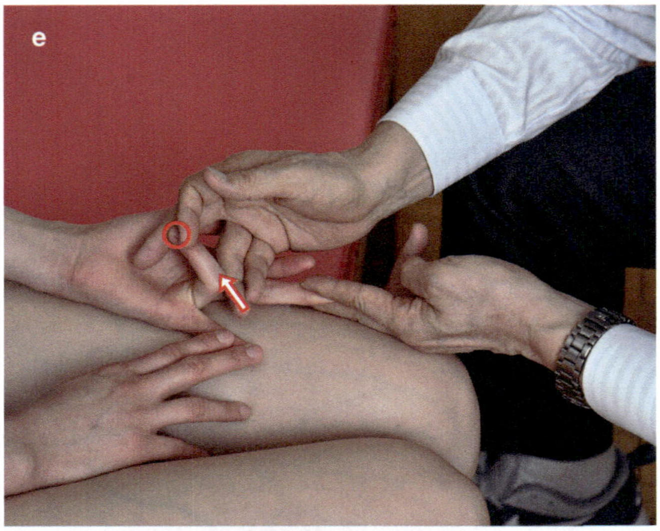

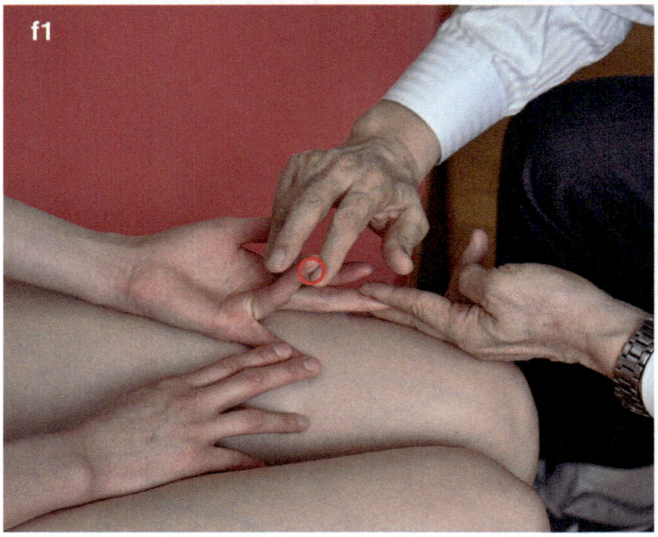

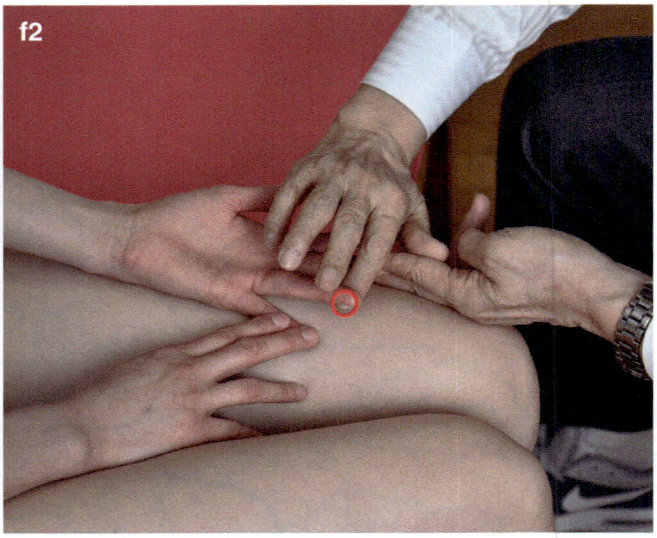

Fig. 4.48 (continued) (**e**) Facilitation to extend the ring finger by flexing the MP joint. (**e**) The therapist uses his ring finger to facilitate extension by flexing the MP joint of the affected hand while maintaining light contact between his index finger and the patient's finger in order to increase the extension movement of the patient's ring finger. The therapist instructs the patient, "Continue extending your ring finger, please."

Additional tips: While performing (d) through (e), the therapist must continuously pronate and supinate his forearm as needed. (**f1**) Continuing extension of the ring finger. The therapist maintains contact between his index finger and the tip (nail) of the patient's ring finger. (**f2**) The end of ring finger extension. When extension of the ring finger reaches its limit, return to (**a**)

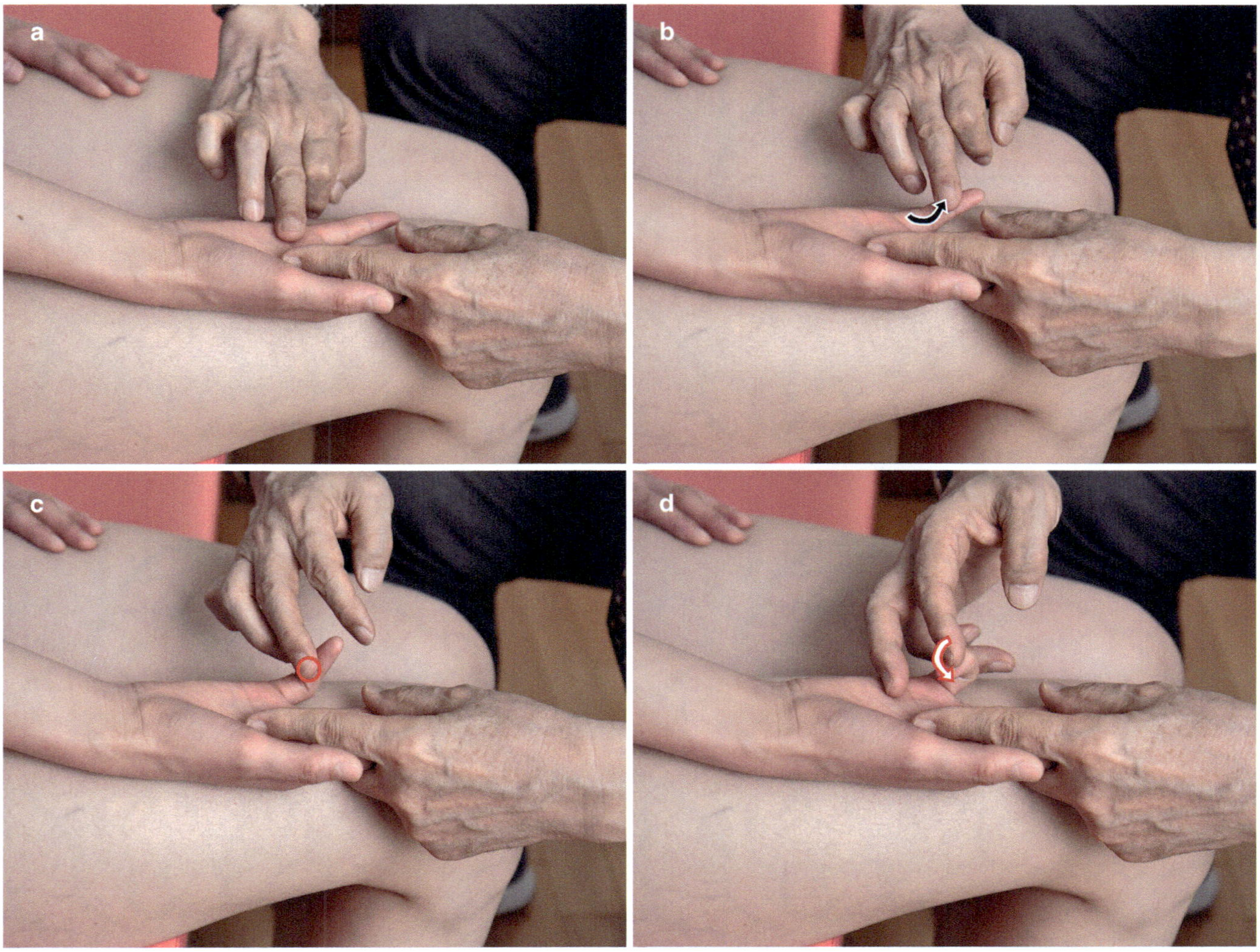

Fig. 4.49 Let's Try: Flexion and extension of the little finger (sitting position, mild paresis) (Video 4.37). (**a**) Support for the little finger as well as the other fingers. (**a**) The therapist gently holds the patient's hand on her thigh and uses his index, middle and ring fingers to hold the patient's index, middle and ring fingers. The therapist then places the middle finger on his other hand near the MP joint of the patient's little finger. In the photograph, the patient's right hand is presented as the affected hand to make it easy to observe the therapist's hand positions. (**b**) Stretching stimulation of the flexor muscle of the little finger. (**b**) The therapist uses his middle finger to rapidly stroke the little finger from the MP joint toward the distal phalanx. The therapist instructs the patient, "Please bend your small finger." (**c**) Flexion of the little finger. (**c**) The therapist uses his middle finger to lightly touch the patient's little finger following the finger's flex movement while instructing the patient, "Continue bending your small finger, please." **Additional tips**: The therapist can perform these finger manipulations smoothly by pronation and supination of his forearm as needed. (**d**) When flexion of the little finger reaches its limit, quickly flex the patient's little finger. (**d**) When flexion in the little finger reaches its limit, the therapist should quickly and gently flex the finger using his index finger. The therapist instructs the patient "Please extend your small finger."

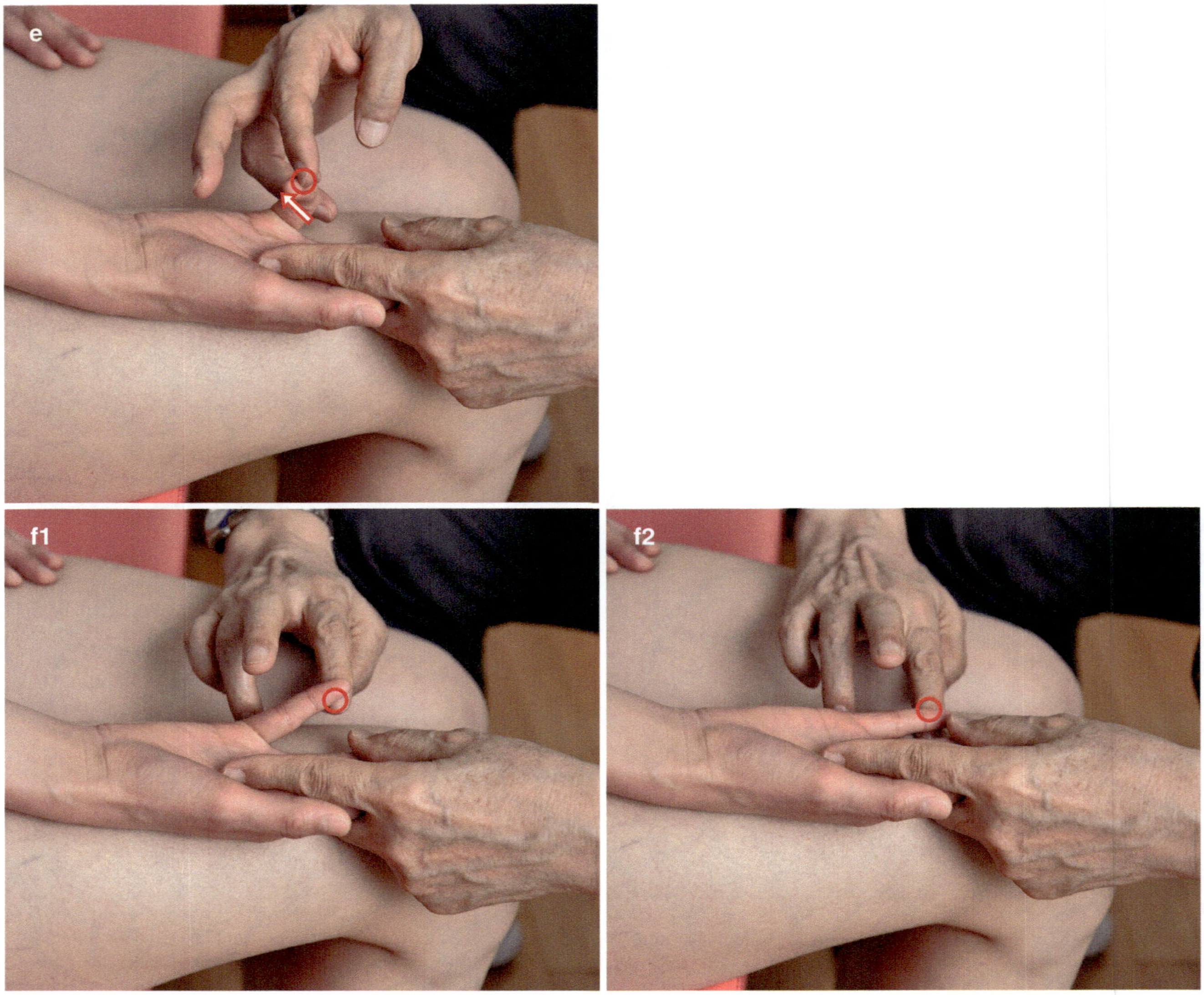

Fig. 4.49 (continued) (**e**) Using the ring finger to facilitate extension of the little finger. (**e**) The therapist uses his ring finger to bend the MP joint of the patient's little finger in order to facilitate extension. "Please extend your small finger." The therapist continues to use his index finger to lightly touch the patient's little finger and induce extension. (**f1**) Continuing extension of the little finger. The therapist maintains contact between his index finger and the distal phalanx (nail) of the patient's little finger. (**f2**) Ending extension of the little finger. When extension of the little finger reaches its limit, return to (**a**)

Repetitive Facilitative Exercise for the Lower Limb

5

Kazumi Kawahira

Many facilitation techniques have been proposed for the treatment of hemiplegic lower limbs, but in spite of numerous scientific studies, none of these methods were proven to be effective in promoting recovery from paralysis.

Many of the techniques of Repetitive Facilitative Exercise described below are based on the PNF and *Brunnstrom* methods. However, we have modified these two methods so that the patient's intended movements can be separated from the synergistic movements, and these intentional movements can be repeated.

5.1 Principles of Physical Therapy for the Lower Limb

For recovery from paralysis of the hemiplegic lower limb, voluntary external rotation, flexion/extension, and abduction/adduction of the hip joints need to be made possible first. Then voluntary movement of the distal lower limb should be induced using movement of the hip joint.

The basic concept of this therapy is to place priority on enabling the movement pattern necessary for walking.

In most cases, the treatment is made more effective by converting the range-of-motion exercise, which is usually performed as a passive movement, to active assistive movement using the Repetitive Facilitative Exercise.

Supplementary Information The online version contains supplementary material available at [https://doi.org/10.1007/978-981-19-0789-0_5].

K. Kawahira (✉)
Laboratory of Repetitive Facilitative Exercise, Kirameki Terrace Health Care Hospital, Kagoshima city, Kagoshima, Japan

5.2 Basic Treatment Techniques

Key Points

1. Prevention of hip joint pain

   Spastic contraction of the adductor muscles in the hip joint often leads to contracture, which restricts abduction and extension. Forceful range-of-motion training for the joint causes pain along the inner thigh (Fig. 5.1a).

   If the patient's paralysis is severe and assistance when walking cannot prevent a clear *Trendelenburg gait*, pain in the area of the hip joint will be caused by elongation of the soft tissue around the joint and stress on the hip joint (Fig. 5.1b).

2. Procedures to reduce spasm of the lower limb extensor muscles

   To manage spasms of the lower limb extensor muscles, it is effective to perform external rotation/flexion of the hip joint, plantar flexion of the ankle joint, and flexion of the toes in the supine position (Fig. 5.2).

   If spasms of the extensor are strong, combinations of these movements can be effective for suppressing spasms.

   If hip joint abduction is limited by spasm of the hip joint adductor muscles, then flexor muscle synergy (external rotation/flexion of the hip joint and flexing of the knee joint) should be performed to suppress spasms (See Fig. 5.10).

   After repeatedly performing the extension/abduction movements of the hip joint while the therapist uses his fingertips to tap on the gluteus medius muscle, knee extension exercise should be performed (Fig. 5.3).

   Since manipulation of the ankle joint suppresses spasms of the extensor muscles, the therapist should hold the patient's ankle with his fingers during Repetitive Facilitative Exercise instead of holding the ankle as it is usually held for range-of-motion exercise (Fig. 5.4).

   When flexion and extension of the knee joint are performed while the therapist inverts the patient's ankle and flexes the patient's toes (negative supporting reflex) (Fig. 5.5), a decrease in spasms was clearly shown by

K. Kawahira et al. (eds.), *Exercise Therapy for Recovery from Hemiplegia*, https://doi.org/10.1007/978-981-19-0789-0_5

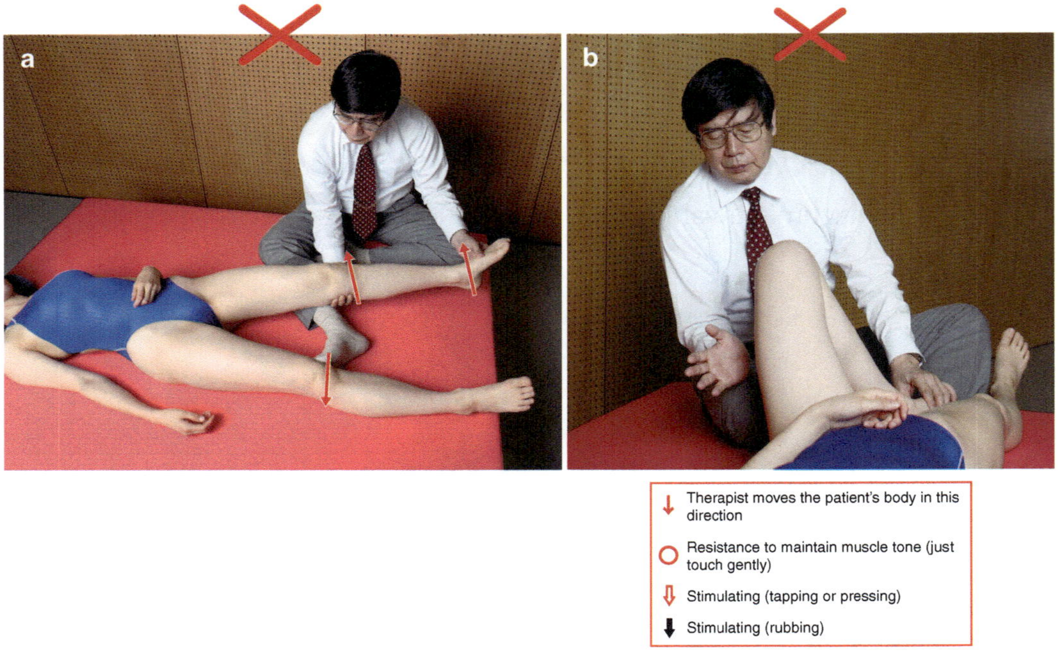

Fig. 5.1 Basic Technique: Prevention of hip joint pain. (**a**) Poor technique. Forceful passive hip abduction. (**a**) Poor technique. In a patient with strong spasms in the adductor muscles, the soft tissue of the hip joint will be damaged if the therapist forcefully abducts the lower limb. The procedure shown in Fig. 5.2 should be used instead. (**b**) Poor technique. Insufficient support during hip abduction. (**b**) Poor technique. It would be better if the therapist didn't reduce resistance or support during the patient's effort to abduct the hip joint. If the patient abducts their hip joint roughly without sufficient support, the soft tissue may be damaged. The stretch reflex in the hip adductor muscles will result in insufficient hip abduction, which is not preferred

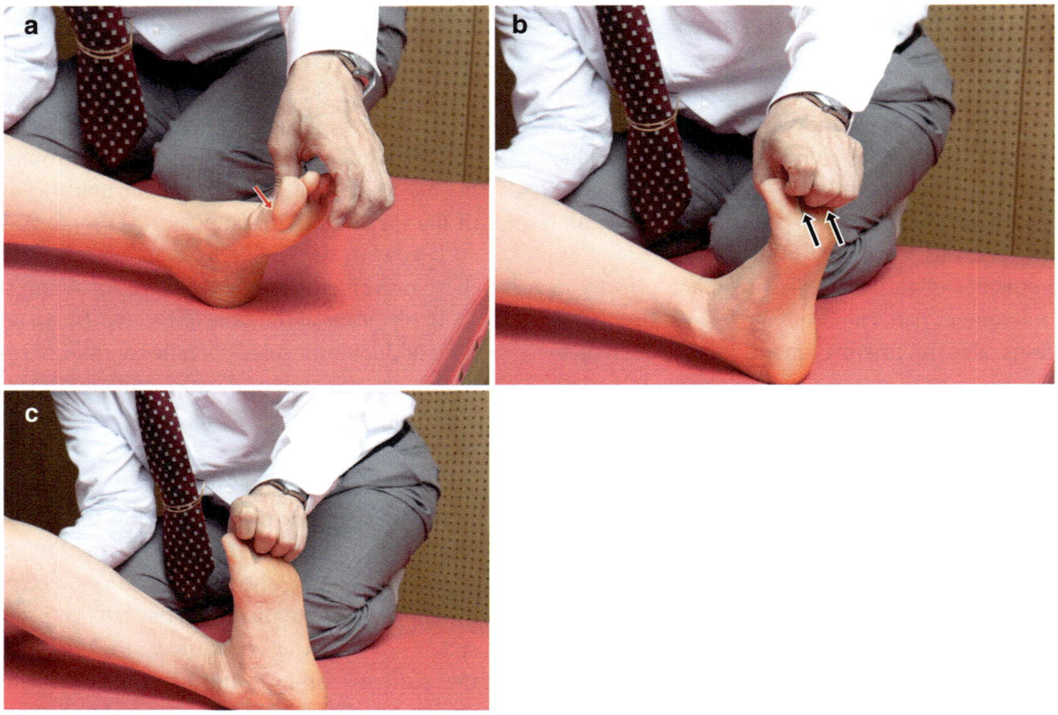

Fig. 5.2 Basic Technique: Dorsiflexion of the ankle by stimulating the toes (Video 5.1). (**a**) The starting position of ankle dorsiflexion (by stimulating the toes). (**a**) The therapist should place his thumb on the dorsal side of the patient's foot and his other four fingers should be placed on the sole side of the toes. Using his thumb, the therapist can hold the ankle in a flexed position as well as adjust the speed of dorsiflexion movement in the ankle joint. (**b**) Stimulation of the patient's toes with the fingertips. (**b**) The therapist stimulates the patient's sole side of the toes as if gently scratching them with his fingers (edge of the fingernail) while simultaneously pressing down on the foot with his thumb. It will be necessary to adjust the speed in order to avoid inducing clonus. During movement, the therapist can adjust the side that he scratches on the patient's toes, since stimulation to the foot interior (big toe) induces inversion of the ankle and stimulation to the foot exterior (little toe) induces ankle eversion. (**c**) When dorsiflexion of the ankle joint and dorsiflexion of the toes induced by stimulation to the toes has reached its limit, return to (**a**). (**c**) Limit of ankle dorsiflexion

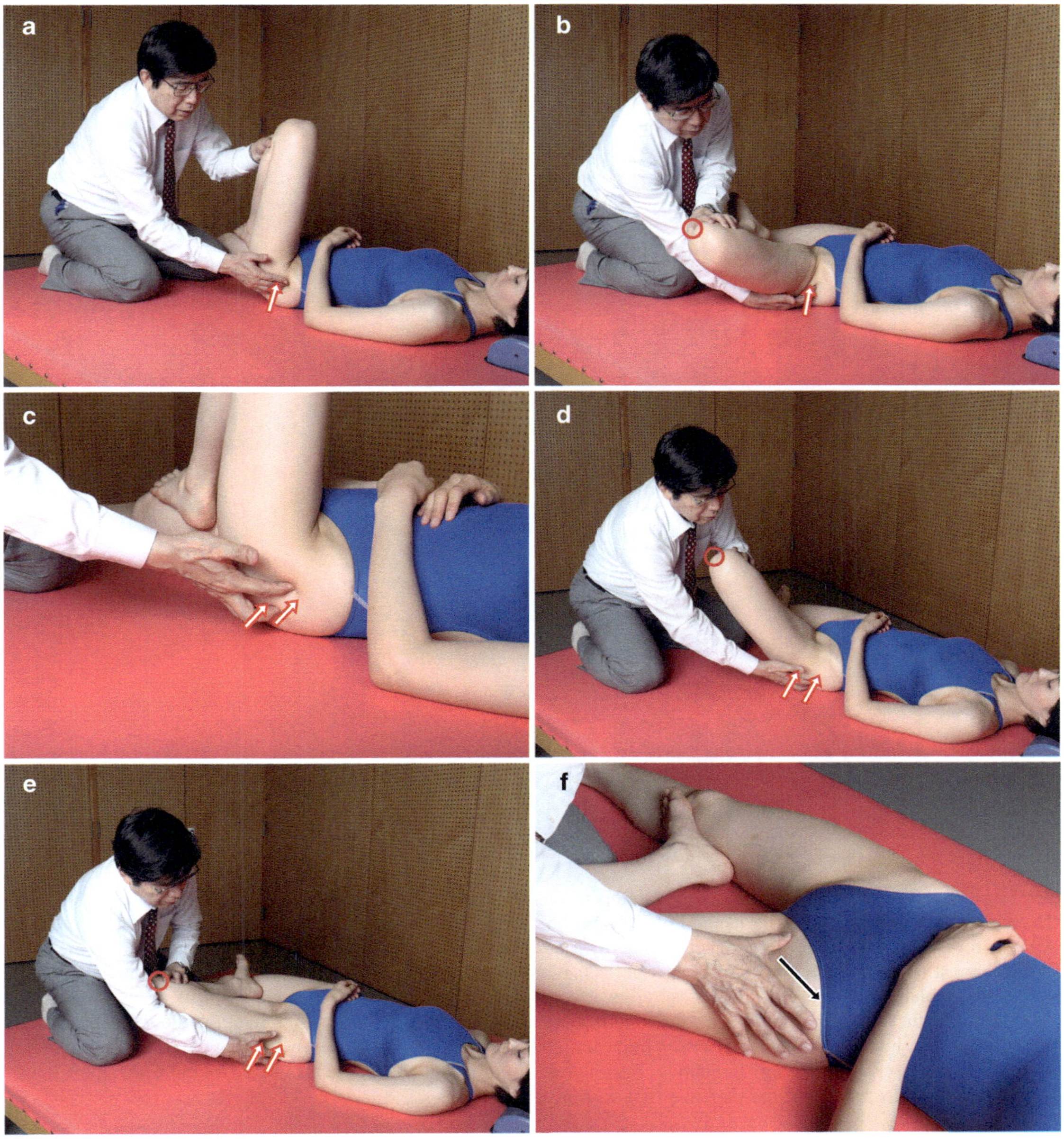

Fig. 5.3 Basic Technique: Suppressing spasms of the adductor muscles of the hip joint (abduction and extension of the hip joint) (Video 5.2). (**a**) Hold the patient's hip in flexion/adduction/external rotation and the knee in flexion while tapping on the abductor muscles of the hip joint. (**a**) In order to suppress spasms of the hip adductor muscles, the therapist should flex and externally rotate the patient's hip as well as flexing the patient's knee. First, the therapist should use his index, middle, and ring fingers to tap the patient's gluteus medius in order to facilitate abduction of the hip joint. Subsequent exercises should be done while the patient's hip joint is externally rotated. (**b**) Hip abduction with knee flexion (while maintaining external rotation). (**b**) The therapist should continuously tap and press the gluteus medius to stimulate the nerve paths. While supporting the patient's limb with his other hand, the therapist should slowly turn the patient's hip joint outward. (**c**) Procedure for extension and abduction of the hip joint. (**c**) Using his index finger, the therapist should tap the patient's gluteus medius, while using his middle and ring fingers to tap the gluteus maximus. This will facilitate extension/abduction of the patient's hip joint. The therapist should instruct the patient, "Turn your knee toward me." During movement, the therapist should also instruct the patient, "Keep your eyes on your knee" in order to encourage her efforts to move the affected limb. (**d**) Extension and abduction (external rotation) of the hip joint while maintaining knee flexion. (**d**) First, the therapist should instruct the patient regarding extension and abduction of the hip joint while maintaining the knee in a flexed position. (**e**) Extension and abduction (external rotation) of the hip joint while maintaining knee flexion. (**e**) The therapist should provide support during extension and abduction of the hip joint until the adductor muscle of the hip joint is sufficiently stretched. If the angle of abduction is insufficient, the therapist should continue slowly stretching the adductor muscle while supporting the patient's knee with his hand. (**f**) Flexion and adduction of the hip joint (external rotation). (**f**) With one hand supporting the patient's knee and keeping the hip joint of the affected limb in an externally rotated position, the therapist should use the thumb of his other hand to rub the inguinal ligament lightly in order to facilitate flexion and adduction of the hip joint. Then the therapist should follow the intended or active flexion and adduction of the hip joint, then continue to provide support until the patient's knee is over her navel. Continue to Fig. 5.11e

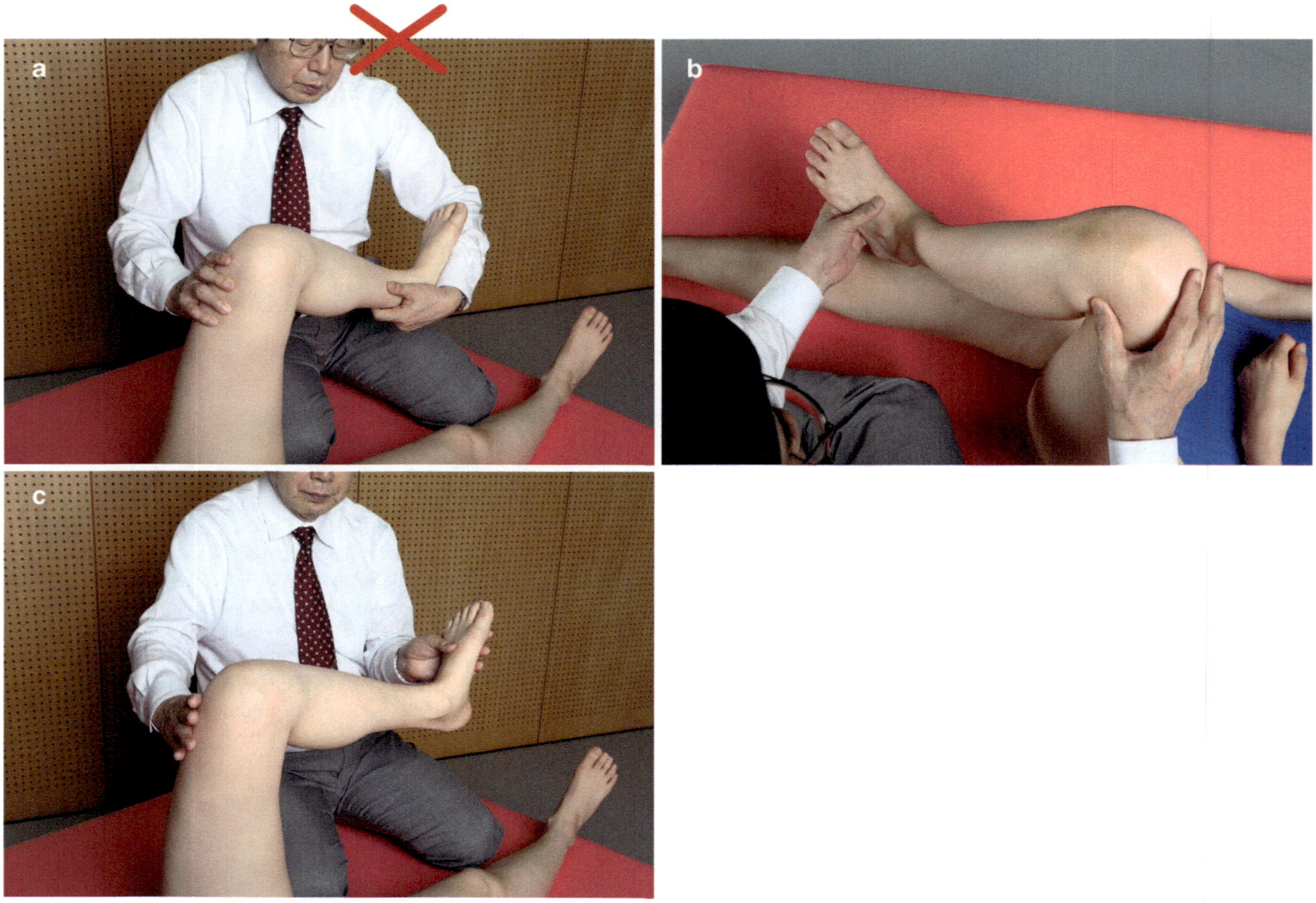

Fig. 5.4 Basic Technique: How to hold the foot and knee for Repetitive Facilitative Exercise. (**a**) Poor technique. The method of holding the patient's lower limb for the standard range of motion exercises. (**a**) Poor technique. If the therapist holds the patient's lower limb with his hand and forearm in the usual manner for range of motion exercises, it will be impossible to perform quick dorsiflexion of the patient's ankle joint or toes. (**b**) The method of holding the patient's lower limb during repetitive facilitative exercise. (**b**) During Repetitive Facilitative Exercise, the therapist should gently hold the patient's foot with his fingers and place his other hand on the patient's knee so that he can control the direction of hip joint movement. In this position, the therapist is ready to use his thumb to apply pressure to the patient's knee in order to manipulate movement of patient's lower limb quickly. (**c**) The method of holding the patient's foot with his fingers, viewed from a different angle. (**c**) It is important to hold the patient's foot gently, not to grip it forcefully. Gently holding the patient's foot makes it possible for the therapist to use his fingers for quick manipulations

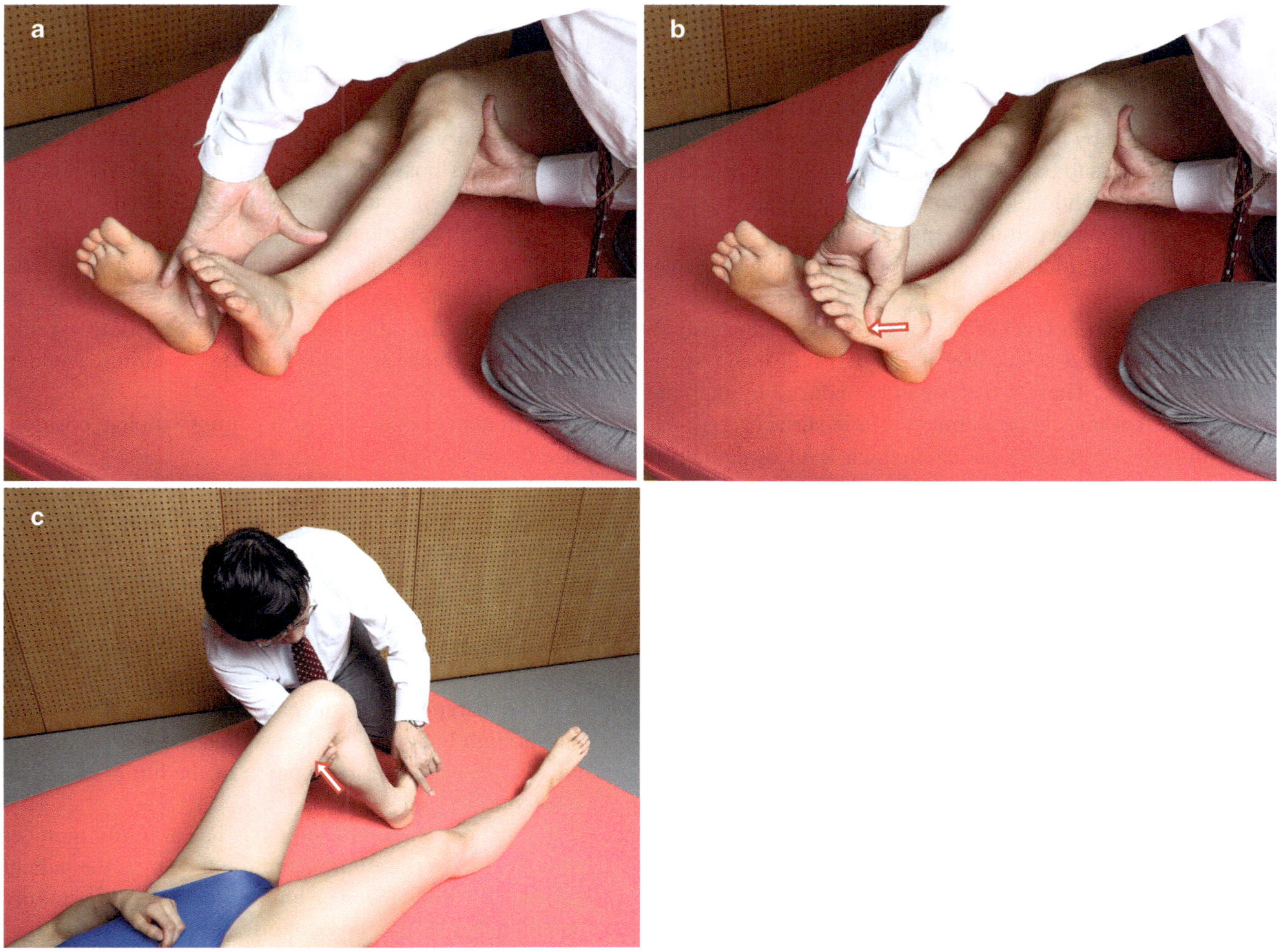

Fig. 5.5 Basic Technique: Suppression of extensor spasticity using negative supporting reflex of the ankle (Videos 5.5 and 5.6). (**a**) Method of supporting the lower limb during the negative supporting reflex and stretch reflex of the extensor digitorum longus muscle. (**a**) The therapist places the fingertips of one hand on the patient's hamstring, at the back of the knee, and the thumb of his other hand on the lateral dorsum of the patient's foot while his other fingers wrap around the sole on the opposite side of the foot. Then the therapist tells the patient to flex the knee while he facilitates knee flexion by inverting the ankle joint and flexing the toes (negative supporting reflex). (**b**) Quick plantar flexion of the ankle and inversion of the foot. (**b**) The therapist should place his middle finger on the patient's sole so that his finger can be used as a fulcrum. Then the therapist should move the patient's ankle into plantar flexion with inversion of the patient's foot while using his thumb to tap the lateral dorsal area of the patient's instep. (**c**) Dorsiflexion and eversion of the ankle joints, and knee flexion. (**c**) If the patient cannot adequately flex her ankle, the therapist should use the fingers of the hand supporting the back of the knee to stimulate the hamstring and guide the patient in flexing, abducting, and externally rotating the hip joint following the synergic movement of the flexor muscles

declines in both the F wave and the F/M ratio, which are two indicators of spasm in the extensor (See Fig. 2.4).

Since supression of spasticity obtained by repetitive facilitative exercise does not persist in the same way as that achieved by hyperthermia, the Repetitive Facilitative Exercise needs to be repeated. Extending the hip and flexing the knee while the patient is standing can suppress spasms of the lower limb extensor muscles.

3. Procedures to increase tension of the target muscle (Fig. 5.10)

   The excitation level of the target nerve tract (muscle) can be increased selectively by stretch reflex and skin-muscle reflex. These reflexes can be induced by tapping on the target muscle or rubbing the skin surface.

   In order to increase the excitement level of the nerve path of the lower limb flexor muscles, passive plantar flexion of the ankle joint and flexing the toes (negative supporting reflex) are effective. To raise the excitement level of the extensor nerve tract, passive ankle dorsiflexion (positive supporting reflex) is effective.

5.3 Procedure for Lower Limb Exercise

The major goal of therapy for the affected lower limb is to re-establish the movement pattern necessary for walking.

Contraction of the hip abductor muscles and quadriceps along with suppression of the triceps surae are important when standing on the affected lower limb. To reproduce this combination of muscle activity appropriately, synergy patterns must be isolated and avoided as much as possible.

During the swing phase of the affected lower limb, it is important to promote pelvic elevation by performing side flexion of the trunk, flexion of the hip joint, and dorsiflexion of the ankle.

5.4 Repetitive Facilitative Exercise for the Hip Joint

Key Points

1. Flexion and extension of the hip joint (while lying on one side) (Figs. 5.6 and 5.7)

   While lying on one side, spasticity of the extensor is reduced and manipulation of the foot becomes easier, which leads to easier induction of hip joint flexion/extension.

   If voluntary extension of the hip joint on the affected side becomes possible, the muscle tonus of the femoris tends to be higher and the knees become less likely to buckle when the patient is standing.

   If voluntary flexion of the hip joint becomes possible, the trunk becomes less likely to lean backward during the swing phase. This allows the center of gravity to transfer smoothly, which is also important during walking training.

2. Abduction/adduction of the hip joint (supine position, knee flexion position) (Fig. 5.8)

   In order to induce abduction/adduction of the hip joint, *Raimiste's phenomenon*, which is one of the associated reactions, can be combined with tapping and fast stretching of the target muscle while the hip joint and knee are flexed.

3. Internal/external rotation of the hip joint (supine position) (Fig. 5.9)

   If the therapist holds the patient's knee joint while the knee is slightly flexed and then stretches the target muscle quickly, internal/external rotation of the hip joint can be induced easily.

   If a patient with mild paralysis has a family member who has received training from the therapist, the patient can place his or her lower limb on the knee of a family member and practice this exercise at home.

4. External rotation, flexion, abduction of the hip joint, and flexion of the knee joint (supine position) (Fig. 5.10)

   Spasticity of the hip adductor muscle can disrupt balance while standing and during the gait cycle. Therefore, hip external rotation, flexion, and abduction and knee flexion are important manipulations to quickly suppress this disruption.

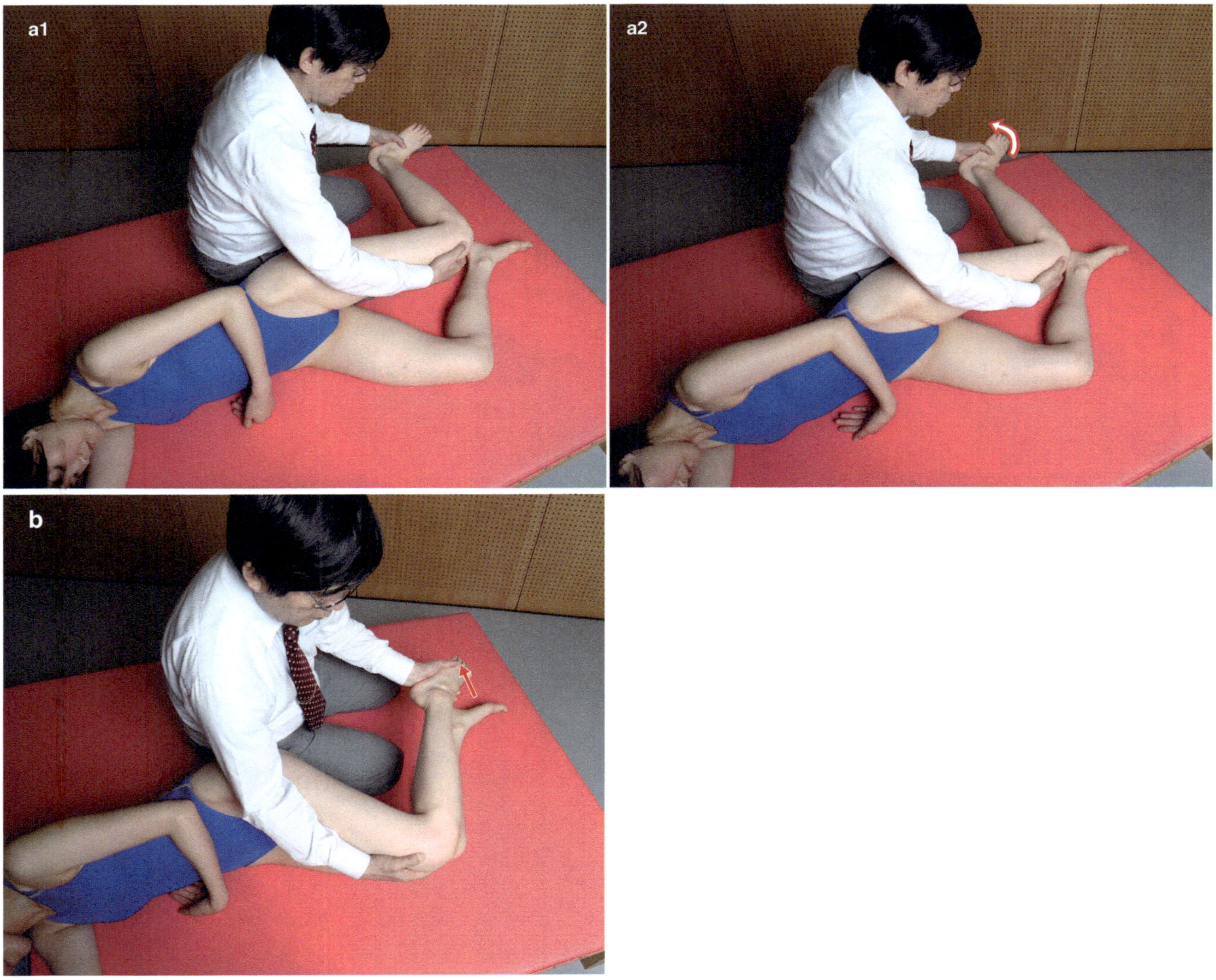

- ↓ Therapist moves the patient's body in this direction
- ○ Resistance to maintain muscle tone (just touch gently)
- ⇩ Stimulating (tapping or pressing)
- ⬇ Stimulating (rubbing)

Fig. 5.6 Let's Try: Hip flexion and extension (lying on the side). (**a1**) Hip positioning together with knee and foot support. (**a1**) The therapist sits beside the patient and places his hip against the patient's hips while lying on one side. The therapist holds the patient's thigh on the affected side (near the knee) and foot, with the patient's knee flexed at 90°. The therapist applies the index finger on the patient's sole and lightly pinches her foot with his thumb and fingers. (**a2**) Starting the hip flexion at the ankle plantar flexed position. (**a2**) The therapist places the hip on the patient's buttocks, then stretches the lower limb in the direction to be extended. The therapist adds ankle planter flexion (inducing a negative supporting reflex), then says, "Bring your knee closer to your forehead," so that the patient flexes the hip joint. (**b**) Continuing hip flexion. (**b**) The therapist maintains the knee angle at the point of ankle plantar flexion and follows the patient's hip flexion.

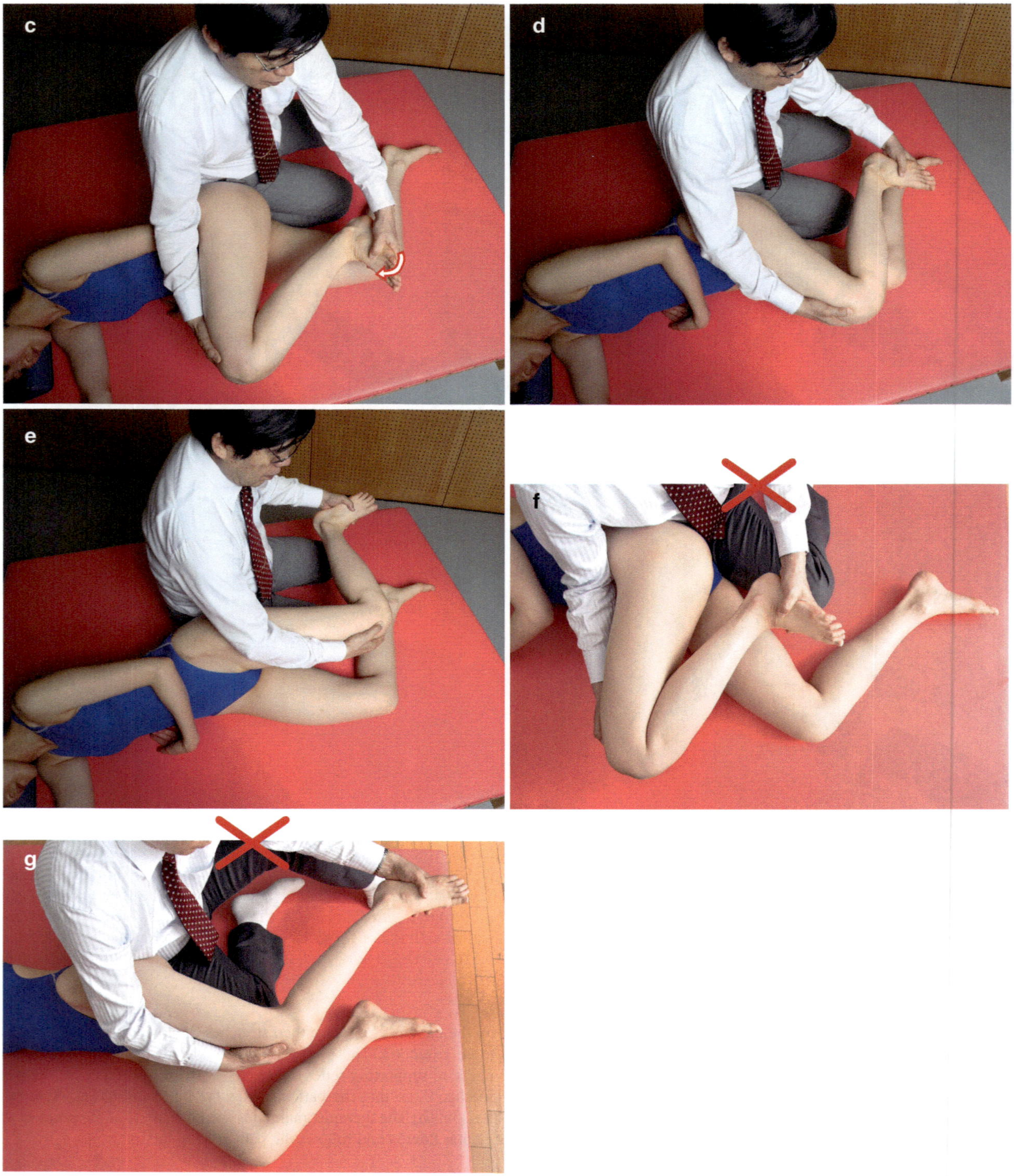

Fig. 5.6 (continued) (**c**) From hip flexion to extension. (**c**) When hip flexion reaches its maximum limit, the therapist adds quick dorsiflexion to the patient's ankle (positive supporting reflex) to increase the tension in the extensor muscles, and tells the patient, "Now extend your hip joint," or "Extend as if you were stretching your hip." (**d**) Continuing hip extension. (**d**) The therapist maintains the patient's ankle dorsiflexion and follows the hip joint extension. (**e**) Hip extension with the knee joint at 90°. (**e**) During this movement, the patient's knee is fixed at 90° and only the hip is extended/flexed. However, if the patient's lower limb is too long to position and it is difficult to keep the knee at 90°, the therapist should hold the knee so that it does not extend at the same time she extends her hip. When knee extension reaches its maximum limit, return to (**b**). If hip extension is insufficient, the therapist should remove their hand from the patient's knee and tap or rub the upper part of the gluteus maximus with one hand. (**f**) Poor practice: Lower limb flexion is performed without separating it from the synergic movement. (**f**) Poor practice: The hip, knee, and ankle joints are extended at the same time. This does not separate the hip joint movements. It is important to keep the knee joint at a fixed angle of 90°. (**g**) Poor practice: The lower limb is extended without being separated from the synergic movement. (**g**) Poor practice: The hip, knee, and ankle joints are extended at the same time. This does not separate the hip joint movements from the synergic movement. The therapist should rub the upper part of the gluteus maximus with one hand when this happens

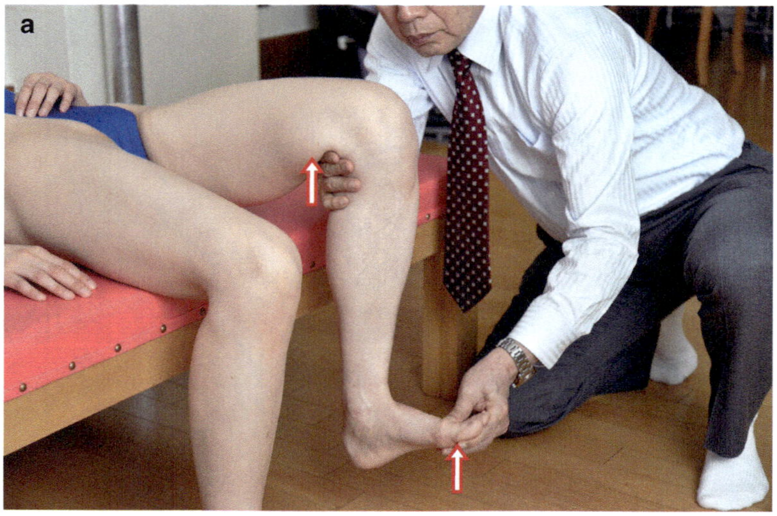

Fig. 5.7 Let's Try: Extension of the hip joint (supine position). (**a**) Hip extension. (**a**) If the patient's hip extension stops before it reaches zero degrees when she lies on one side, allow the patient to lie on her back and have the lower leg drop from the bed. The therapist flexes the patient's hip joint on the affected side and the foot quickly, saying, "Step on the floor." This should be more effective when the hip joint of the unaffected lower limb is fully flexed, with the patient holding it with her arm, allowing the pelvis to be raised and offering the affected lower limb more possibility for movement

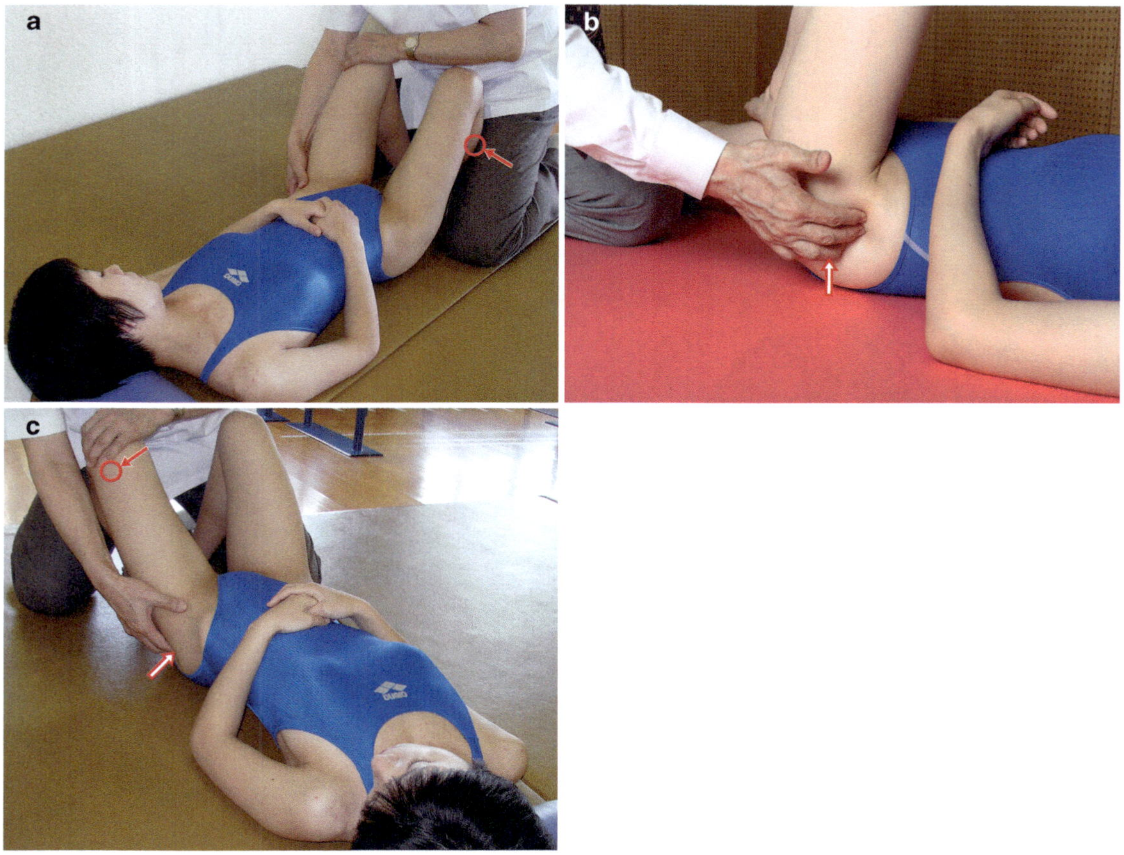

Fig. 5.8 Let's Try: Hip abduction and adduction (supine position, knee flexed). (**a**) Positioning both lower limbs and supporting the buttocks and knee. (**a**) The therapist holds both the patient's knees flexed at 120° and positions the unaffected lower limb, which the therapist's thigh locks from the outer side. The therapist places one hand on the hip abductor muscles (between the great trochanter and iliac crest) and the other hand on the knee on the affected side. (**b**) Extension stimulation and tapping on the hip abductor muscles. (**b**) The therapist stretches the patient's hip quickly towards the adduction direction with one hand on her knee, saying, "Stretch both your lower limbs outwards" in order to induce hip adduction, and taps the gluteus medius. If hip abduction is insufficient, the therapist should say, "Spread your lower limbs" and then limit the abduction in the unaffected lower limb using his own lower limbs. To assist abduction from the patient's affected hip, the therapist applies resistance with the fingertips supporting the patient's knee. (**c**) From hip abduction to adduction. (**c**) When hip abduction reaches its maximum limit, the therapist adds a quick stretch in the direction of the hip abduction and says, "Bring your lower limb inward to the center." If hip abduction is insufficient, the therapist should instruct the patient to "Close your both lower limbs together." Although not shown in the figure, the induction of the hip adduction becomes easier if the therapist taps on the upper part of the hip adductor muscles (inner side of the thigh, near the inguinal part) with his fingers

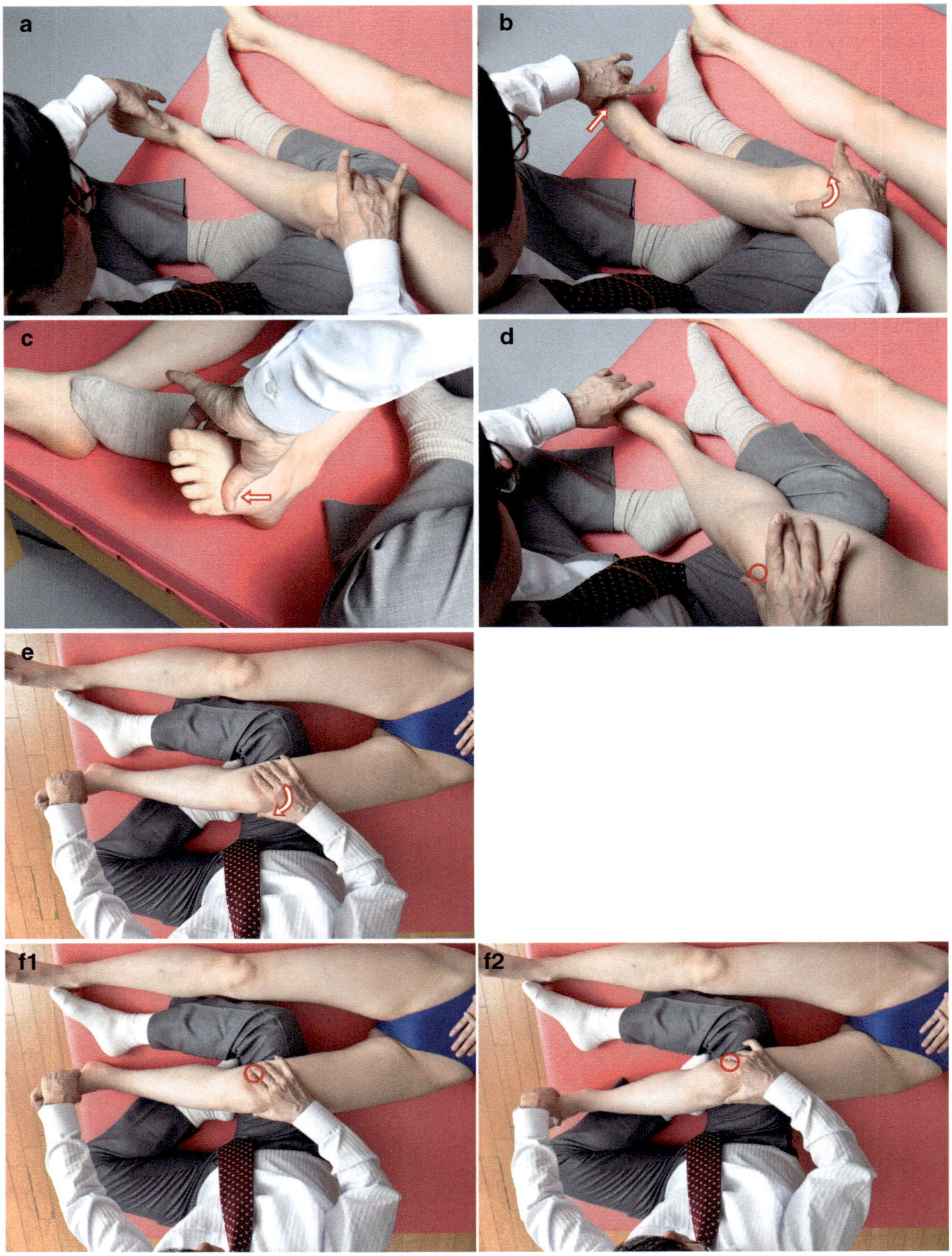

Fig. 5.9 Let's Try: Internal/External hip joint rotation (supine position). (**a**) Supporting the knee and foot. (**a**) The therapist sits with the patient's affected knee on his thigh, holding the patient's patella with the fingertips of one hand, and the patient's foot with the other hand. (**b**) Stretching the hip's external rotator muscles and foot inversion. (**b**) The therapist moves the hand on the patella and rotates the patient's hip joint quickly. Next he instructs the patient to "Turn your foot outwards," and then, using his thumb, taps the patient's dorsum of foot (dorsum pedis). **Tips**: The therapist should use his fingers for quick stretching and should apply gentle pressure. (**c**) Tapping on the foot, viewed from a lower angle. (**c**) Placing the therapist's middle finger on the patient's sole as the fulcrum, he taps the patient's instep (the part closest to her little toe) with his thumb. (**d**) From external hip rotation to internal hip rotation. (**d**) The therapist applies gentle resistance to the external hip rotation. Once the movement reaches its maximum limit, the therapist externally rotates the patient's hip quickly, then says "Turn your foot inwards" so that the patient performs an internal rotation of the hip joint. (**e**) Quick external hip joint rotation to facilitate internal rotation of the hip. (**e**) The therapist holds the patient's knee with his fingers as if pinching the knee with his thumb and fingers, then rotates the knee quickly outwards. (**f1**) Continuing internal rotation of the hip joint. (**f1**) The therapist applies gentle resistance to the patient's knee with the fingers that hold the knee joint, which challenges the patient's internal rotation of the hip joint, and this maintain muscle activity. (**f2**) From internal to external hip rotation. (**f2**) Once the internal rotation of the hip reaches its maximum limit, return to (**b**)

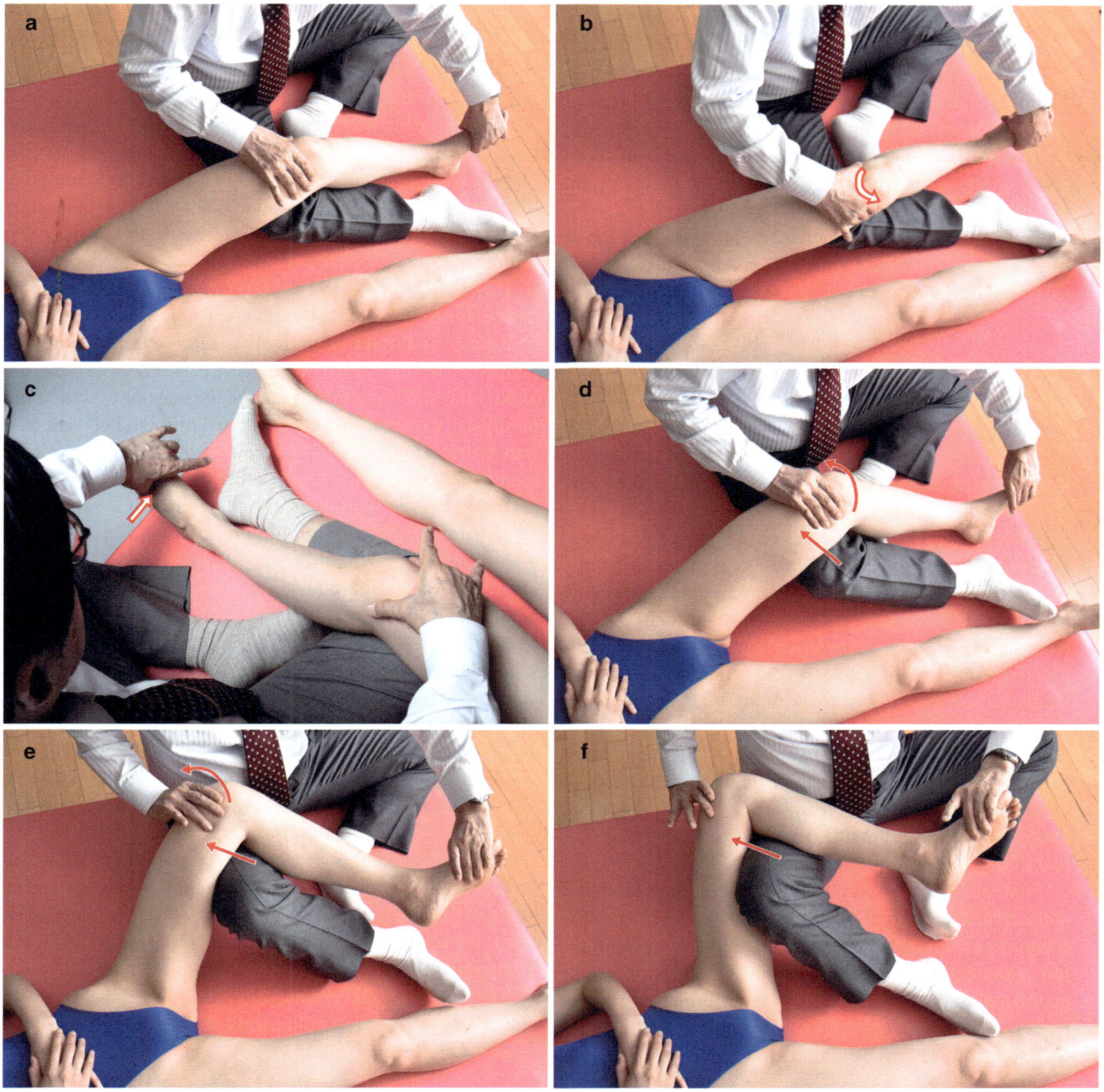

Fig. 5.10 How to suppress spasticity in the hip adductor muscles (Synergic movement: external rotation, flexion, and hip joint abduction, and knee flexion) (supine position) (Video 5.7). (**a**) Supporting the knee and foot. (**a**) The therapist sits with the patient's affected knee on his thigh, positioning the patient's knee at a slightly flexed angle. He places one hand lightly on the patient's patella as if gently pinching the patella with the fingertips, then holds the patient's foot while placing his thumb on her dorsum of foot (dorsum pedis). This positioning and manipulation make it easier to induce outer hip rotation in the pattern (Fig. 5.12): The patterns are hip extension/abduction (in the external rotation position), knee extension, and ankle fixed at 0° (no dorsiflexion no extension) ⇔ hip flexion/adduction (in the external rotation position), and knee flexion. (**b**) Stretch reflex induction in external rotator muscles by quick internal rotation of the hip joint. (**b**) The therapist holds the patella with the fingertips and rotates the patient's hip joint quickly to induce external rotation. (**c**) Quick foot inversion to induce stretch reflex in the foot dorsiflexor muscles. (**c**) The therapist then taps the dorsal side of the patient's foot with his thumb, which inverts the patient's foot (to induce eversion). (**d**) Assisting flexion/external rotation in the hip and knee joints. (**d**) The therapist uses the thigh to assist the patient's hip external rotation/abduction/flexion and knee flexion. (**e**) Continuing hip external rotation/flexion/abduction and knee flexion. (**e**) The therapist uses the thigh to fully flex the patient's hip and knee joints. If the patient's hip movement is on the point of stopping, the therapist should apply gentle assistance to the hip's external rotation/flexion/abduction. The therapist should not apply passive movement quickly, as it may cause pain. (**f**) Stretching the hip adductor muscles with a passive movement. (**f**) The therapist stretches the patient's hip adductor muscles to their maximum limit with the thigh and both hands, but only to the extent that the patient does not feel pain. As a result, spasticity can be suppressed within a short time

5.5 Repetitive Facilitative Exercise for the Lower Limb

The primary goal of this exercise is to achieve the movement patterns necessary for walking. For example, hip extension and abduction (maintaining an externally rotated position) and knee extension (maintaining the ankle angle at 0°) are necessary while standing on the affected lower limb.

A combination of hip flexion/adduction (while the hip is externally rotated) and knee joint flexion (ankle dorsiflexion) is necessary during the swing phase of the affected lower limb.

Key Points

1. Hip extension/abduction and flexion/adduction (supine position, knee flexed position) (Fig. 5.11)

   This manipulation is very important for separating synergistic movements and for the learning/acquisition of new movements.

   The movement patterns necessary for walking. Extension/abduction of the hip joint is induced by the stretch reflex of the gluteus maximus and gluteus medius and dermal reflex while suppressing spasticity of the extensor muscles by hip external rotation and knee flexion. Even if paralysis is quite severe and the patient seems unable to move his or her lower limbs at all, this manipulation will be effective.
2. Hip extension/abduction/external rotation (maintaining external rotation position), knee extension, ankle joint at 0° fixation ⇆ hip flexion/adduction/external rotation, knee flexion, ankle dorsiflexion (supine position: modified PNF pattern) (Fig. 5.12)

   First, at a knee flexed position, perform hip extension, abduction, and external rotation (maintaining the external rotation position), then repeat, alternating with hip flexion, adduction, and external rotation. Next, hip extension, abduction and external rotation are performed in combination with knee extension, and hip flexion, adduction and external rotation in combination with knee flexion.

   In order to suppress contraction of the hip adductor muscles associated with hip extension, hip abduction/adduction is prioritized over hip flexion/extension.

   When the hip joint starts flexing, quickly promote internal rotation of the hip joint in order to induce that joint's external rotation. Hip joint and knee joint extension should be performed within the range and load that can suppress internal rotation of the hip joint and ankle planter flexion caused by spasticity, which are increased by extension of the lower limb (Fig. 5.12).

   If the therapist follows the conventional PNF patterns and performs these exercises with the patient's hip extended, abducted, and internally rotated, contraction of the hip adductor muscles will occur in the early stage and hip abduction will not be facilitated. For patients with mild paralysis, exercises using the conventional PNF method will work. However, for patients with synergistic movements or spasticity in the hip adductor muscles and triceps of the calf, it is better to use the Repetitive Facilitative Exercise while the patient's hip joint is held in extension/abduction/external rotation and the ankle joint is held at zero degrees. Furthermore, it is easier for the hip joint to adduct/abduct on the horizontal surface if hip flexion/extension is avoided.

   If PNF exercises are being performed as preparation for walking pattern training, this method includes several movements that the hemiplegic patient will not be able to do, such as hip extension/abduction/internal rotation, ankle dorsiflexion and subsequent quick plantar flexion (push-off). These movements are all observed from the mid-stance to toe-off phase during normal walking.

   If the patient requires a lower limb brace because of equinus foot, drop-foot, or back knee, then hip and knee extension along with plantar flexion of the ankle will prevent the patient from smoothly transferring the center of gravity forward. Furthermore, these PNF exercises will worsen the back knee.

   The conventional PNF method can be applied only to patients who have very mild paralysis without spasms of the extensor muscles.

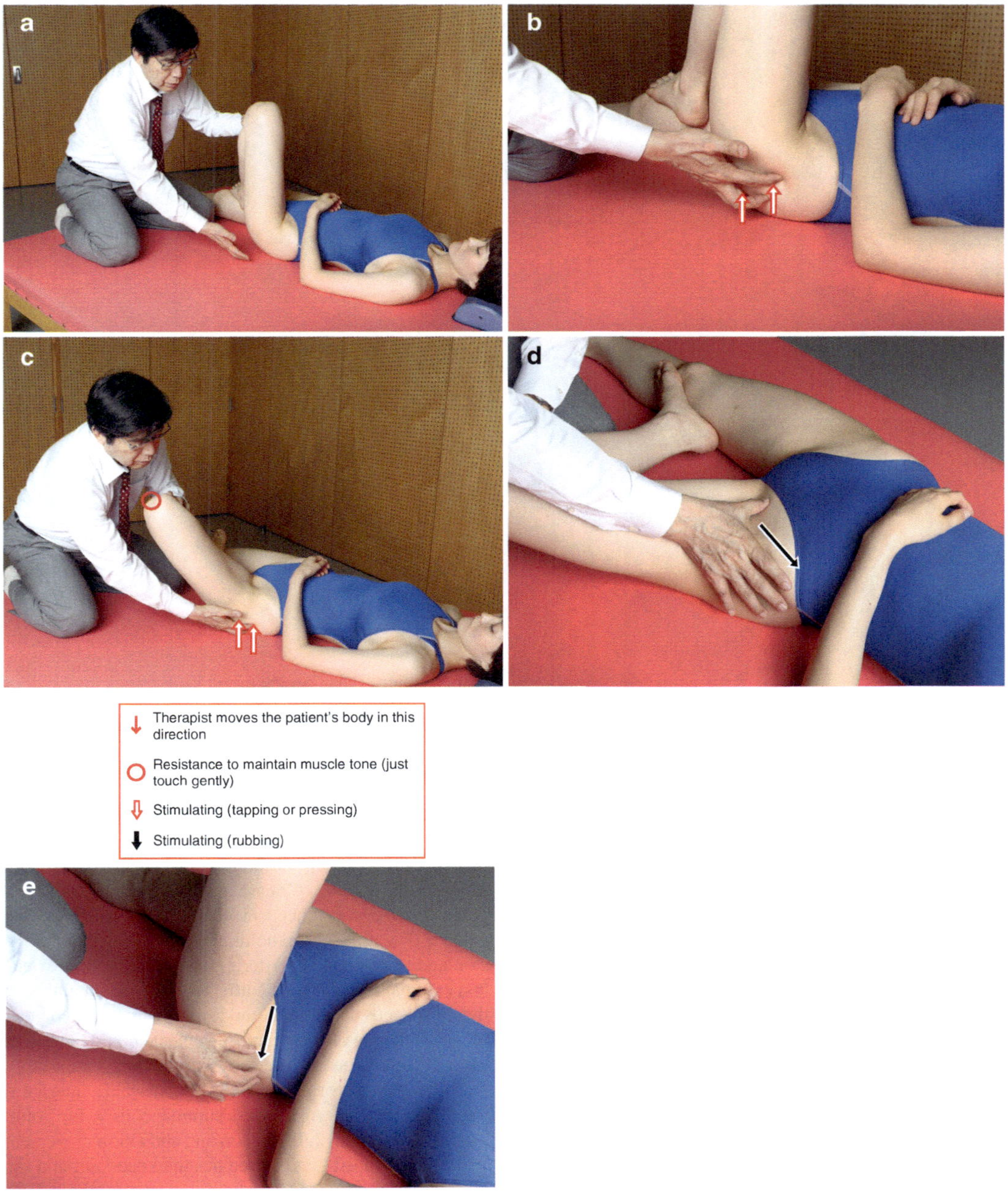

Fig. 5.11 Let's Try: Hip extension, abduction/flexion, and adduction/extension (supine position, knee flexion position) (Videos 5.3 and 5.4). (**a**) Supporting the lower limb and hip. (**a**) The therapist sits on the patient's affected side, grasping the proximal third of the lower leg from the front with one hand, and places the patient's hip joint in a flexed position with the knee fully flexed. The therapist opens out the fingers for tapping and places them near the gluteus maximus and gluteus medius. (**b**) Stimulating extension and abduction in the hip joint (side view). (**b**) To stimulate abduction in the hip joint, the therapist uses an index finger to tap the gluteus medius muscle located between the affected limb's greater trochanter and the iliac crest. To facilitate hip joint extension, the therapist taps the gluteus maximus muscle using the tips of the middle, ring, and small fingers. **Key point**: The therapist achieves the greatest effect by spreading the hand wide before tapping the gluteus medius with the index finger and gluteus maximus with the other fingers. (**c**) Continuing hip extension, abduction, and external rotation. (**c**) Support the lower legs during hip extension, abduction, and external rotation (hold secure in order to keep the knee turned out and downward). In order to suppress spasticity in the hip adductor muscle, it is important to extend and abduct the hip joint, keeping it in an external rotating position and not attempting internal rotation. (**d**) From hip extension and abduction to hip flexion and adduction. (**d**) When the extension, abduction, and external rotation of the hip joint reach their maximum limit, (with the knee held in a position turned out and downward), the therapist instructs the patient: "Please move your knee up toward your navel." The therapist then rubs the inguinal ligament with his thumb. (**e**) Continuing hip flexion and adduction (externally rotated position). (**e**) If the movement to flexion, adduction, and external rotation of the hip joint (while still holding) stops after the therapist has rubbed the inguinal ligament with his thumb, return to step (**a**)

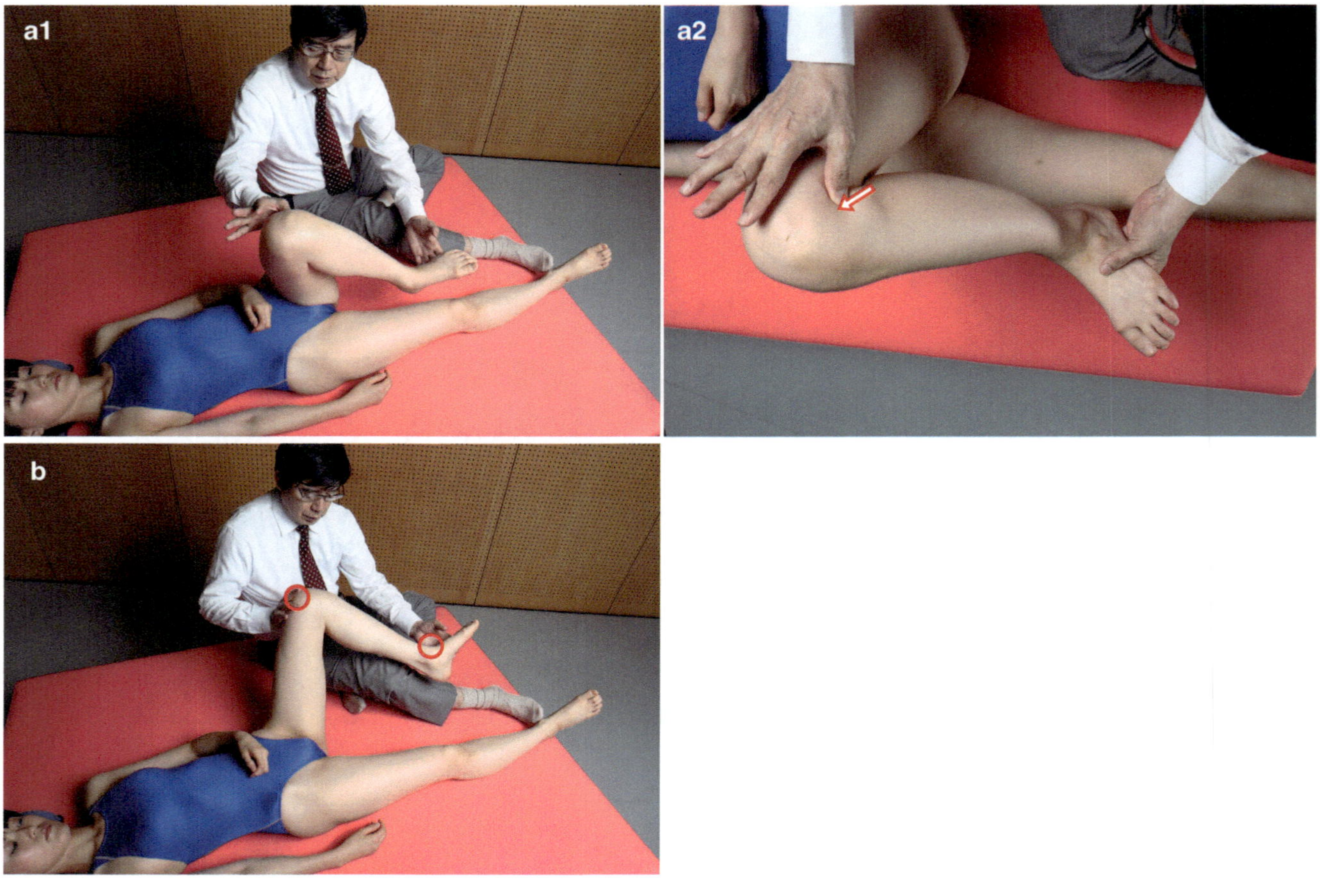

Fig. 5.12 Let's Try: Hip joint extension, abduction, and external rotation (retention); knee joint extension with ankle joint fixed at 0° ⇔ hip flexion, adduction, and external rotation knee flexion as well as ankle dorsiflexion (supine position: modified PNF pattern). (**a1**) Supporting the knee and foot. (**a1**) The therapist sits by the patient's affected side and places his lower limb at position (**c**). Alternatively, he places his lower limb at the patient's knee position when she moves her hip joint to perform extension, abduction, and external rotation (retention) as well as knee joint extension. The therapist holds the patient's knee, supporting it with his thumb. The therapist grasps the patient's foot on the affected limb with the thumb and all four fingers of his other hand, telling the patient, "Please leave the bending and stretching of the knee to me." (**a2**) Supporting the foot as seen from above. (**a2**) The therapist holds the patient's hip joint flexion, adduction, and external rotation while keeping the knee on the affected limb at an angle. The therapist pushes the patient's knee quickly in the direction of the navel with his thumb and instructs the patient, "Please move your knee outward/downward" or "Please move your knee toward me." The therapist keeps the patient's ankle at 0° with the hand holding the foot. (**b**) Continuing hip joint extension and abduction. (**b**) After performing hip joint extension, abduction, and external rotation on the affected limb (maintaining the position), the therapist keeps the patient's ankle at 0° with his left hand. Only in the case of a high degree of voluntary movement should the therapist say, "Please extend your knee a little" to induce knee extension. **Key points 1.** According to the original PNF pattern, where internal rotation and hip joint extension is emphasized simultaneously, hip joint abduction will likely be restricted. **2.** Extension and abduction movements are facilitated when the therapist gently stretches the knee joint and asks the patient to perform extension and abduction movements (about 30°).

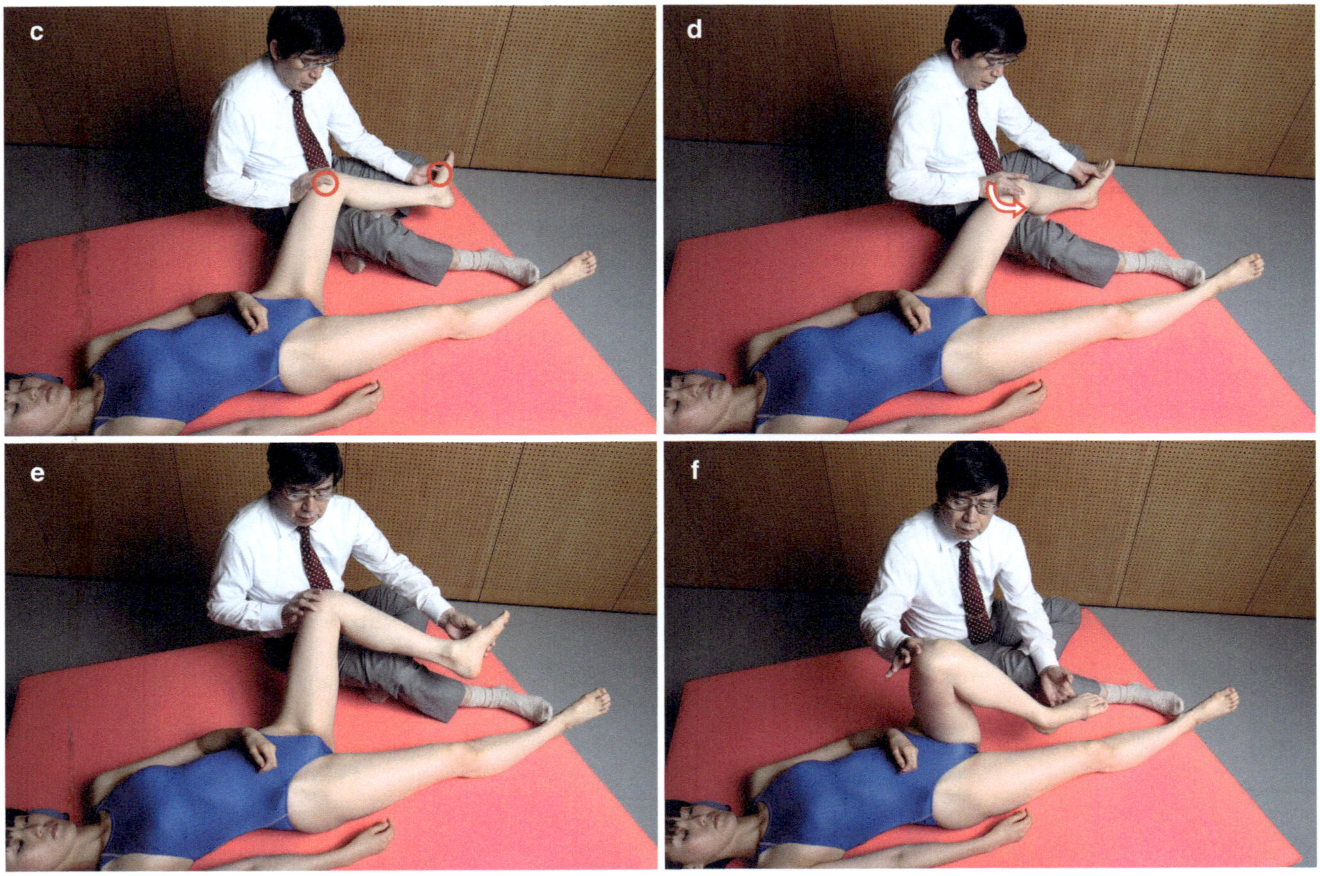

Fig. 5.12 (continued) (**c**) Continuing hip joint extension/abduction (keeping an external rotation position). (**c**) The therapist applies slight resistance with his hand on the patient's knee, maintaining an external rotation position, and then the patient continues to perform an unassisted hip abduction. Although the knee joint extension is performed by the therapist, once spasticity in the extensor has increased and hip joint adduction and ankle joint flexion have begun, the therapist instructs the patient to stop. Then the knee joint extension is stopped on the therapist's knee. (**d**) When extension and abduction in the hip joint cease, the therapist facilitates a combination of flexion, adduction, and external rotation. (**d**) If the affected limb's extension has stopped, the therapist instructs the patient: "Please relax." Once the extensor muscles are relaxed, the therapist quickly rotates the hip in an internal direction with his hand holding the patient's knee. The therapist instructs the patient, "Pull your knee over your navel," then indicates external hip joint rotation, flexion, and adduction. (**e**) Hip joint flexion/adduction/external rotation and ankle dorsiflexion. (**e**) Emphasizing hip joint flexion and external rotation. If quick-stretching stimulation to the hip's external rotation muscles is insufficient, hip joint adduction and internal rotation occurs and hip joint flexion becomes inadequate. **Key point**: To promote ankle dorsiflexion, the therapist scratches the patient's sole with the index finger of the hand while holding the patient's foot. (**f**) Continuing hip joint flexion, adduction/external rotation, and ankle dorsiflexion. (**f**) When flexion, adduction, and external rotation of the affected hip joint stop (where the knee stops above the navel), return to (**a**). **Key point**: In the case of contraction (spasticity) in the hip adductor muscle, the therapist emphasizes hip extension and abduction. The knee extension is controlled by the therapist, who changes the direction from "knee to navel" to "knee to forehead"

5.6 Repetitive Facilitative Exercise for the Knee

Knee flexion exercise can be performed more easily while the patient is seated or lying on his or her side. If the patient has no shoulder pain and can flex the knee slightly, this exercise can be performed in the prone position. However, the prone position is not so practical because the patient may experience pain in the shoulder and cannot watch his or her own lower limb move during the exercise.

Key Points

1. Flexion of the hip and knee joints, and ankle dorsiflexion (supine position) (Fig. 5.13)

   To facilitate knee flexion, spasticity of the lower limb extensor muscles needs to be suppressed.
2. Knee flexion/extension (in sitting) (Fig. 5.14)

   As flexion of the knee joint while sitting is performed with the hip flexed, the flexion of the knee can be induced easily. If the patient cannot extend the knee joint while the hip is flexed, the therapist can apply stimulation to the skin overlying the quadriceps. Furthermore, passive ankle dorsiflexion will also improve the patient's ability to perform voluntary knee extension.

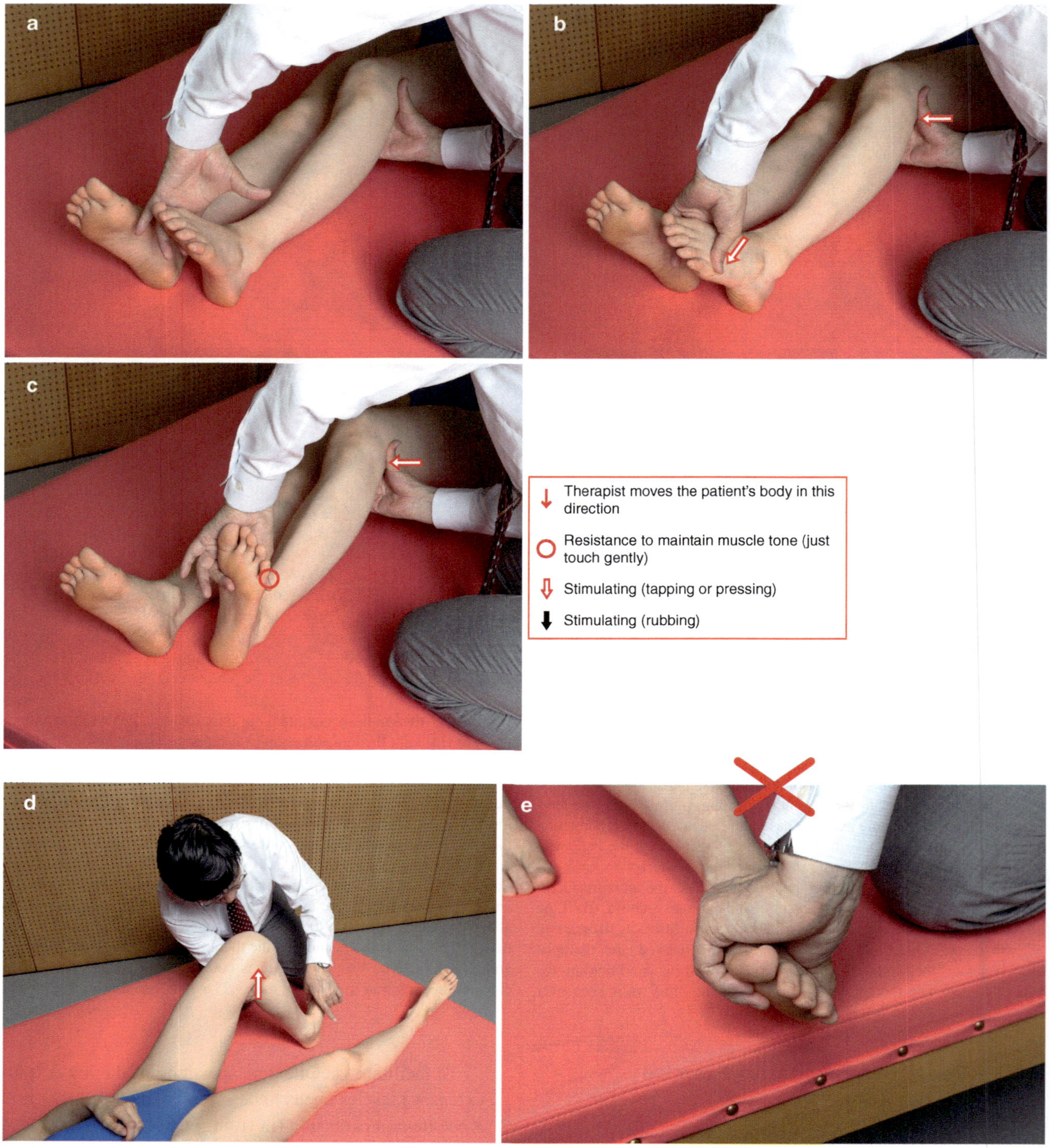

Fig. 5.13 Let's Try: Hip and knee flexion and foot dorsiflexion (supine position). (**a**) Supporting the knee and foot. (**a**) The therapist holds the back of the knee with one hand and the foot with the other hand. The therapist places his thumb on the dorsal side of the patient's foot towards the patient's small toe, then places the middle finger on the sole of the patient's foot, holding it with the fingertips. (**b**) Negative supporting reflex, stretch reflex of the extensor digitorum longus muscle, and stimulation for the knee's flexor muscles. (**b**) The therapist quickly inverts the patient's foot and flexes her knee while tapping the patient's hamstrings with a pinching movement using the thumb, index finger, and middle finger. (**c**) Continuing knee flexion and ankle dorsiflexion. (**c**) The therapist uses his thumb and palm to reduce the speed of ankle joint dorsiflexion, adding light resistance as appropriate. If clonus is induced, the therapist reduces the speed of dorsiflexion by placing his palm on the dorsal side of the patient's foot. (**d**) Continuing knee flexion and ankle dorsiflexion. (**d**) When the muscle for knee flexion is weak, the therapist increases knee flexion gently. (**e**) Poor example (foot support). (**e**) Poor technique: The therapist is holding the patient's foot from his index finger to his little finger and this tightness makes it difficult for the patient's foot to move quickly

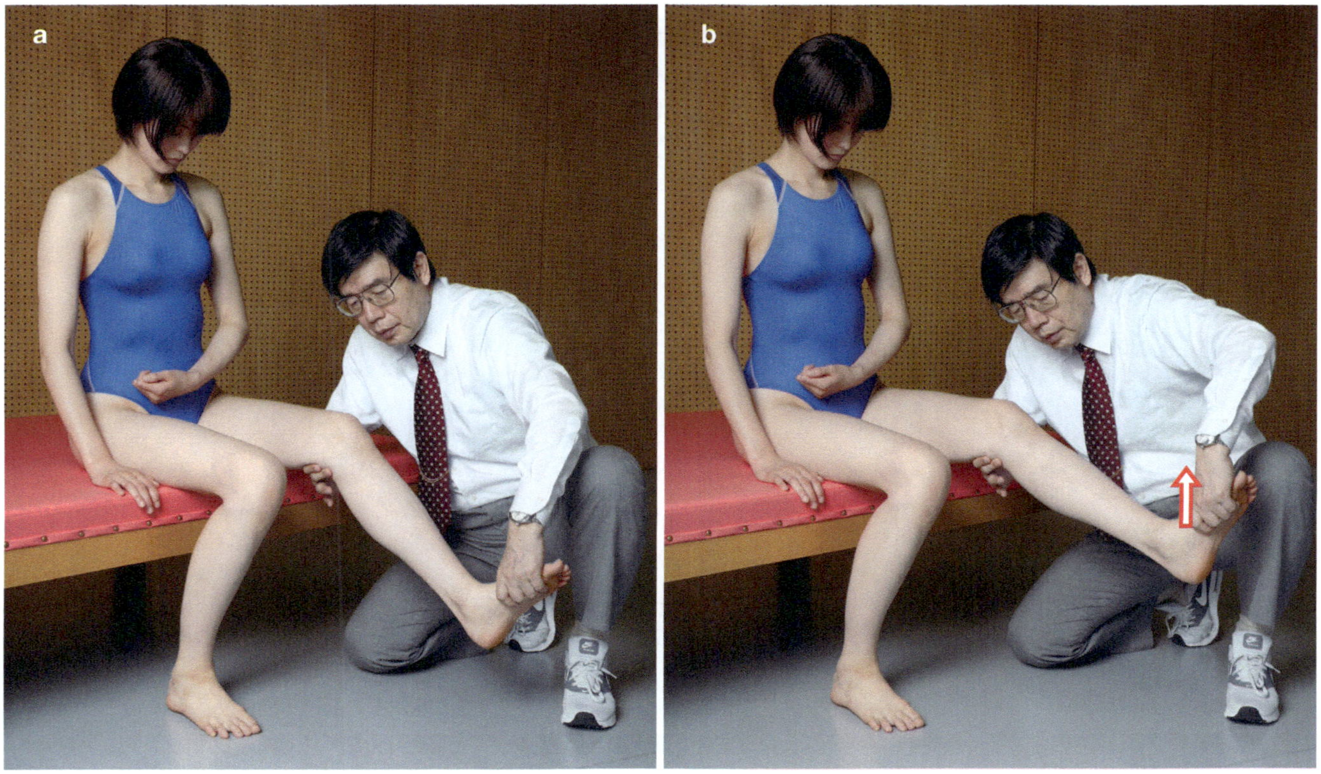

Fig. 5.14 Let's Try: Flexion and knee joint extension (seated position) (Videos 5.8 and 5.9). (**a**) Supporting the thigh and foot. (**a**) The therapist sits beside the patient's affected limb, grasps the hamstring from below with one hand while holding the foot from the dorsal side with the other, and sets the knee joint position ready for flexion. (**b**) Stimulating hamstring stretching to flex the knee joint and negative supporting reflex. (**b**) The therapist says, "Please bend your knee," then pulls the knee joint quickly with his hand holding the patient's foot until resistance is felt. Apply stretching stimulation to the knee's flexor muscles while simultaneously performing a quick inversion on the foot. **Key point**: The therapist performs a quick stretch using his fingers but never adding too much force.

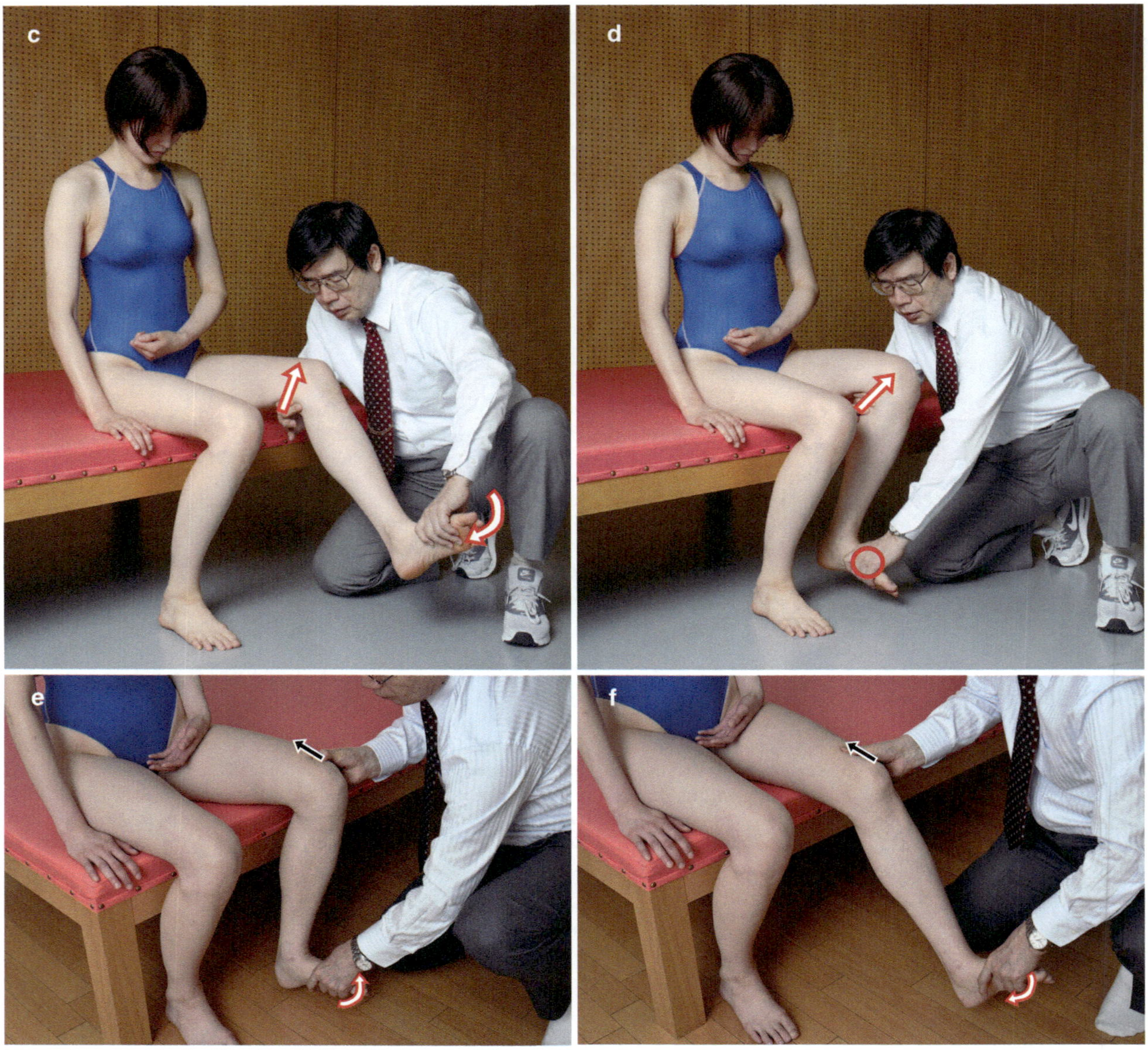

Fig. 5.14 (continued) (**c**) Tapping on the knee flexor muscle. (**c**) The therapist taps the patient's hamstrings gently with one hand and applies slight resistance to knee flexion with the other hand holding the foot. (**d**) Continuing knee flexion. (**d**) The therapist adds slight resistance to the patient's knee joint flexion while maintaining ankle plantar flexion. (**e**) From knee flexion to extension. (**e**) When knee joint flexion stops, the therapist quickly performs ankle dorsiflexion and knee joint flexion for the patient's knee extension, then rubs the patient's patella tendon with his thumb in order to increase tension in the knee extensor muscle. The therapist instructs the patient: "Please extend your knee." **Key point**: In the case of strong spasticity in the extensor muscles, such as clonus occurring in ankle dorsiflexion, the therapist reduces the speed and magnitude of ankle dorsiflexion to the extent that clonus no longer occurs. (**f**) Continuing knee extensions. (**f**) Follow to knee extensions

5.7 Repetitive Facilitative Exercise for the Ankle

There are several techniques that will help induce ankle dorsiflexion. These methods include adding resistance when the hip joint flexes or holding and quickly inverting the ankle while internally rotating the hip joint (facilitating external rotation of the hip) as if the therapist is pulling the whole lower limb downward. If the patient tries to pull her knee to the chest and hold that position, her knee will be flexed and spasticity in the extensor muscles will be suppressed, making ankle eversion easier.

Key Points

1. Ankle dorsiflexion (supine position) (Fig. 5.15)

   The therapist should use both hands to facilitate dorsiflexion and eversion of the ankle.
2. Ankle dorsiflexion (sitting position) (Fig. 5.16)

   Since the hip and knee joints are flexed, dorsiflexion of the ankle can be facilitated easily.
3. Dorsiflexion of the toes and ankle (supine position) (Fig. 5.17)

   Many patients have difficulty in recovering voluntary movement of the toes. However, if dorsiflexion of the ankle can be induced at the synergistic level, it will be possible for the therapist to induce dorsiflexion of the ankle joint and toes by performing a tickling movement on the patient's sole (from the metatarsal head to the tip of the toe). In order to prevent plantar flexion due to clonus, the therapist should carefully consider which technique would apply the optimal level of stimulation at the most appropriate site.

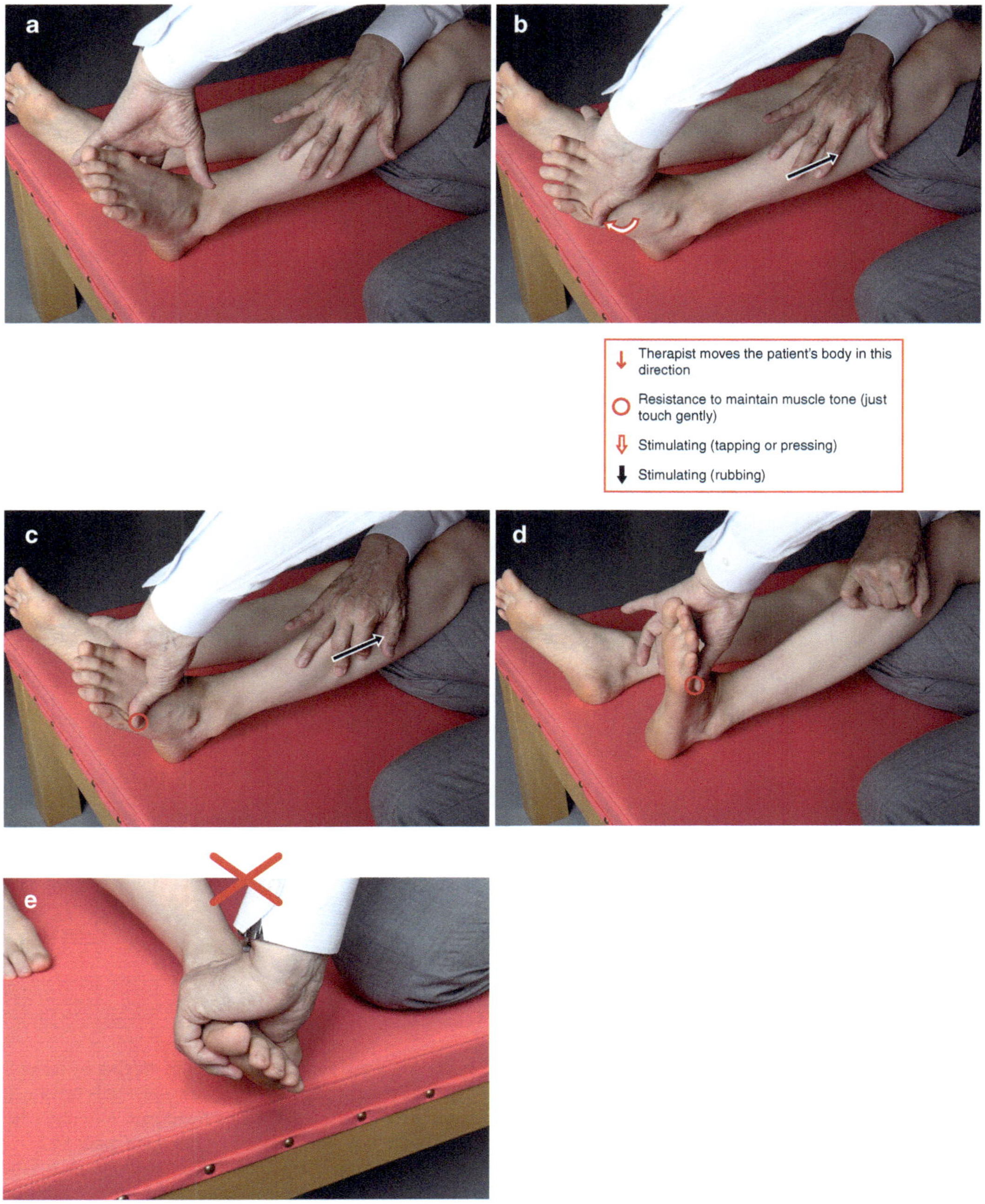

Fig. 5.15 Let's Try: Ankle joint dorsiflexion (supine position) (Video 5.10). (**a**) Supporting the lower leg and foot. (**a**) The patient's head is placed on a high pillow in a position to see her lower limbs. The therapist places his knee under the patient's affected limb, positioning her knee for flexion. The therapist then places one hand on the affected limb's extensor digitorum longus muscle while placing the other on the dorsal side of the foot, which he holds with his thumb and middle finger. As a preparatory exercise, while knee joint flexion exercises can be performed passively on the therapist's knee, the therapist can also perform dorsal flexion for the ankle and repeat dorsiflexion movement through synergistic movement. (**b**) Giving stretching stimulation to the ankle joint dorsiflexion muscle group. (**b**) The therapist uses his thumb to tap and induce the ankle's plantar flexion in the patient, quickly performs an inversion, then instructs the patient: "Please bring your knee close towards your forehead and lift your leg." (**c**) Rubbing the top of the ankle joint dorsiflexion muscle group. (**c**) While stimulating the extensor digitorum longus muscle or by rubbing the more fibula side with one hand, the therapist uses the other hand to apply slight resistance to ankle joint dorsiflexion/eversion. (**d**) Continuing ankle joint dorsiflexion. (**d**) When the ankle dorsiflexion stops, return to (**a**). (**e**) Poor example. (**e**) Poor technique: If the therapist taps his thumb using a style that is accompanied by forearm pronation, it will result in a quick inversion. However, should the therapist tap as if pushing the palm on the patient's foot, this will slow the inversion process, as inducted ankle dorsiflexion will be reduced

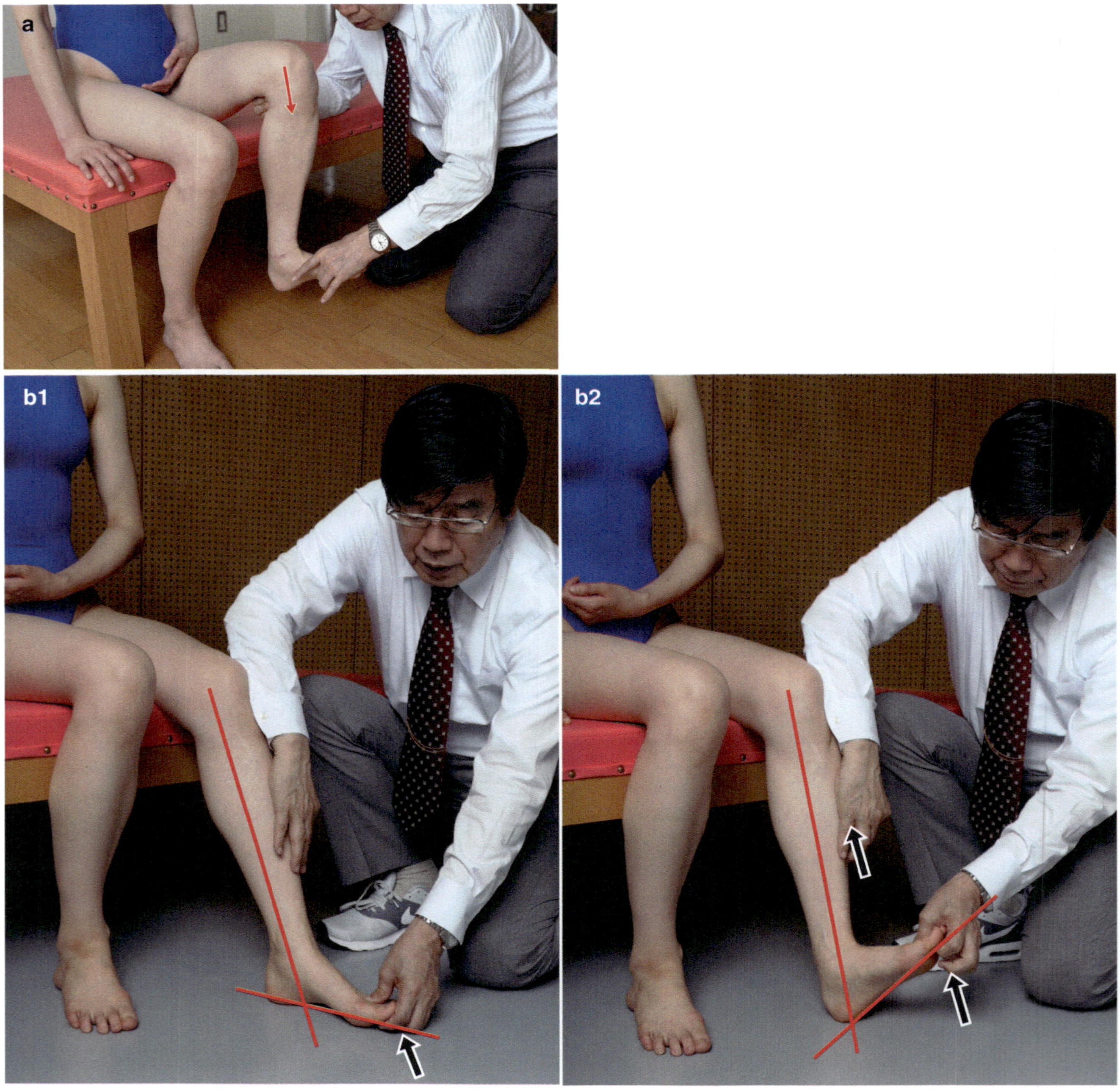

Fig. 5.16 Let's Try: Ankle Dorsiflexion (sitting position) (Video 5.11). (**a**) Procedure to increase the lower limb flexor muscle tension. In order to induce dorsiflexion in the ankle joint, the therapist instructs the patient to flex the hip joint and says, "Please move your knee close to your forehead." The therapist increases the tension in the ankle's dorsiflexion muscle group by applying resistance to the hip joint flexing through holding the patient's knee with one hand. The therapist can easily apply resistance to the hip joint flexing by using his hand to hold down the patient's knee. (**b1**) Using fingernails to scratch the underside/sole side of the toes. (**b1**) Using fingernails to scratch the underside/sole side of the toes makes ankle dorsiflexion easier. (**b2**) Using fingernails to scratch the underside/sole side of the toes. (**b2**) When the patient's knee is flexed and the lateral side of the lower leg is rubbed, ankle dorsiflexion becomes easy.

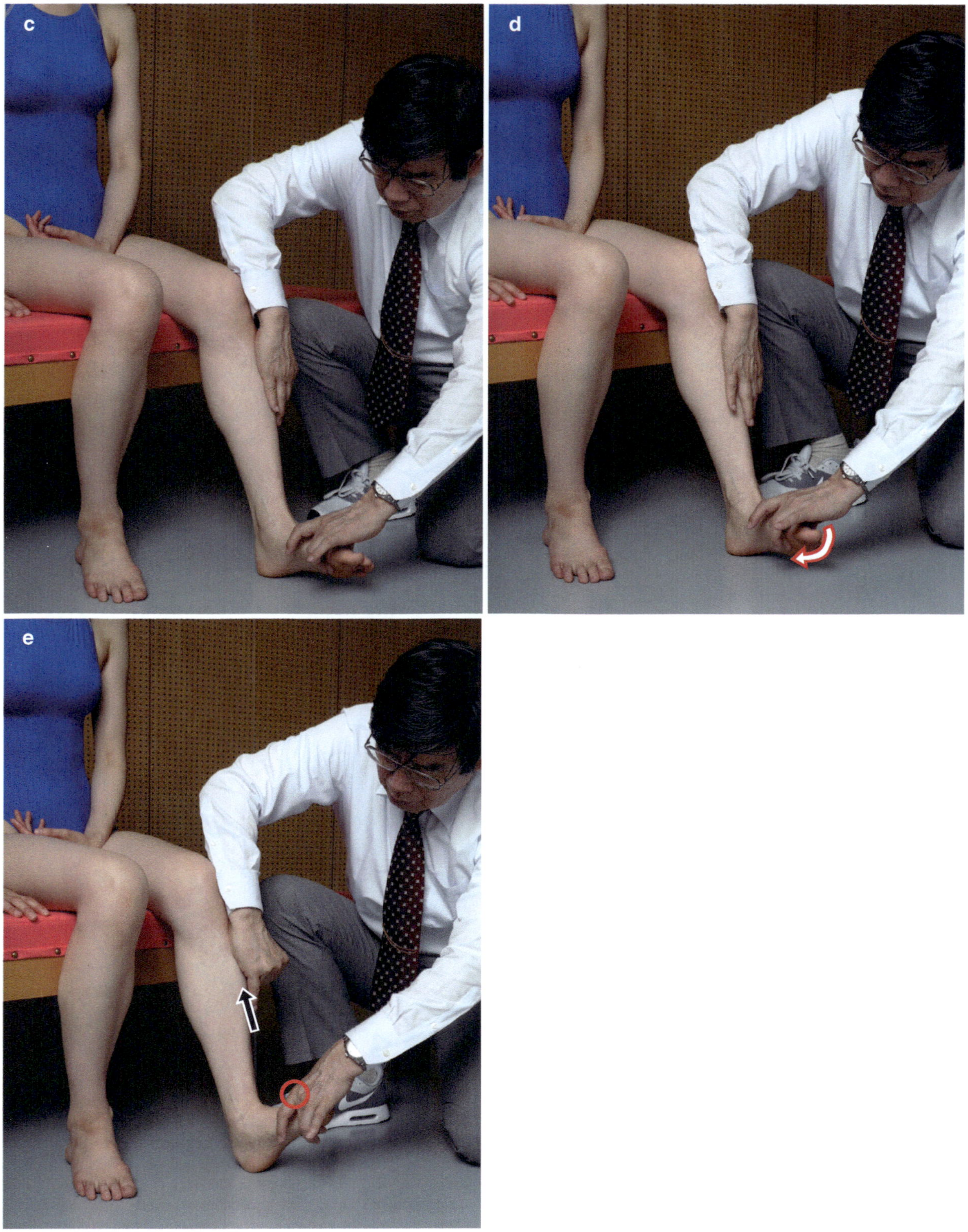

Fig. 5.16 (continued) (**c**) Inverting the procedure. (**c**) The therapist gently holds the patient's knee on the affected lower limb with one hand, keeping the hip joint at 90° and the knee joint at 60°, while holding the foot with the other hand and positioning the ankle joint at 0°. (**d**) Stretching stimulation for the ankle dorsiflexion muscle. (**d**) The therapist quickly makes ankle joint flexion/inversion with his thumb and instructs the patient: "Please raise your toe while keeping your heel rested on the floor." If the patient's ankle eversion on the affected lower limb is insufficient, the therapist presses the patient's knee with his elbow and uses the other hand to hold the foot. Right after performing planter flexion with the hand holding the foot, the therapist stimulates (by rubbing) the extensor digitorum longus muscle with the hand on the upper arm placed on her knee. (**e**) Continuing ankle dorsiflexion. (**e**) When ankle dorsiflexion stops, return to (**c**). **Key point**: When the patient is in a reclined position (slouching backward) the hip and knee joints are in extensive flexion to hold the affected limbs, the extensor spasticity is suppressed, and the gastrocnemius muscles, which are biarticular muscles are then relaxed. Therefore, ankle joint dorsiflexion is induced easily

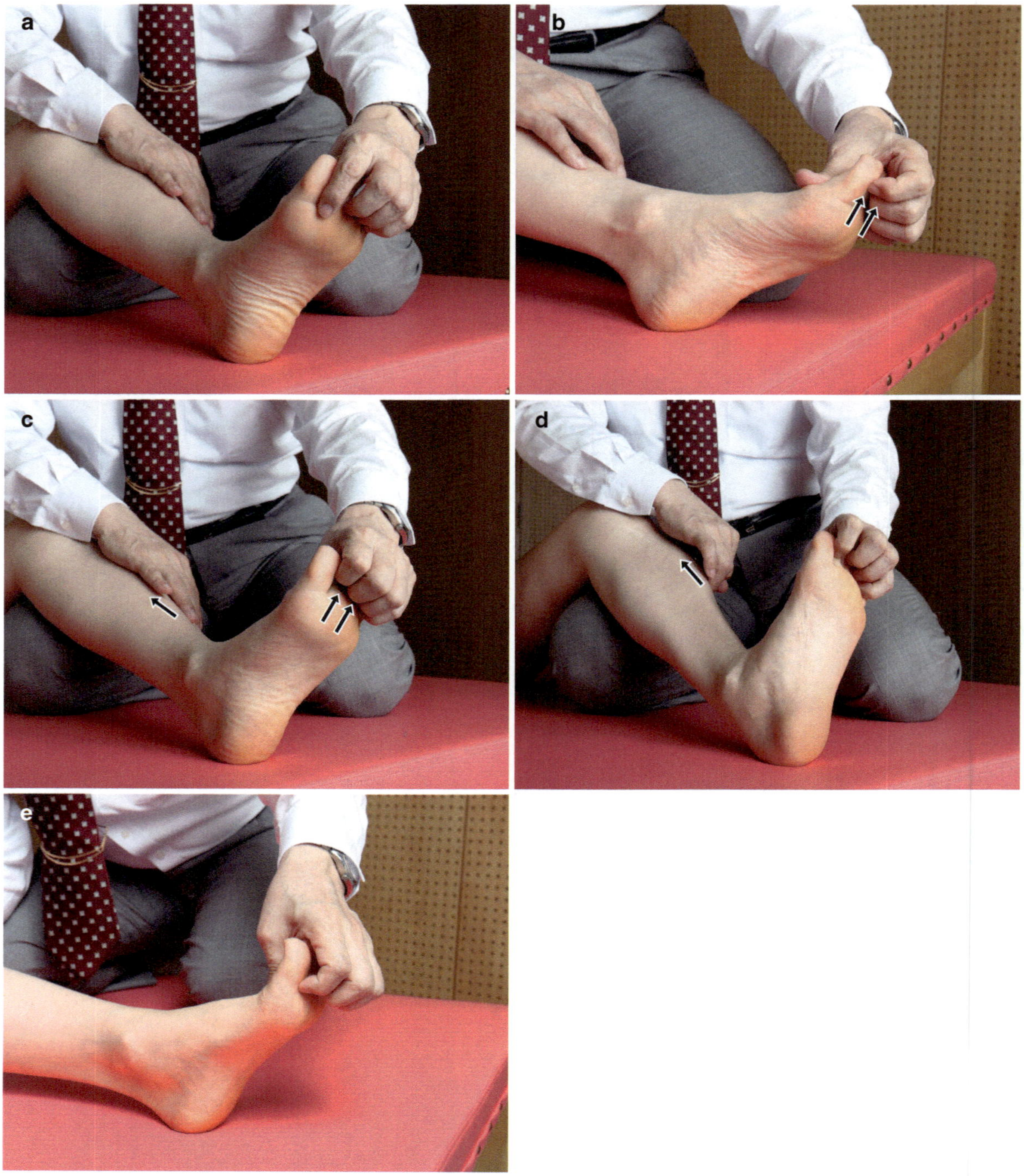

Fig. 5.17 Let's Try: Toe and Ankle dorsiflexion (supine position). (**a**) Supporting the lower leg and foot. (**a**) The therapist holds the patient's ankle gently in a plantar flexed position, with the thumb covering the patient's toes, and with the fingers allowing the toes and the fingertips to touch the sole (metatarso phalangeal joint), and the fingers of the other hand placed on the extensor digitorum longus muscle. **Key point**: When the objective is dorsiflexion in the toes, the therapist holds the patient's toes with his fingers and keeps the ankle plantar flexed position by using his thumb. (**b**) Ankle and toe plantar flexion. (**b**) The therapist facilitates a plantar flex on the patient's ankle joint by pushing the dorsal side of the patient's foot with the thumb and at the same time facilitates dorsiflexion in the toes using his fingers in a scratching motion. In order to adjust the speed of ankle dorsiflexion, the therapist applies slight resistance to the patient's foot with his thumb holding the dorsal side of the patient's foot. (**c**) Scratching the toe. (**c**) The therapist scratches the toes from the sole (metatarso phalangeal joint) by using his fingertips or nails and induce dorsiflexion in the toes and ankle. (**d**) Ankle joint dorsiflexion and continuing toe dorsiflexion. (**d**) As the patient's ankle joint is dorsiflexed and the toes are stretched, the therapist moves the stimulation point to the tip of the toe. As the stimulation to the sole is repeated, the reaction becomes dull, so if the therapist repeats these ten or more times, it becomes necessary to have a break of 10 seconds or thereabouts. Return to (**a**). (**e**) Stimulation to the toes. (**e**) Shows how the therapist stimulates the patient's toes through light scratching using his fingernail tips

Walking Training to Utilize the Function of the Affected Lower Limb

6

Kazumi Kawahira

An ideal walking gait is: (1) highly stable, (2) fast enough for daily use, (3) minimally abnormal in appearance, and (4) not harmful to the muscle spasticity from the hemiplegia or resulting in deformation of the joints.

Some effective measures to aim for with these outcomes are strengthening the unaffected lower limb standing firmly on the unaffected lower limb, and using a lower limb brace and cane. If the therapist holds the patient's body and moves the patient's lower limb, we cannot expect to benefit from this training. The aims of the exercises are to obtain smooth movement of the center of gravity, minimize the difference in the walking rhythm between the lower limbs, and to bring down the affected foot evenly and gently onto the floor during the initial contact phase of walking.

When starting training for standing and walking, we need to emphasize (1) improved balance while standing on the unaffected lower limb and (2) enabling the patient to quickly move their center of gravity from the unaffected to the affected lower limb and back again. The patient needs to know the maximum load that can be applied to the trunk and lower limb on the affected side, thereby bringing the center of gravity back to the unaffected lower limb and standing firmly with the unaffected lower limb at the end of the movement. The affected lower limb should leave the floor slightly, making it easier for the lower limb to take a step.

The most important advice the therapist must give during walking training is to repeat, "Stand firmly on the unaffected side," but <u>not</u> to say, "Put your weight onto the affected side." The instruction "Load your weight onto the affected side" is harmful because it could lead to increased spasticity in the affected lower limb as well as a leaning body axis (trunk and head) toward the affected side, a lowered shoulder girdle muscle, difficulty moving the center of gravity to the unaffected side, and less foot clearance for the affected lower limb.

It is not effective to first begin with a three-point gait pattern using a quad cane. Instead, if prediction for substantial recovery is possible, level up to the two-point gait pattern using a T-cane.

It is usually effective to first begin with the two-point gait pattern using a parallel bar and a T-cane.

Figure 6.1 shows how to maintain balance when standing on the unaffected lower limb, the walking pattern we should aim for, and the points we should look out for. In order to increase walking with a good pattern, the patient should use the parallel bars often for support and wear a brace when walking outside of the bars so that the therapist can evaluate progress by comparing the walking patterns inside and outside of the bars.

Check Points

1. If the patient cannot stand evenly on the unaffected lower limb due to old age or moderate-to-severe paralysis

   If the patient cannot stand stably on the unaffected lower limb in the early stage of rehabilitation, they should walk in a 3-point gait pattern while moving (hand or cane → affected lower → unaffected lower limb). This gait pattern has two of three support points (the affected lower limb, unaffected lower limb, and a cane) on the ground at all times. The patient needs to place the unaffected lower limb a little ahead of the affected lower limb, then repeat the standing-up movement to strengthen the unaffected lower limb. The patient will then aim to achieve the 2-point gait pattern, supporting the cane with the affected and unaffected lower limb only. The patient will use a T-cane or Lofstrand crutch, positioning it slightly ahead of the unaffected side. During the phase when both lower limbs stand side by side, the patient may find it difficult to step out with the affected lower limb and find that the upper body is forced backward the moment he/she tries to swing the affected lower limb forward.

Supplementary Information The online version contains supplementary material available at [https://doi.org/10.1007/978-981-19-0789-0_6].

K. Kawahira (✉)
Laboratory of Repetitive Facilitative Exercise, Kirameki Terrace Health Care Hospital, Kagoshima city, Kagoshima, Japan

K. Kawahira et al. (eds.), *Exercise Therapy for Recovery from Hemiplegia*, https://doi.org/10.1007/978-981-19-0789-0_6

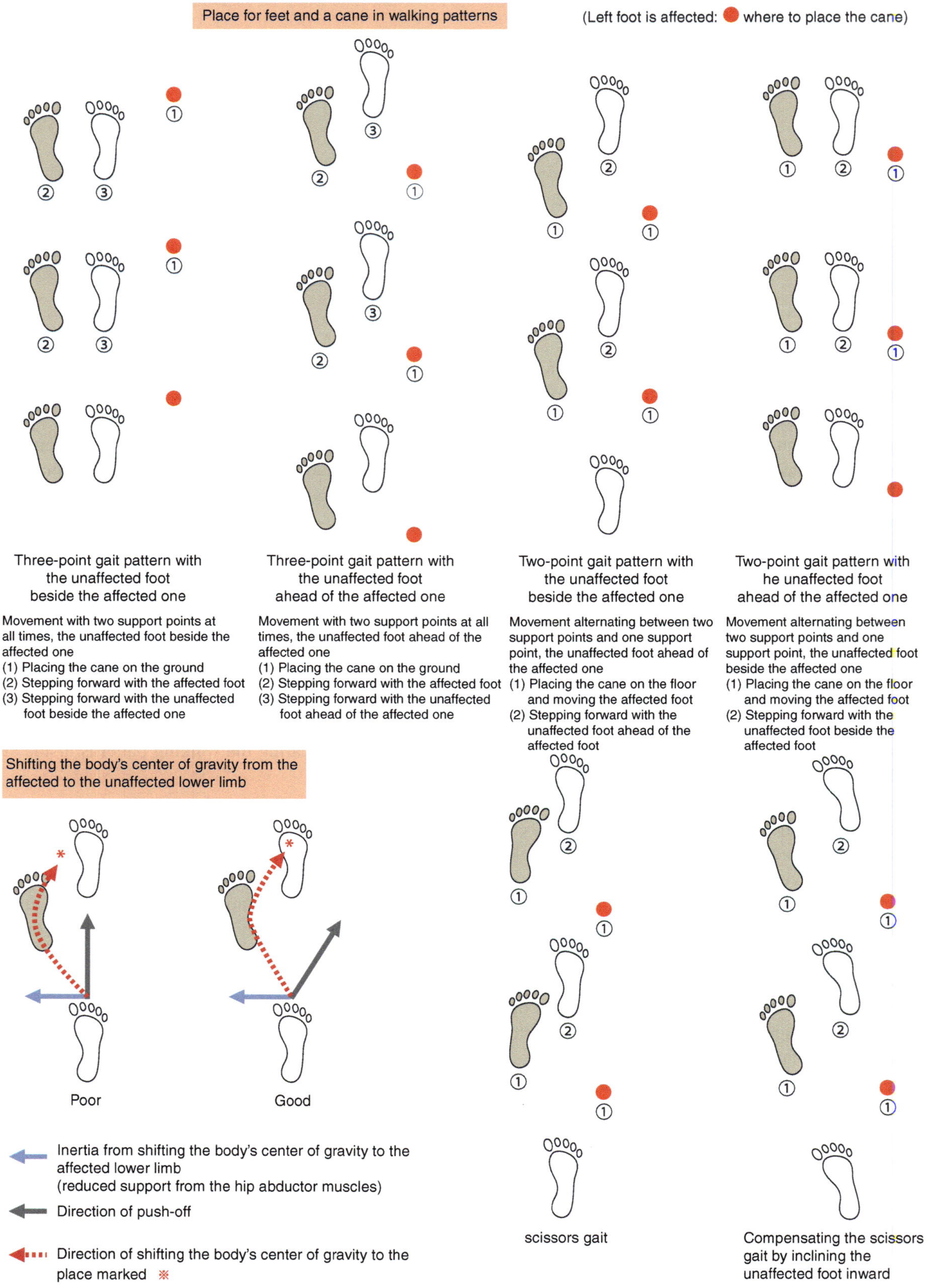

Fig. 6.1 Places for the feet and a cane for walking patterns. One objective in walking training is to shift the body's center of gravity smoothly. Another is to reduce the contracture of extensor muscles before and after stepping on the floor and later standing on the unaffected lower limb. To achieve this using the cane, the hip joint and knee joint should be slightly bent. For the body's center of gravity to shift smoothly during clearance and mid-stance of the affected lower limb, the patient should point the toes on the affected side slightly outwards

2. If the patient is young but unstable when standing on the unaffected lower limb due to mild-to-moderate paralysis
   The patient should train by a 2-point gait pattern on the parallel bars and strengthen standing on the unaffected lower limb. This improves standing balance so that the patient can do a 2-point gait pattern with a T-cane and place the unaffected lower limb ahead. If wearing a lower limb brace to improve the clearance of the toes, the patient should avoid stamping forcefully on the floor and trying to raise the toes. Then it is likely the patient will learn to walk without the brace.
3. If able to stand stably on the unaffected lower limb, the patient should walk in a 2-point gait pattern with a T-cane placing the unaffected lower limb ahead.
4. If the patient cannot maintain balance while standing due to ataxia or hemiplegia on both sides.

Patients who cannot expect a drastic improvement should begin with smaller steps, such as a 3-point gait pattern using a walker, Lofstrand crutch, or quad cane.

We need to evaluate balance while the patient is standing and estimate the walking pattern the patient may be able to achieve. Timing for placing the crutch and foot and the rhythm of shifting the body's center of gravity vary depending on walking patterns of a 2-point or a 3-point gait pattern. Therefore, the patient needs to repeat the same walking pattern that could be obtained from the exercise.

Lower Limb Brace and Stick
Rehabilitation therapists ask whether using a brace and a cane or crutch might slow the patient's recovery from paralysis and make the patient dependent on such devices in order to walk, but numerous studies have indicated that this is not the correct. Whether the patient is in an acute, recovery, or chronic stage, use of these devices facilitates recovery from paralysis and restores the ability to walk, which also prevents problems arising from disuse.

6.1 Training for Standing Balance

Standing balance affects the ability to walk and transfers skills from a bed to a wheelchair. The goal of this exercise is to stand on the unaffected lower limb. The exercise should be performed using parallel bars or handrails for effectiveness and the patient's safety.

The purpose of the training exercises and precautions taken during the training are stated as follows. The patient needs to repeat the standing-up movement 100 times in order to keep a stable standing position and to learn to stand firmly on the unaffected side and maintain stability. Even if the heel of the affected lower limb does not maintain contact with the floor, stability on a single lower limb while standing is important for controlling the affected lower limb.

6.1.1 Purpose

1. Achieving firmly standing balance on the unaffected lower limb (Fig. 6.2):
   The patient should place the center of gravity onto the unaffected lower limb without bending the body to the affected side.
2. Moving the center of gravity smoothly from the affected to the unaffected lower limb (Figs. 6.3 and 6.4)
   The patient should move his/her center of gravity slightly onto the affected lower limb for a short time, extend the affected lower limb to support the body weight, and then move the center of gravity completely onto the unaffected lower limb. By widening the step width, the center of gravity quickly returns from the affected lower limb to the unaffected lower limb.

Key Point
The ability to walk relates to moving the center of gravity between the two lower limbs. It is unrelated to the ability to support the whole bodyweight on the affected lower limb or to support it equally on both lower limbs.

Key Point
The affected lower limb cannot make any small adjustments to its movements. Exercise in standing on the affected lower limb for the sole purpose of balance training will bring no benefits.

3. The patient should keep her eyes ahead of the toes so that the affected lower limb is within sight. Visual feedback will partially compensate for the paralysis and sensory disturbance. You should not instruct the patient by saying, "Raise your chin and push out your chest." Such a posture will increase the extensor muscle spasticity and make it difficult to control the affected lower limb.

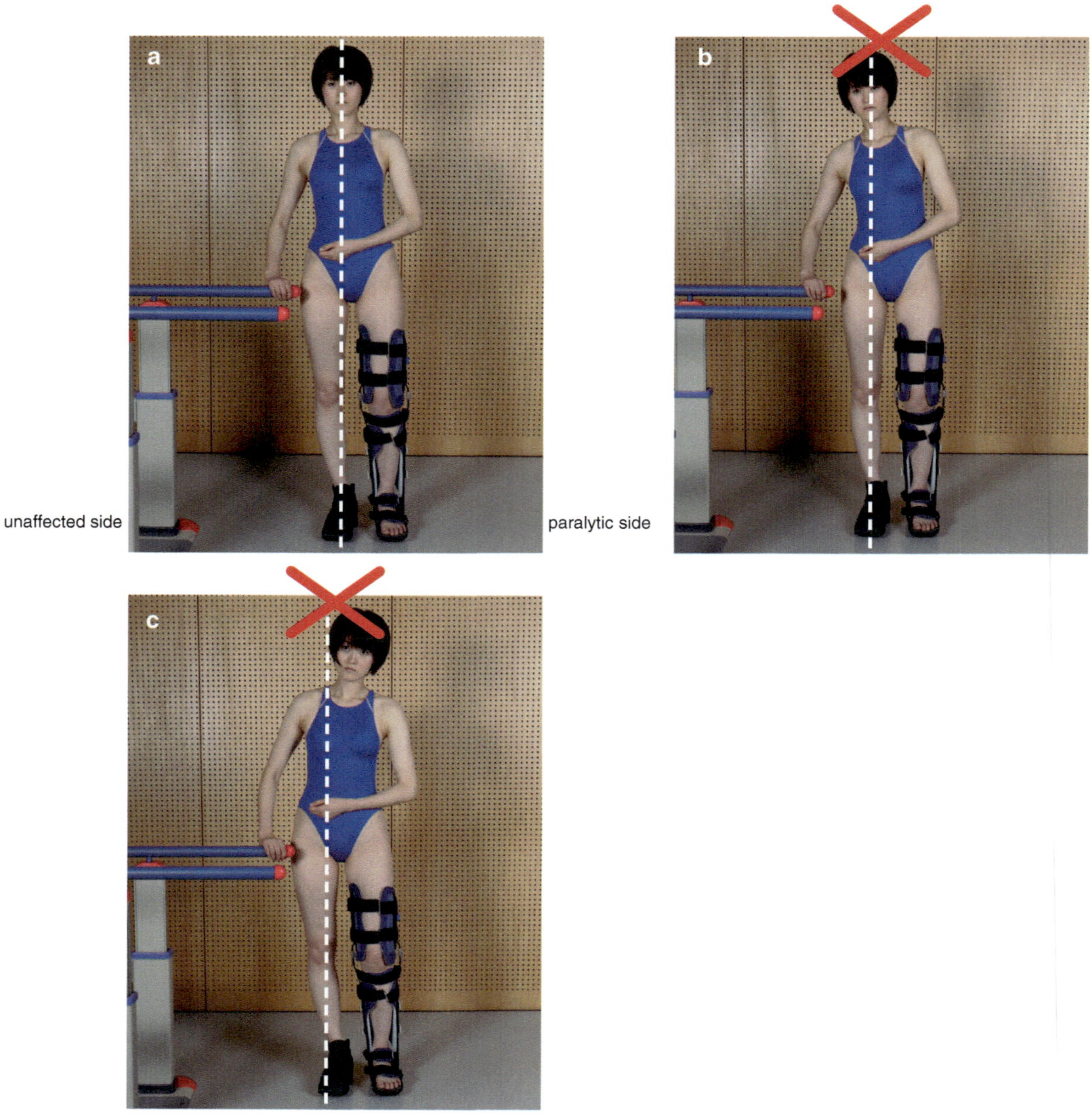

Fig. 6.2 Basic Technique: Posture to be taken when standing. Note: The procedure here is in condition with knee locking of LLB. (**a**) The posture the patient should take. (**a**) The sole of the shoe (heel) on the affected side's lower limb hovers off the floor. The trunk should not bend laterally towards the unaffected side. The patient can lift the affected side's heel from the floor naturally when she transfers her weight onto the unaffected lower limb, as her head aligns with the extension line for the unaffected lower limb. Do not raise the pelvis on the affected side by bending the trunk laterally. The center of gravity is on the unaffected lower limb. (**b**) Poor example: Standing on the unaffected lower limb and bending the trunk laterally towards the unaffected side. (**b**) The patient's trunk leans onto the unaffected side. The affected sole of the shoe (heel, left foot) does not touch the floor. The center of gravity is on the unaffected lower limb, which is good, but the patient cannot achieve toe-clearance. It is important that the patient raises the pelvis when stepping out onto the affected lower limb and place the foot gently on the floor. If the patient's paralysis is severe, the therapist should say "Straighten your lower back," which makes it easier for the patient to extend the lower limb. The patient should place the center of gravity onto the unaffected lower limb so that the sole of the shoe (the heel) of the affected side leaves the floor slightly. During the gait motion cycle, the patient's foot can lift from the floor (achieving toe-clearance) by first loading her body weight properly onto the affected lower limb. Loading (onto a leg) leads to propulsion for walking. If the sole of the patient's shoe (the heel) on the affected side does not leave the floor, the therapist should advise her to "Stand firmly on your unaffected limb." (**c**) This is incorrect: Standing on the unaffected lower limb but with the trunk lateral bending towards the affected side. This is the posture that we should avoid by any means. (**c**) Here the patient is lateral bending the trunk towards the affected side, putting the center of gravity onto the affected lower limb. Even if the patient swings the affected lower limb, the toe will drag, leading to a circumduction gait. Placing the center of gravity onto the affected lower limb during gait leads to Trendelenburg's walking pattern or a fall. With this posture, the patient cannot improve the gait speed. An instruction to "Place your center of gravity onto your affected leg" leads to this posture, which is incorrect

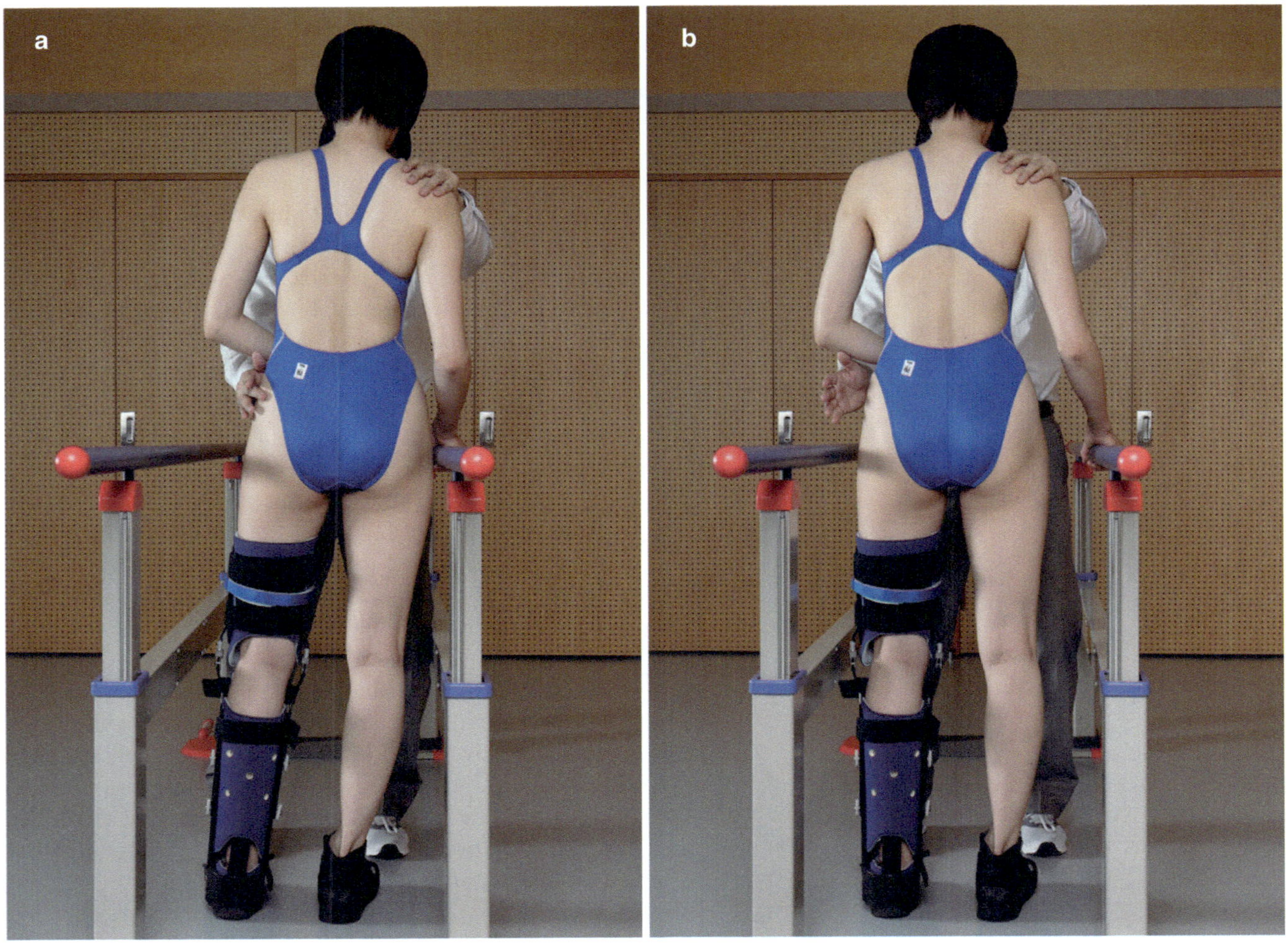

Fig. 6.3 Let's try: Assisting the movement of the center of gravity between the affected and unaffected lower limb (Videos 6.1 and 6.2). (**a**) Moving the center of gravity from the unaffected to the affected side with the therapist supporting the movement. (**a**) The patient has both feet opens to shoulder-width and places the center of gravity onto the unaffected lower limb. **Key point**: To prevent the knee giving way and to assist the extension/flexion of the knee, fixation from the inner side of the knee works better. See Fig. 6.4a and b. (**b**) Shifting the center of gravity onto the affected lower limb. (**b**) The therapist says, "Place your weight slightly onto the affected lower limb." The shift in the center of gravity should be small so that the patient can extend the affected lower limb and move the center of gravity back onto the unaffected lower limb. Asking the patient to stand with the center of gravity solely on the affected lower limb has no benefit. The patient should place the weight slightly on the affected side, flexing the knee a little, and right after that she should bring the weight back onto the unaffected lower limb. The therapist needs to ask the patient to repeat this movement many times and have the distance in the shifting center of gravity between the affected and unaffected lower limb, gradually increase. Note: The picture shows that the therapist's fingers do not rest on the patient's hip. The picture was taken at the moment the therapist is tapping his fingers.

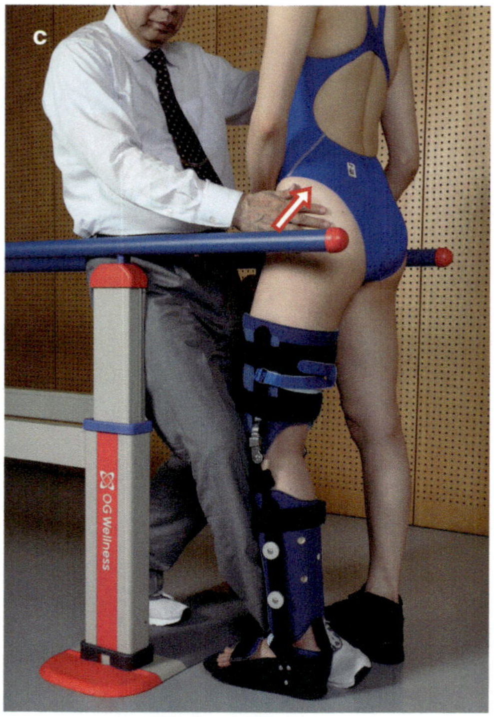

↓ Therapist moves the patient's body in this direction

○ Resistance to maintain muscle tone (just by touching gently)

⇩ Stimulating (tapping or pressing)

⬇ Stimulating (rubbing)

Fig. 6.3 (continued) (**c**) Shift in the center of gravity from the affected to the unaffected lower limb and how to stimulate the lower limb. (**c**) If the knee joint of the brace is not fixed, the therapist should use his lower limb to support the patient's lower limb in order to prevent the knee giving way. The therapist should also tap the hip or the gluteus maximus muscle. This support makes it easier for the patient to shift the center of gravity using the affected lower limb

6.1.2 Exercise for a Patient with Unstable Standing Balance

For the patient who can't control a knee hyperextension (back knee) due to the spasticity of gastrocnemius, we need to keep the hip joint flexed forward slightly to suppress involuntary spasms of extensor muscles or else use a short leg brace. For an unstable knee, due to flaccid paralysis, the patient should use a long leg brace. To raise the tonus level in the extensor muscles, we apply vibratory stimulation to the whole of the affected lower limb, electrical stimulation to the gluteus medius and quadriceps, or rapid stretching (stretch reflex) to the triceps muscle in the calf. Then we let the patient stand and say, "Stretch your lower back," to make it easier to extend the hip and knee joints.

Even if asked to straighten the upper body, the patient may have to rely on using the unaffected arm on the parallel bar and will be unable to stop lateral bending towards the affected side. In such a case, the patient is likely to have an impairment in recognizing the spatial axis or lose trunk stability on the affected side. To correct recognition of the spatial axis, the patient should try to straighten her body by looking at a pillar or other vertically aligned object after (guiding) passive movement of the body to the left and right. To strengthen the trunk, the patient's training needs to include a lateral bend between the affected and unaffected sides, turning over while lying down, and raising the pelvis on the affected side.

Even if asked to stand straight, a patient with cognitive impairment will lean to the affected side within a short time. We need to let the patient stand on the unaffected lower limb, significantly bend their body laterally towards the unaffected side, then return to a straight position so as to stand completely on the unaffected side. By repeating these moves, the patient's balance while standing will be improved.

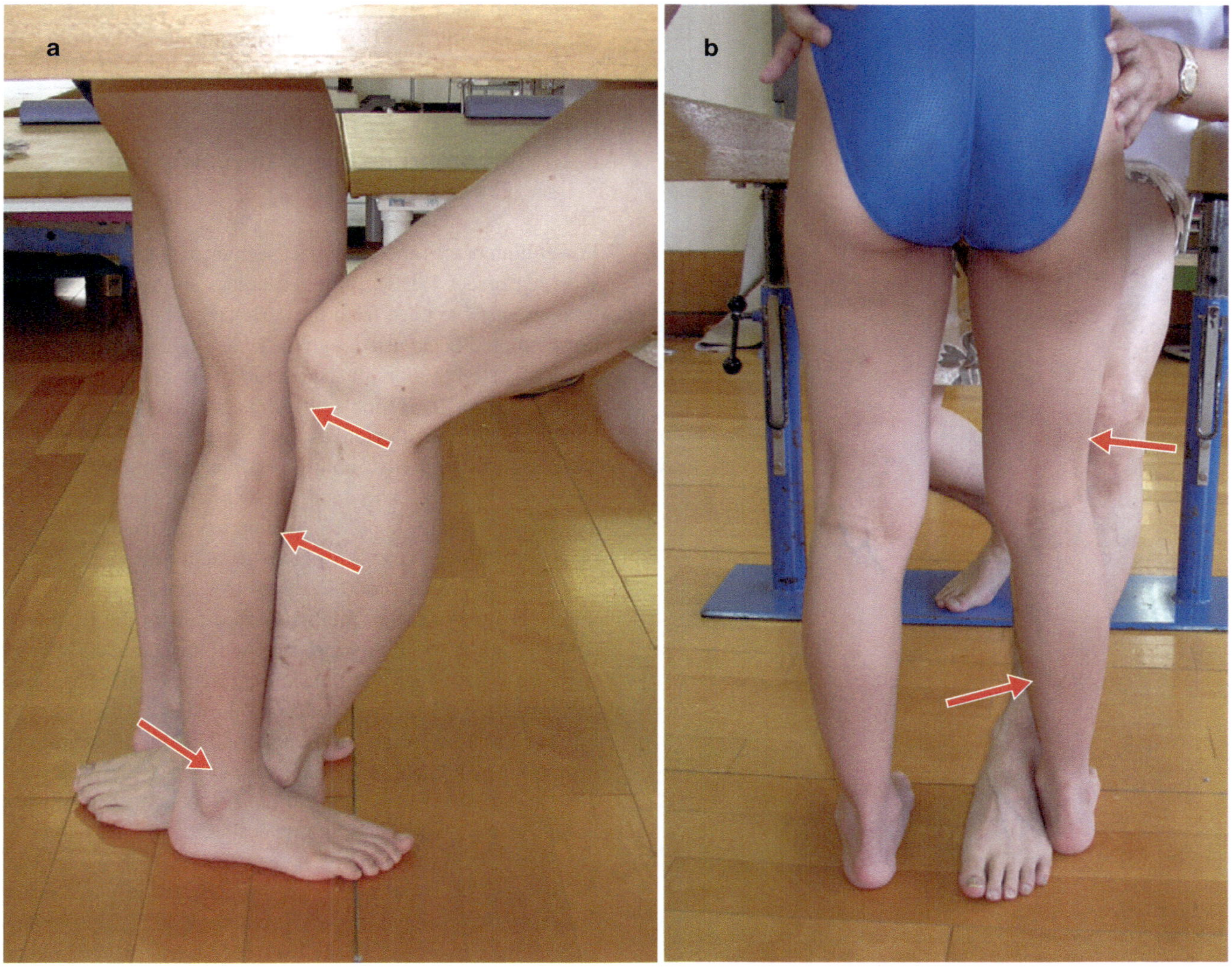

Fig. 6.4 Let's try: How to assist the patient's affected lower limb. (**a**) Fixing the patient's knee and shin (side view). (**a**) This supporting method is used when we need to prevent the patient's knee giving way and when assisting the patient to extend the knee in order to shift the center of gravity towards the unaffected lower limb. The therapist needs to press both the knee and shin against those of the patient, not just at the knee. If the therapist presses only at the knee, typically he will fail to prevent the patient's knee giving way. The therapist should carry out this procedure near the handrail or within the parallel bars to maintain his posture and balance. (**b**) Fixing the patient's knee and shin (back view). (**b**) Allowing the patient to extend the hip joint on the affected side, the therapist should stimulate the gluteus maximus and gluteus medius. To further allow the patient to keep extending the knee, the therapist should hold the knee and shin against his own. If the patient wears a long leg brace, this exercise will be easier, which will lead to voluntary movement of the affected lower limb.

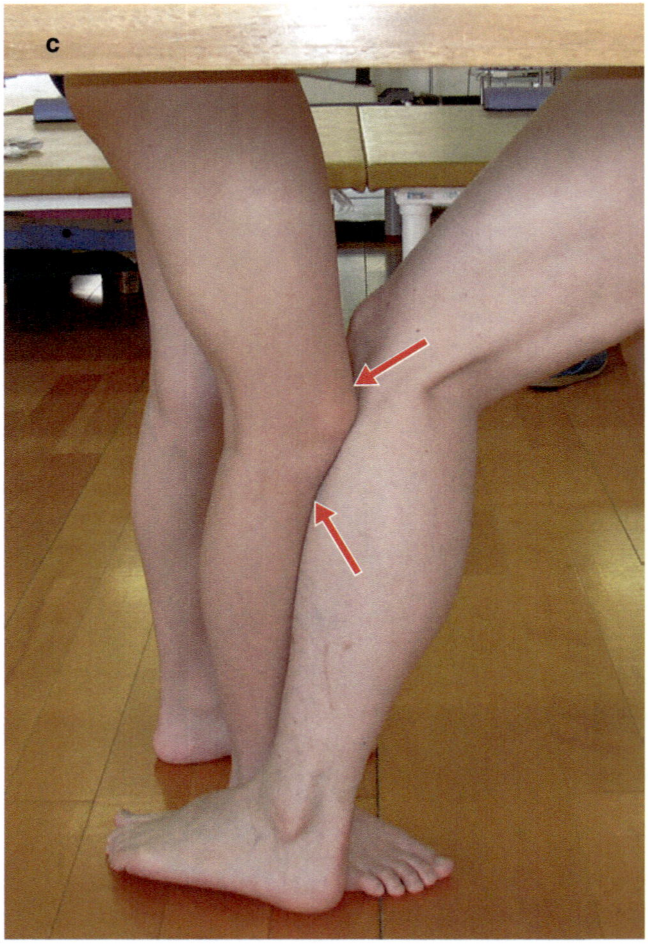

Fig. 6.4 (continued) (**c**) Supporting the lateral side of the patient's shin. (**c**) This supporting method does not work well if the patient's paralysis is too severe to remain sufficiently stable. For patients with light paralysis, the therapist can support the knee from the lateral side, as shown in picture **c** (this is different from the method shown in picture **a**). To prevent the knee from giving way and to move the center of gravity onto the affected lower limb, the therapist should support the patient's knee. If the therapist just places their own knee on the patient' s knees, then the patient' shin will not be stable, so it is important the shin fronts are pressed close together

6.2 Lower Limb Brace and Cane

The benefits of using a lower limb brace or cane include preventing falling, improving posture when walking, suppressing spasticity, and increasing the amount of exercise a patient can perform. The patient should be encouraged to conduct the exercises on parallel bars or a lower limb brace and continue exercising outside the rehabilitation facility. Walking without a lower limb brace leads to reduced stability, gait, and speed and should be limited to situations such as bathing or walking for short distances at home.

Although the patient can dorsal flex the ankle while seated, if the heel fails to make contact with the floor while walking or if the foot-flat phase begins too soon after heel contact is made, the patient should use the parallel bars and walk with small steps, focusing on the heel contact phase with knee and hip joints slightly flexed. By holding onto the parallel bars before placing a foot on the floor, the patient can rest the heel on the floor with less difficulty.

6.2.1 Appropriate Use of the Lower Limb Brace

Lower limb braces are categorized as long leg braces (LLB) and short leg braces (SLB). Long leg braces are used for patients with an unstable knee from the early stages of treatment. The braces should be adjusted with the knee slightly flexed and the ankle at a slightly dorsiflexed position with restriction against plantar flexion (posterior stop), thus limiting plantar flexion while allowing dorsiflexion. In the recovery stage, depending on the muscles' supportive ability and the patients' capacity to walk, the range of knee and ankle joint motion should be expanded. Patients should try to keep toe off the floor and do plenty of walking exercise using the parallel bars.

Short leg braces should be used to suppress spasticity in the triceps calf muscle, facilitate toe off from the floor, and shift the center of gravity after the heel-strike makes contact with the floor. Since the braces help achieve good walking patterns, they should be used from the start of the patient's rehabilitation program. To judge whether the ankle joint of orthosis is necessary, the patient should be asked to walk and observed to see if the ankle joint of orthosis turns backward in later phases when standing up from the edge of a chair. If the dorsiflexion of the ankle is insufficient, then the heel on the affected side does not touch the floor, the affected lower limb is pulled backward, and the patient is likely to lose balance. If the patient has a back knee before leaving the heel on the affected side, we should evaluate a 5–10 mm additional sole to supplement the flexion in the ankle joint.

6.2.2 Choice and Usage of Canes (Fig. 6.5)

A T-cane is suitable providing the patient can stand stably on the unaffected lower limb. A Lofstrand crutch is recommended when the patient's stance on the unaffected lower limb is unstable. Use a walker cane if the patient's stance is very unstable.

During the two-point walking phase using a T-cane, the patient should place the cane on the outer side of the unaffected foot at a near-vertical angle parallel to a position between the metatarsal and heel of the affected foot. The cane should then be placed on the floor just before the affected heel touches the floor. The moment the heel on the affected side touches the floor, the center of gravity moves away from the unaffected lower limb, causing the unaffected heel to step off the floor and leading to maximum instability in the patient's stance. The patient should place the cane gently to support stability and make it easier to control the spasticity in the extensor muscles of the affected lower limb.

Patients with ataxia or low muscular force show reduced balance when standing on the unaffected lower limb. The

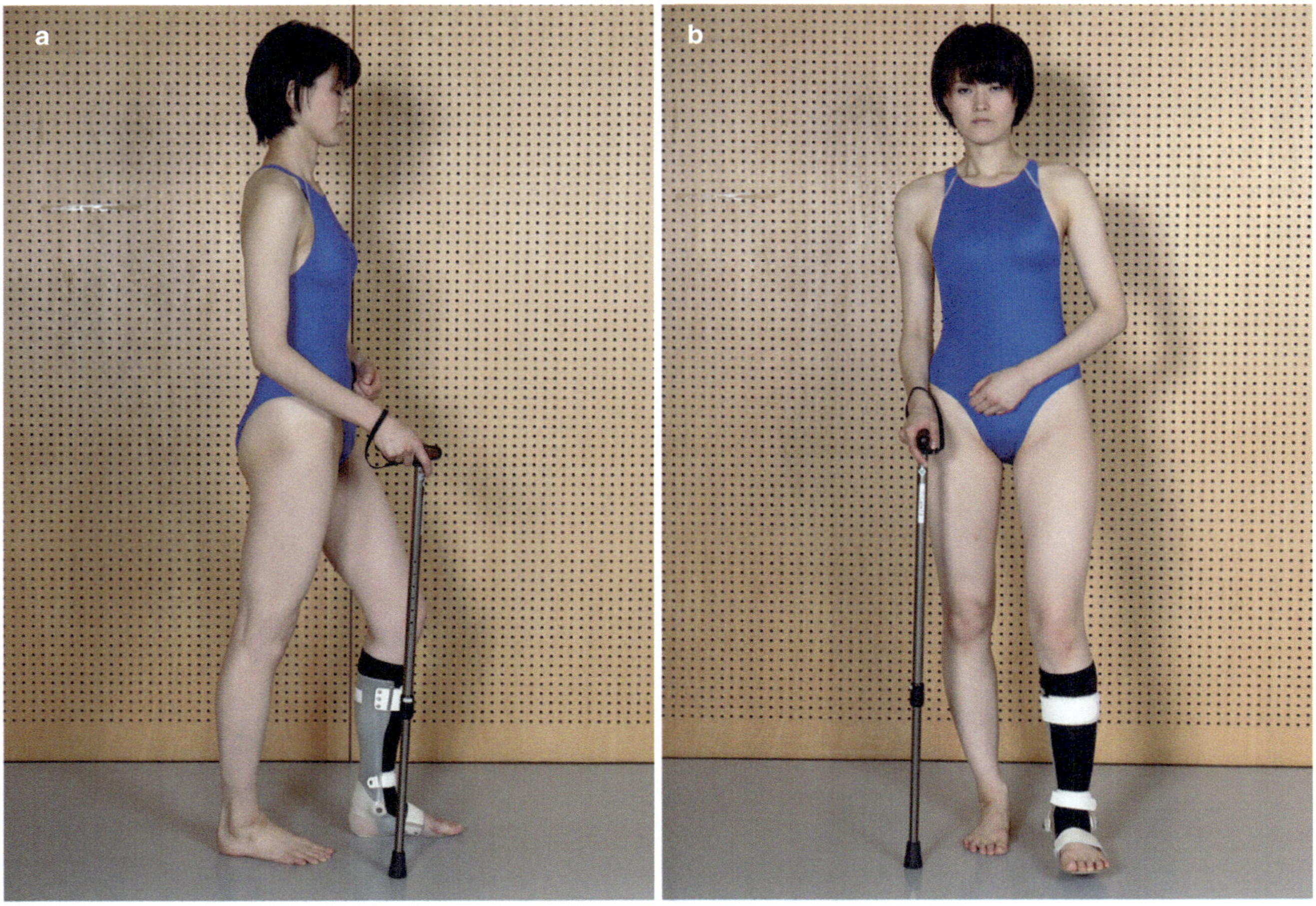

Fig. 6.5 Let's try: Using the parallel bars or a cane during gait training (Video 6.4). (**a**) How to hold the cane (side view). The basic use of the T-cane is shown here. With a balanced stance on the unaffected lower limb, the patient should place the cane in the position of a heel, supporting the affected side, slightly before the affected foot touches the floor. The patient should keep the cane almost vertical to prevent the affected knee from turning inward or outward and to achieve a rhythmical and appropriate walking speed. (**a**) Position of the T-cane and affected lower limb (side view). The T-cane is placed beside the heel on the affected side. The patient's elbow that holds the stick should flex slightly and touch her trunk. The wrist is flexed so that the cane and the forearm are not aligned. This make it possible to hold the cane with stability. (**b**) Position of the cane and the affected lower limb (front view). (**b**) How to hold the cane (front view). The cane is almost vertically straight. The patient's elbow flexes slightly and touches the body. The wrist is dorsi-flexed. The cane and the arm are not aligned, which allows stability. The patient places the elbow against the body to stabilize the T-cane, which makes it easier to align with the walking rhythm. The patient should not press the T-cane hard against the floor but use it to touch the floor gently while supporting unstable steps with the affected lower limb.

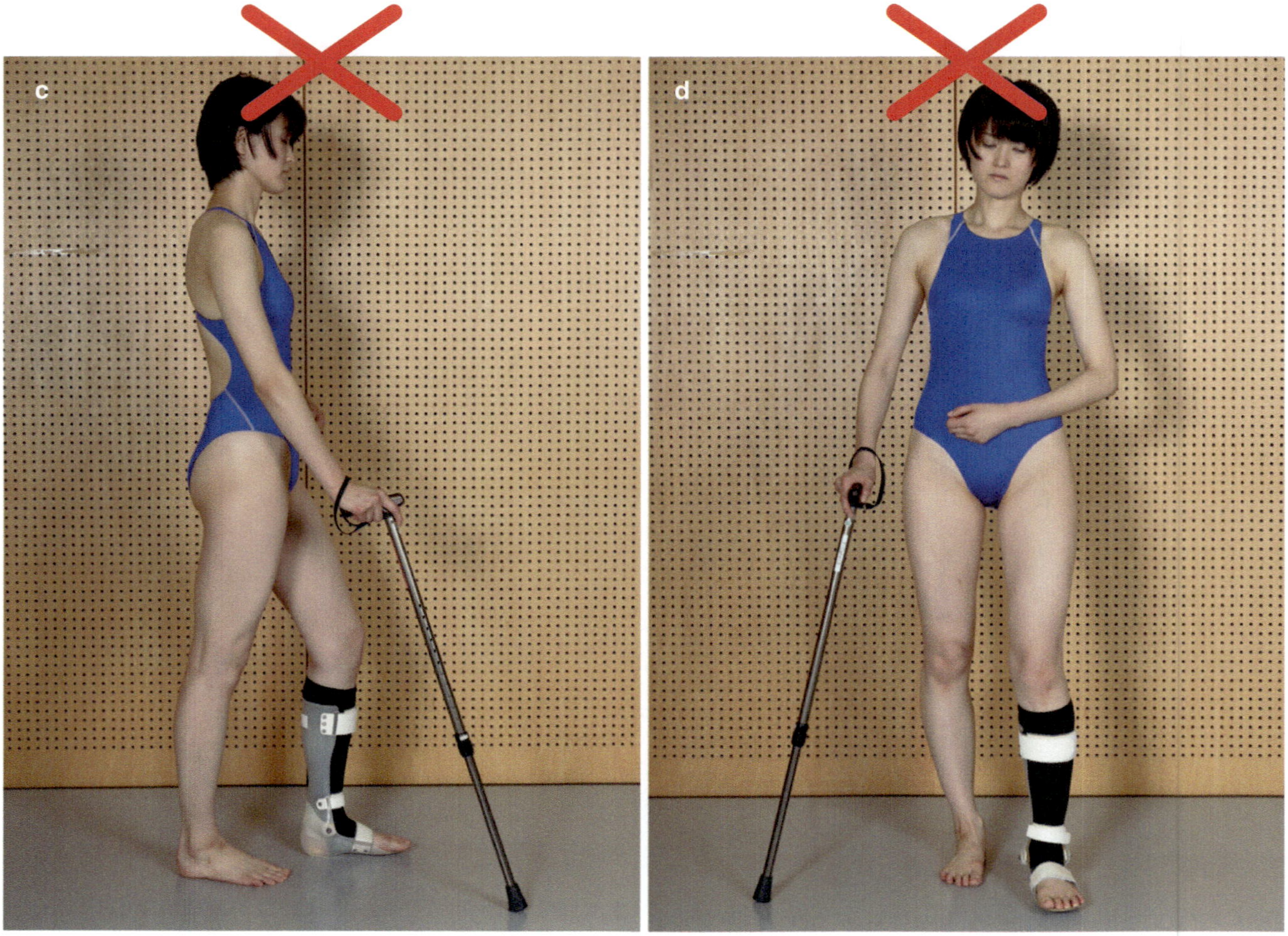

Fig. 6.5 (continued) (**c**) Bad example: Inappropriate position of the T-cane (side view). (**c**) Bad example: Poor positioning of the T-cane and the affected lower limb (side view). The cane is placed too far ahead of the affected heel. The arm holding the T-cane is extended and does not touch her body. The patient places the T-cane ahead of the patient's lower limb, so the patient cannot keep balance while standing. Also, having the elbow away from the body makes the cane unstable and results in poor support. When the T-cane is not vertical, the patient cannot move her center of gravity well, which prevents her from increasing her walking speed. (**d**) Bad example: Poor positioning of the T-cane and the affected lower limb (front view). (**d**) The T-cane is not placed vertically. The arm holding the T-cane is extended and is away from the body. The more the cane sways, the more the trunk and affected lower limb are pushed toward the affected side. This escalates Trendelenburg gait pattern, worsening the clearance of the toe, and leading to deformation in the affected knee. Therefore, the patient must (1) be advised during walking exercises that standing firmly on the unaffected lower limb avoids this type of incorrect cane usage, and (2) told, "Stand firmly on the unaffected lower limb to avoid this type of incorrect cane usage." Most patients with this poor technique grip the cane tightly with stress in the forearm to shoulder, moving the center of gravity onto the affected lower limb extremely slowly, or else holding the center of gravity in the middle of the standing phase, which leaves the unaffected foot higher off the floor, and places it down slowly beside the affected one. Shifting the center of gravity by pushing off is insufficient, so the patient flexes the hip joint and leans her upper body forward, raising the possibility of losing balance and slanting backwards. **Key point**: It is important for patients with hemiplegia in the walking training to have a smooth shift in the center of gravity during the stance phase onto the affected lower limb. Making the unaffected lower limb able to "push-off" allows for a smooth shift in the center of gravity. If the therapist repeats the instruction, "Transfer your weight more on the affected lower limb," the training will go into a negative spiral.

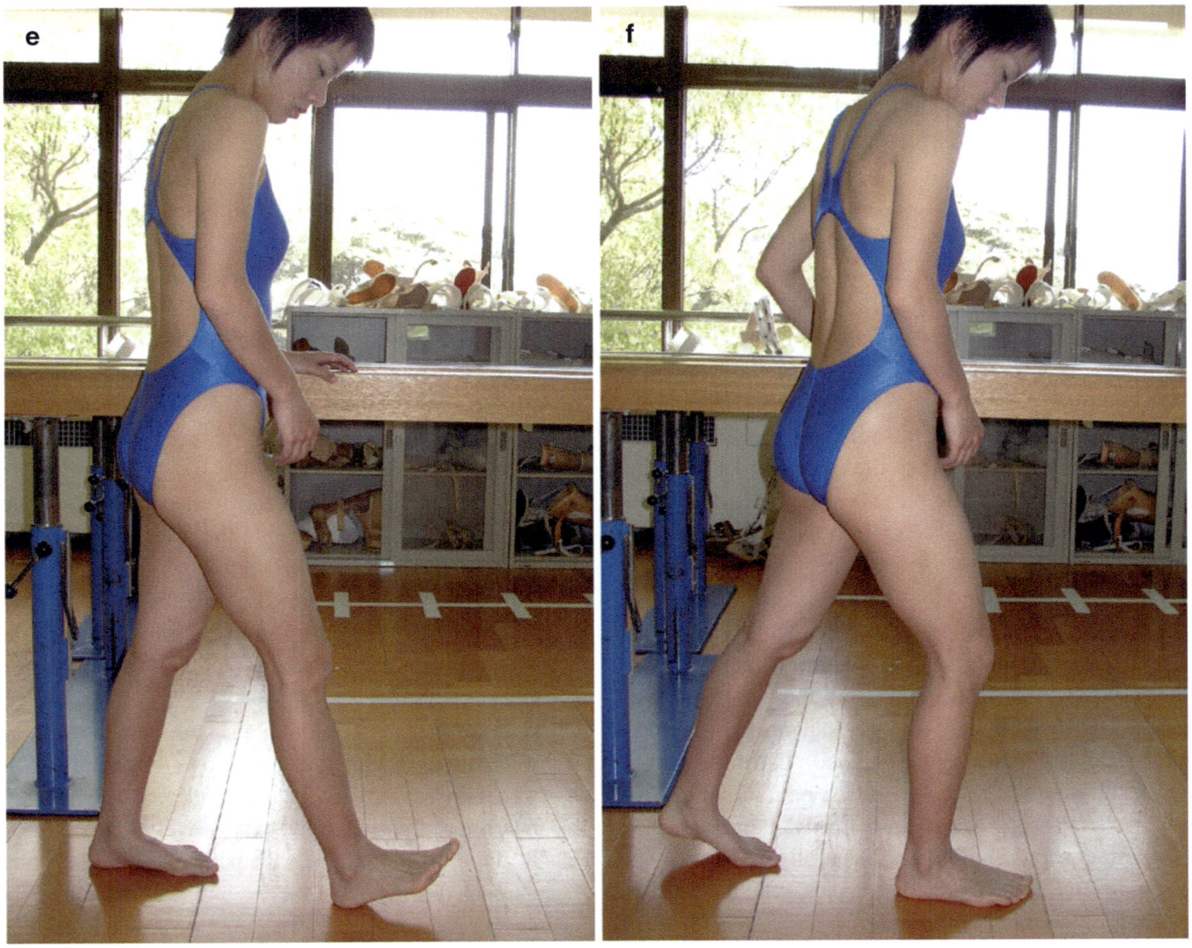

Fig. 6.5 (continued) (**e**) Training to place the foot down, starting from the heel contact, using the parallel bars. (**f**) Training to place the heel contact, and then the foot-flat. (**e**, **f**) Training heel contact on the affected side. Even if the patient's paralysis is mild, the patient should repeat the training either by touching the ground with the heel first or touching with the heel and then placing a full "foot-flat" onto the floor. The patient should flex the hip and knee joints slightly. Training should be started by walking with small steps such as two-point gait. Note: The stroke at the training start should be one third of the stride shown in these pictures

cane should be placed slightly away from the patient's body and extended so that the patient can keep the elbow close to the body for better support.

6.2.3 Raised Shoe Insole for the Unaffected Heel

When the patient uses a brace on one lower limb, an insole is required for the foot on the other side to offset the difference in lower limb length. The insole should have a height of 1–2 cm to cover the thickness of the brace sole and an additional buffer space between the brace and shoe. The shoe lift should also be high enough to obtain easier clearance for the affected foot. If the patient's abductor muscles in the hip joint or trunk muscles are too weak, the pelvis on the affected side drops by a length corresponding to the thickness of the insole on the unaffected foot, thus failing to provide easy clearance for the toes.

An abduction gait occurs due to either of the following reasons:

1. An abduction or circumduction gait develops to compensate for the difference in lower limb length.
2. An abduction gait develops due to a patient's habitual attempts to adjust lower limb movements.

For a patient with problem (1), additional height for the unaffected foot helps to improve walking. However, for (2), an insole for the unaffected foot will cause the patient's head to be involuntarily raised and lowered, with the head raised and lowered interchangeably when the patient stands on the affected lower limb.

If the patient is strong enough to stand on the unaffected lower limb properly, it becomes possible to raise the pelvis and keep the affected lower limb off the floor. Supplementing the difference in height from the lower limb brace should be sufficient.

- Examining the difference in lower limb length

  Allow the patient to be seated with the knees bent at a 90° angle and the heels placed together. If the height of the knee on the affected side is higher, prepare an insole for the unaffected heel to keep both knee heights level.

6.3 Walking Training

During the standing phase of the affected lower limb, walking training requires a smooth shift in the center of gravity. To achieve this, planter flexion of the unaffected lower limb needs to be suppressed (using the lower limb brace in most cases) and the push-off from the affected lower limb proceeding diagonally forward from the unaffected lower limb is needed. This will allow the patient to move forward; otherwise, the patient will stumble or step backwards to maintain stability.

If the patient's unaffected lower limb is raised and swings slowly, it means the unaffected push-off capability or force is insufficient. The following shift in weight to the unaffected lower limb will also be insufficient, and the patient is likely to lean towards the affected side or backwards, making it easy to lose stability (causes poor balance). During the swing phase with the affected lower limb, the patient should first stand on the unaffected lower limb firmly, before taking a step. The patient should hold onto the parallel bars or cane to suppress spasticity in the extensor muscles, which is likely to increase, before stepping onto the floor.

The patient should hold the parallel bars or cane gently with the hand on the unaffected side just before the affected foot lands on the floor. This stabilizes the affected lower limb, induces maximum voluntary movement of the affected lower limb, and makes it possible to achieve a walking pattern with a smooth movement in the center of gravity.

If the therapist places emphasis on saying, "Place your weight on the affected lower limb," then excessive weight will be imposed on the affected lower limb while the patient tries to stand. This makes it difficult to shift the center of gravity smoothly, extends the time for standing on the affected lower limb, and reduces the push-off using the unaffected foot.

This will also lead to Trendelenburg gait pattern, loss of balance while standing, lowered walking speed, dragging the affected lower limb, more obvious abnormal walking, and a deformed knee joint. The instruction, "Place more weight on the affected lower limb," does not bring any benefits. The therapist should say, "Stand firmly on the unaffected lower limb."

1. Training on the walking pattern (Figs. 6.6, 6.7, 6.8, 6.9, 6.10, 6.11, 6.12)

   Repeating the target walking pattern strengthens the nerve path which achieves the pattern and will result in acquiring an ideal walking pattern. Facilitative therapy should be continued during the walking pattern training as it is important to achieve the target pattern (Figs. 6.8 and 6.9). Parallel bars, canes, and lower limb braces help to achieve this. Training without them does not deliver any good results.

2. Assisting the patients with Trendelenburg gait pattern, scissors gait, etc. (Figs. 6.8 and 6.9)

   Walking with the affected foot turned outwards (an abduction gait or a circumduction gait) and having the affected lower limb crossed over the unaffected lower limb (scissors gait) can cause knocked-knees or bowed-legs.

   To prevent a knee deformity, the unaffected lower limb should move straight beside the affected lower limb and land beside the cane. Sufficient push-off allows the center of gravity to move from the affected lower limb side to the unaffected lower limb, and does not allow the patient's weight to stay long on the affected lower limb, leading to smooth walking.

   If the patient walks with a scissors gait or inner rotation of the hip joint with the toe turned to the front or inwards, or even if the patient walks with Trendelenburg gait pattern and the center of gravity moves excessively to the affected side, the therapist should guide the patient to slant the body to the unaffected side so that the toe on the affected side turn outwards. This suppresses dragging the toe and placing excessive weight on the affected lower limb, which makes it easier to shift the center of gravity to the unaffected lower limb and improve walking stability and speed.

3. Assistance walking upstairs/downstairs (Figs. 6.13 and 6.14)

   During the voluntary movements, either the dorsal premotor area is strongly involved when responding to outer stimulation or the inner side of premotor area (supplementary motor area) is strongly involved when there is no outer stimulation. Patients with damage including the supplemental motor area and cingulate gyrus often have mild-to-moderate paralysis with complications of difficulty beginning a movement, grasp reflex, and alien hand syndrome.

   There are some tips for letting these patients move their limbs. Even if the patient cannot move the arm despite the instruction "Extend your elbow," the arm and fingers may extend if an object is placed in front of the patient. Even if the patient cannot flex the knee despite giving the instruction "Raise your lower limb," the patient can raise the lower limb if a bar is placed in front of the patient or the patient is allowed to stand in front of the stairs. These tips should be utilized for the actual treatment [1].

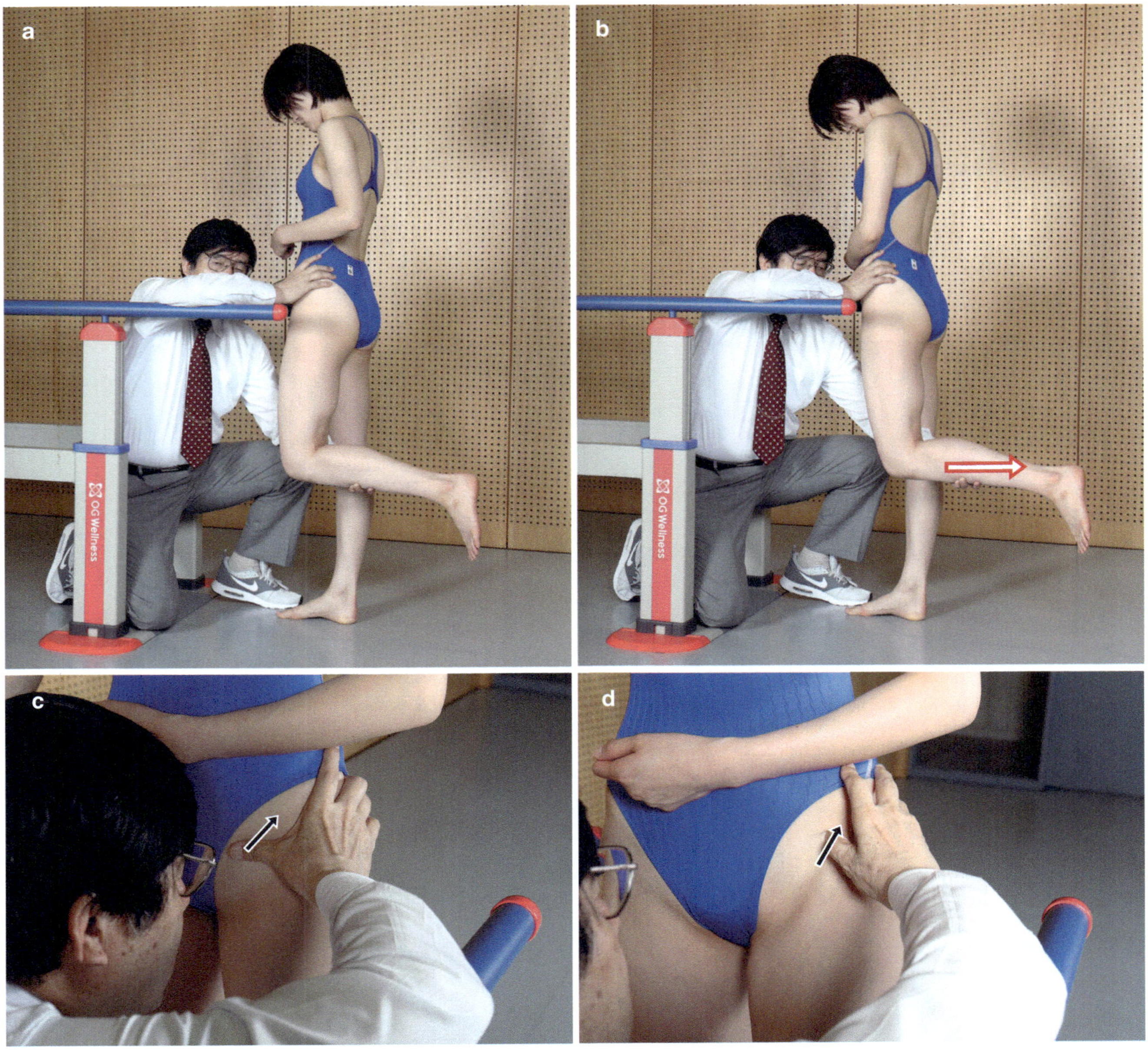

↓ Therapist moves the patient's body in this direction

○ Resistance to maintain muscle tone (just by touching gently)

⇩ Stimulating (tapping or pressing)

⬇ Stimulating (rubbing)

Fig. 6.6 Let's try: Training and assisting the walking pattern (Video 6.3). (**a**) Supporting the hip and lower limb. Techniques to induce stepping forward movement through hip flexion (**a** to **d**). (**a**) The therapist holds the patient's affected knee at 90°. The patient repeats flexing the hip joint. The therapist should apply their hand to the affected side of the hip to support the patient's movement. **Key point:** To prevent falling while facilitating a hip joint flex, the therapist should place the elbow on the parallel bars, hold the Iliac crest, and support the patient's affected lower limb with the other hand. (**b**) Quick extension of the hip joint. (**b**) The therapist extends the patient's hip joint quickly with a hand on the affected lower limb and instructs the patient to "Move your knee close towards your forehead," to facilitate the hip joint flexing. (**c**) Stimulation for the inguinal ligament. (**c**) The therapist rubs gently above the inguinal ligament with the thumb, which is placed on the patient's Iliac crest. (**d**) Continuous stimulation to the inguinal ligament. (**d**) The therapist rubs the inguinal ligament with his thumb up to the Iliac crest.

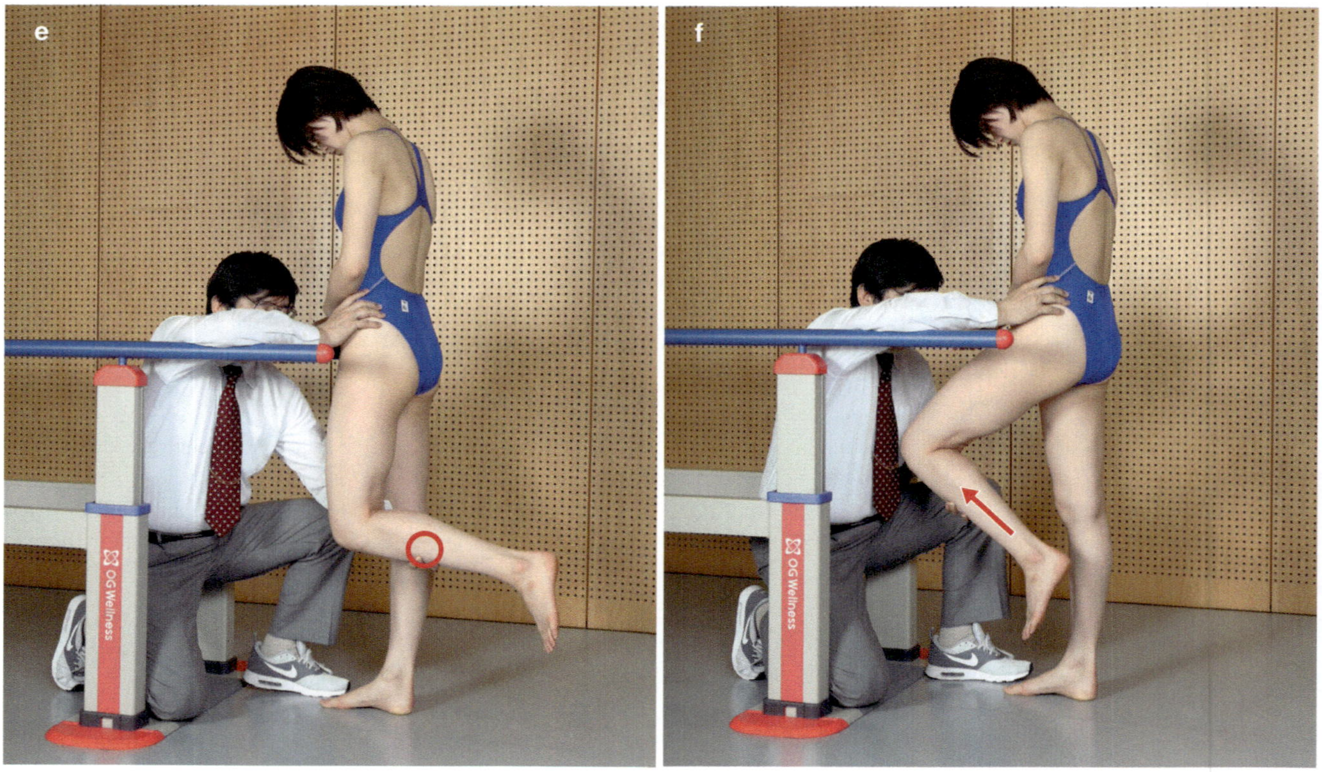

Fig. 6.6 (continued) (**e**) Flex the hip joint. (**e**) The therapist lets the patient keep flexing her knee at 90° and continues bending her hip joint. (**f**) Continuing to flex the hip joint and preventing the upper body turning backwards. (**f**) The patient repeats raising the lower limb by flexing the hip joint and tries not to turn the upper body backwards. To accomplish this, the therapist should instruct the patient to "Move your knee close towards your forehead," then "Try to bend your body"

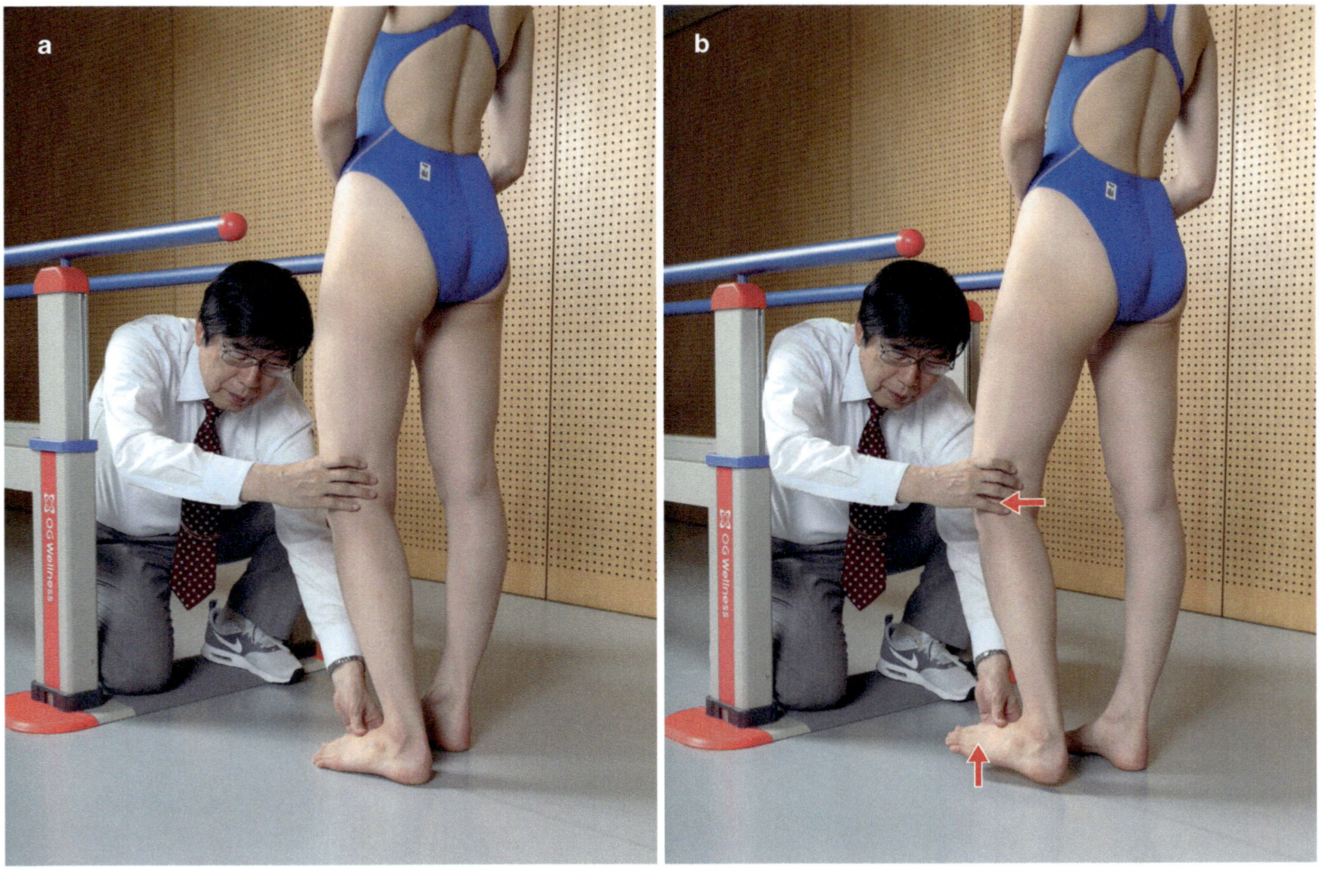

Fig. 6.7 Let's try: Training for the walking pattern-Two-point gait pattern and a stepping-out movement (Video 6.3). (**a**) Holding the affected knee and foot. (**a**) The therapist holds the patient's lower limb, helps raise the foot off the floor, and instructs her to "Stand firmly on the unaffected lower limb." The patient repeats the stepping out movement. It does not matter if the patient drags the toe. The arm of the unaffected side should move in synchronization with the affected lower limb when stepping out. If the patient cannot step out by flexing the hip joint while standing, the stepping forward exercise by flexing the knee is effective (see Fig. 6.6). If the patient cannot dorsi-flex the ankle, a short leg brace should be used. (**b**) Holding the foot during the swing phase. (**b**) The therapist holds the affected foot and repeats the movement between the mid-swing and the heel strike. To suppress extensor muscle spasticity, which becomes more likely to happen before the foot touches the floor, the patient should not try to raise the lower limb too high.

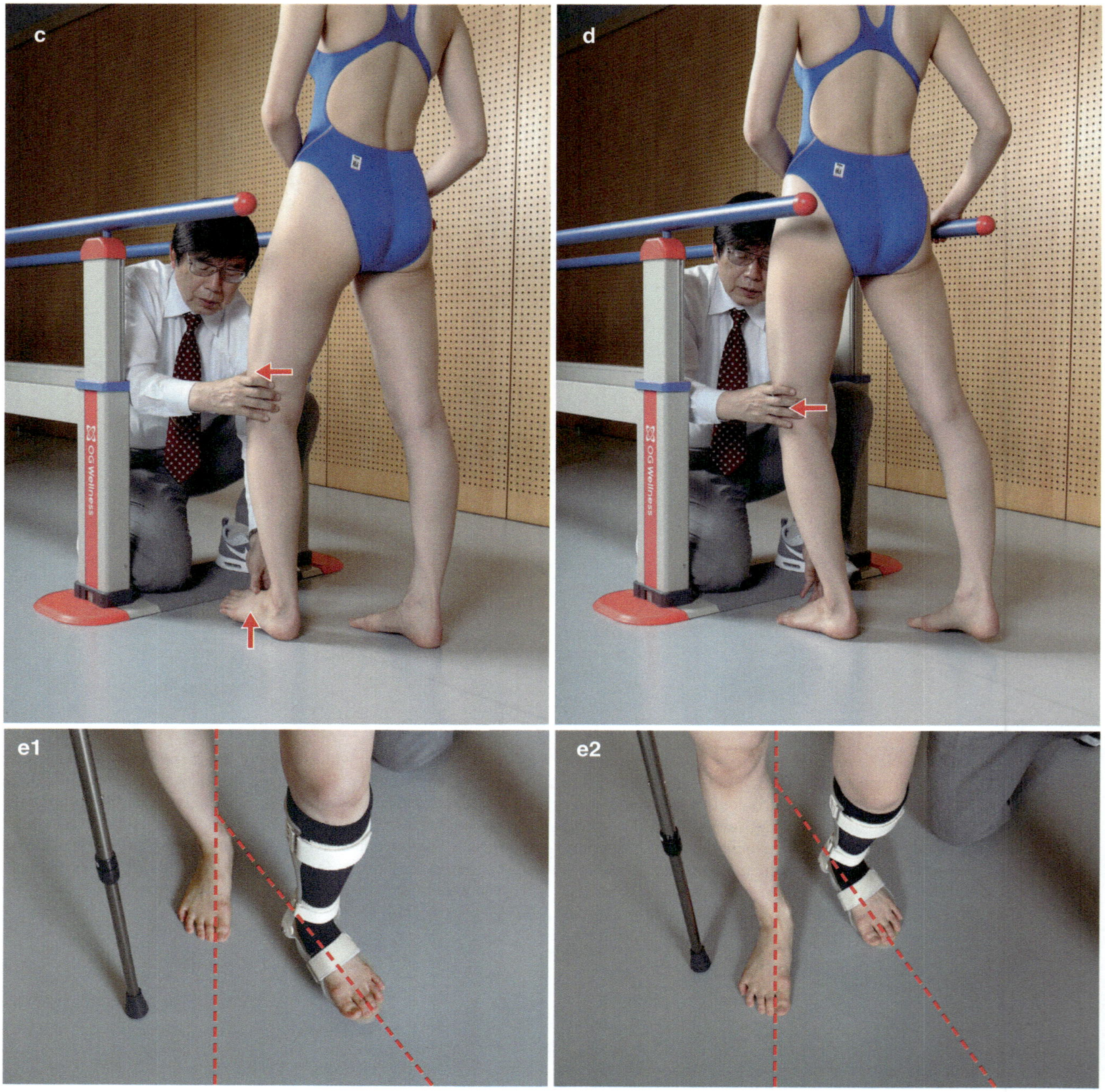

Fig. 6.7 (continued) (**c**) Holding the knee and foot around heel contact. (**c**) To suppress extensor muscle spasticity when the heel lands on the floor, the patient's knee should be kept slightly flexed. (**d**) Holding the knee and foot when placing the heel on the floor (at the moment of heel contact). (**d**) Right before the affected foot lands on the floor with the knee slightly flexed, put the unaffected hand on the parallel bar. This makes it easier to control the movement of the center of gravity. The therapist holds the foot but releases it when the heel touches the floor. (**e1**, **e2**) From the heel contact to the loading on the lower limb. After the heel makes contact, the patient moves the center of gravity forwards, keeping the knee flexed slightly. The therapist should repeat the instruction to the patient, "Stand firmly on the unaffected lower limb," to avoid making too much effort to load on the affected lower limb. **Key Point**: Training to push off and step off the unaffected lower limb is done with the condition shown in "**e1** and **e2**." The patient places the affected foot ahead of the other and moves the center of gravity smoothly before and after mid-stance of the affected lower limb. After repeating a smooth shift in the center of gravity, the patient moves on to training strong push-offs as well as steps diagonally forward with the unaffected foot. The point is to use the push-off with the unaffected lower limb, shifting the center of gravity away from the affected to the unaffected lower limb completely, and achieve standing firmly on the unaffected lower limb. Stepping out onto the affected lower limb from standing on the unaffected lower limb should be repeated until they become a continual series of movements

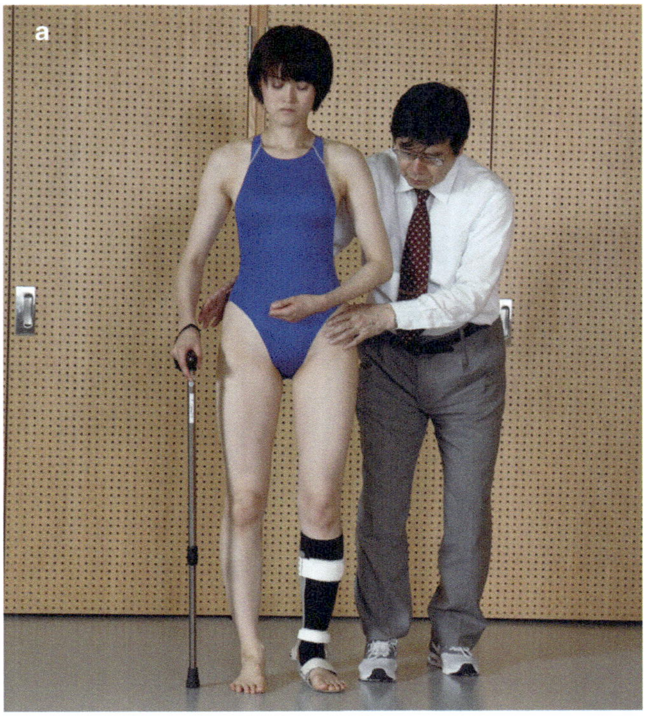

Fig. 6.8 Let's try: Facilitating two-point walking (Video 6.5). (**a**) Positions for Repetitive Facilitative Exercise for gait training. (**a**) The therapist stands beside the patient's unaffected side, and applies the shoulder gently to the patient's back. The therapist applies the ball of the thumb on the Iliac crest of the unaffected side and middle and ring fingers on the gluteus medius of affected side. The therapist then applies the tip of the middle finger onto the inguinal ligament on the affected side. The therapist should instruct the patient to "Stand on the unaffected lower limb firmly, and swing the hand holding the cane and affected lower limb together," then "Hold the cane gently, let the cane hang straight down, and do not move it forcefully."

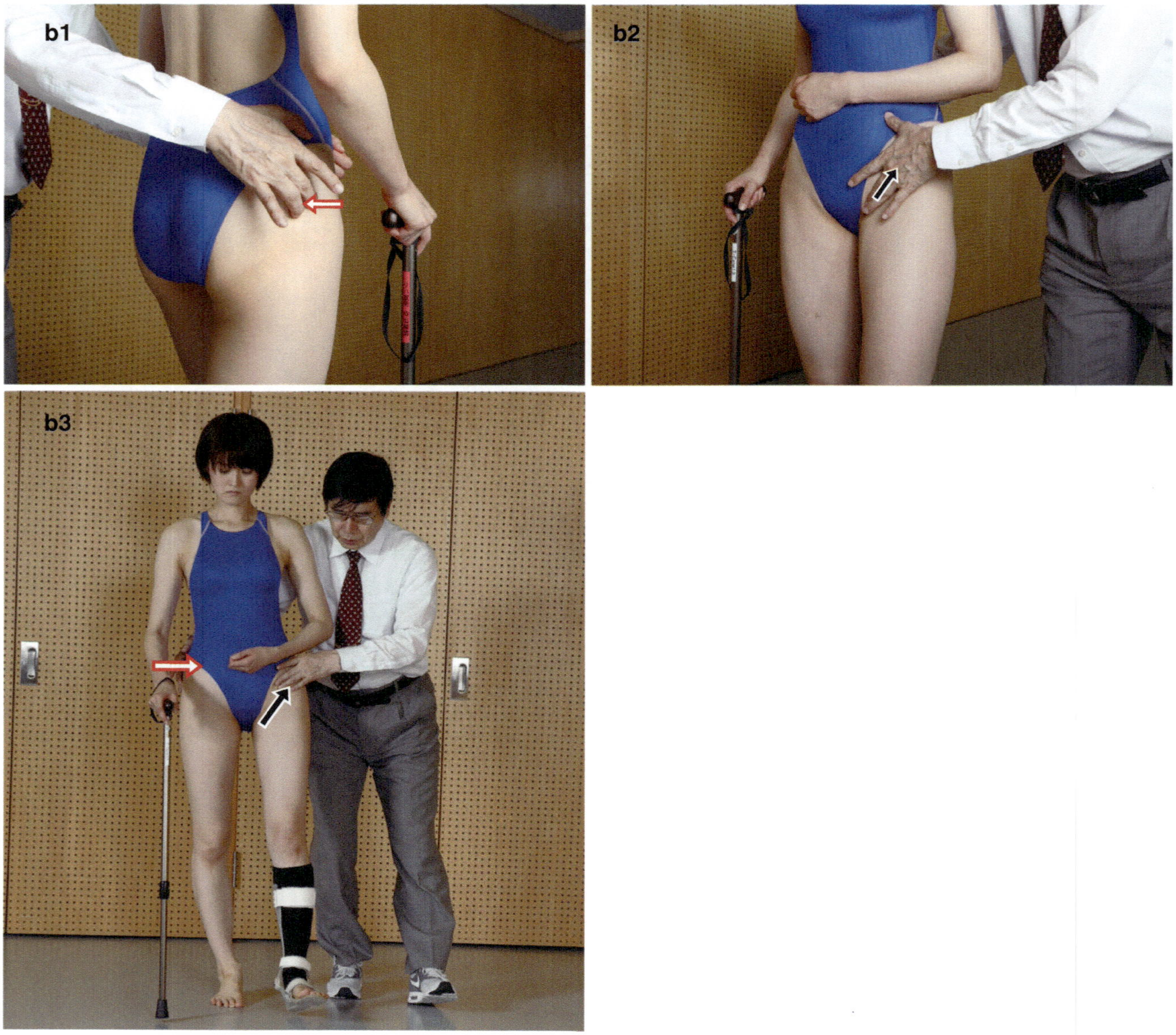

Fig. 6.8 (continued) (**b1**) Tapping the gluteus medius on the unaffected side. (**b2**) Stimulating the inguinal part of affected lower limb. (**b3**) Swinging the affected lower limb forward. (**b1-3**) The therapist instructs the patient to "Swing your affected lower limb forward," then taps the gluteus medius on patient's unaffected side with therapist's middle and ring fingers. At the same time, therapist uses the tip of the middle finger to rub the patient's inguinal ligament. The therapist should apply his shoulder to prevent the patient from leaning backwards. **Key point**: The therapist should continue counting "One, two…" in words while the patient walks to maintain a walking rhythm.

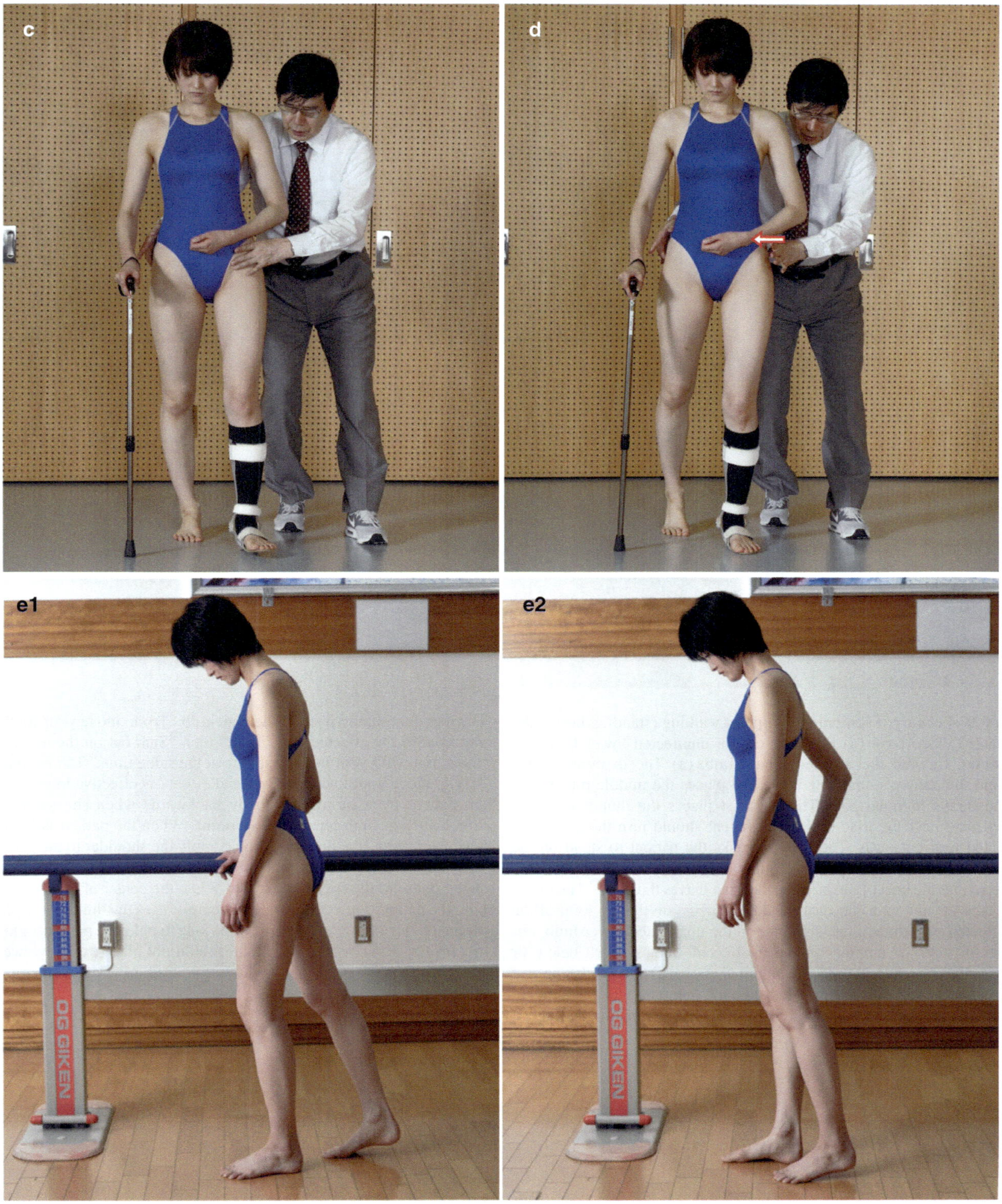

Fig. 6.8 (continued) (**c**) Affected lower limb control at heel contact. (**c**) The therapist instructs the patient to "Place your foot on the floor gently" to keep control over the affected lower limb's movements. When the patient's foot steps on the floor, the therapist leaves the middle finger off the inguinal ligament and places the tip of the thumb on the gluteus medius. (**d**) Poking the gluteus medius with the thumb when the patient stands on the affected lower limb. (**d**) To let the patient push off with the unaffected lower limb quickly and move the center of gravity to the affected lower limb and back to the unaffected lower limb smoothly, the therapist instructs the patient to "Stand on the affected lower limb gently and return your weight to the unaffected lower limb." Then the therapist pokes the patient's gluteus medius with the thumb when the patient stands on the affected lower limb. (**e1**) Push-off with the unaffected lower limb. (**e2**) Standing on the unaffected lower limb firmly. (**e1** and **e2**) The patient should push off with the unaffected lower limb firmly and quickly so that the unaffected foot lands ahead of the affected foot. The therapist should guide the patient to make a series of movements: standing on the unaffected lower limb firmly, the unaffected foot leaving the floor naturally, and proceeding to the next step out with the unaffected lower limb

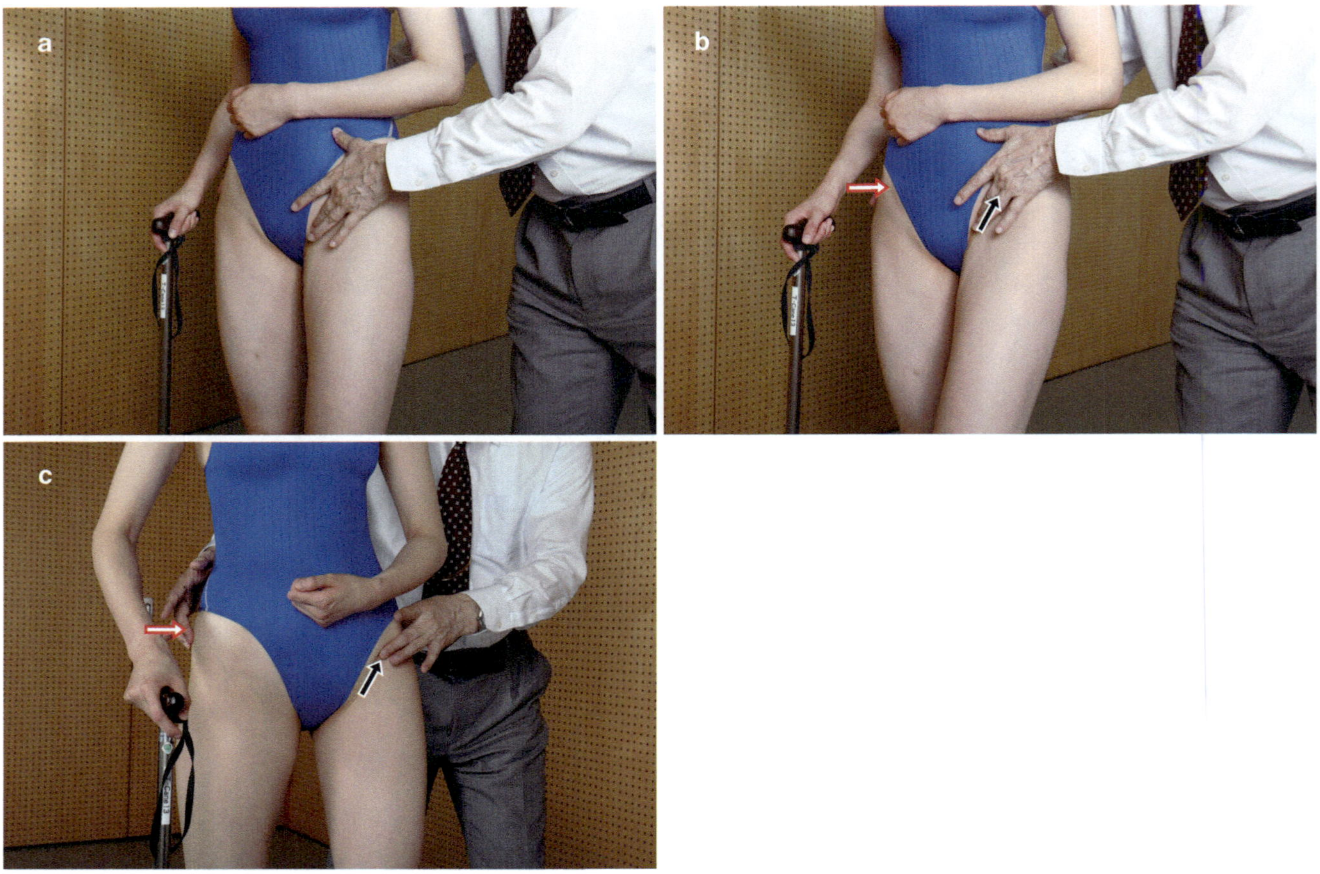

Fig. 6.9 Let's try: Facilitating two-point walking (standing beside the patient) (Video 6.6). (**a**) Standing on the unaffected lower limb and swinging forward the affected lower limb. (**a**) The therapist stands behind the patient's unaffected side and places the middle finger near the patient's inguinal part. The therapist places the thumb from the other hand near the Iliac crest. The patient should turn the toe of the affected foot outwards. The therapist asks the patient to stand on the unaffected lower limb and says, "Stand firmly on the unaffected lower limb so that the heel of the affected side can leaves the floor." The therapist taps the abductor muscles (Middle gluteus) on the hip joint of the unaffected side to facilitate standing on the unaffected lower limb. For better pictures of tapping on the unaffected side, see "**c**" below or Fig. 6.8b1. (**b**) Stimulating the inguinal part. (**b**) While the patient moves the weight to the unaffected side, the therapist rubs the inguinal ligament of the affected side with long strokes using the middle fingers. The therapist should instruct the patient to "Try to move your forehead and knee of the affected side closer" then "Step out on the unaffected lower limb with your hand swinging at the same time." The instruction "Try to move your knee and forehead closer" is effective in preventing the patient from turning the body backwards when she swings the affected lower limb forward. **Key point**: When the patient is likely to bend backwards, the therapist should apply the shoulder to the patient's back and hold. (**c**) How to manipulate the affected lower limb from the swing phase to the stance phase. (**c**) When the center of gravity moves onto the patient's affected foot, the therapist uses the thumb of the hand placed on the iliac crest on the affected side to poke the patient's gluteus medius for stimulation. The patient should not place excessive weight on the affected lower limb, which can lead to Trendelenburg gait pattern

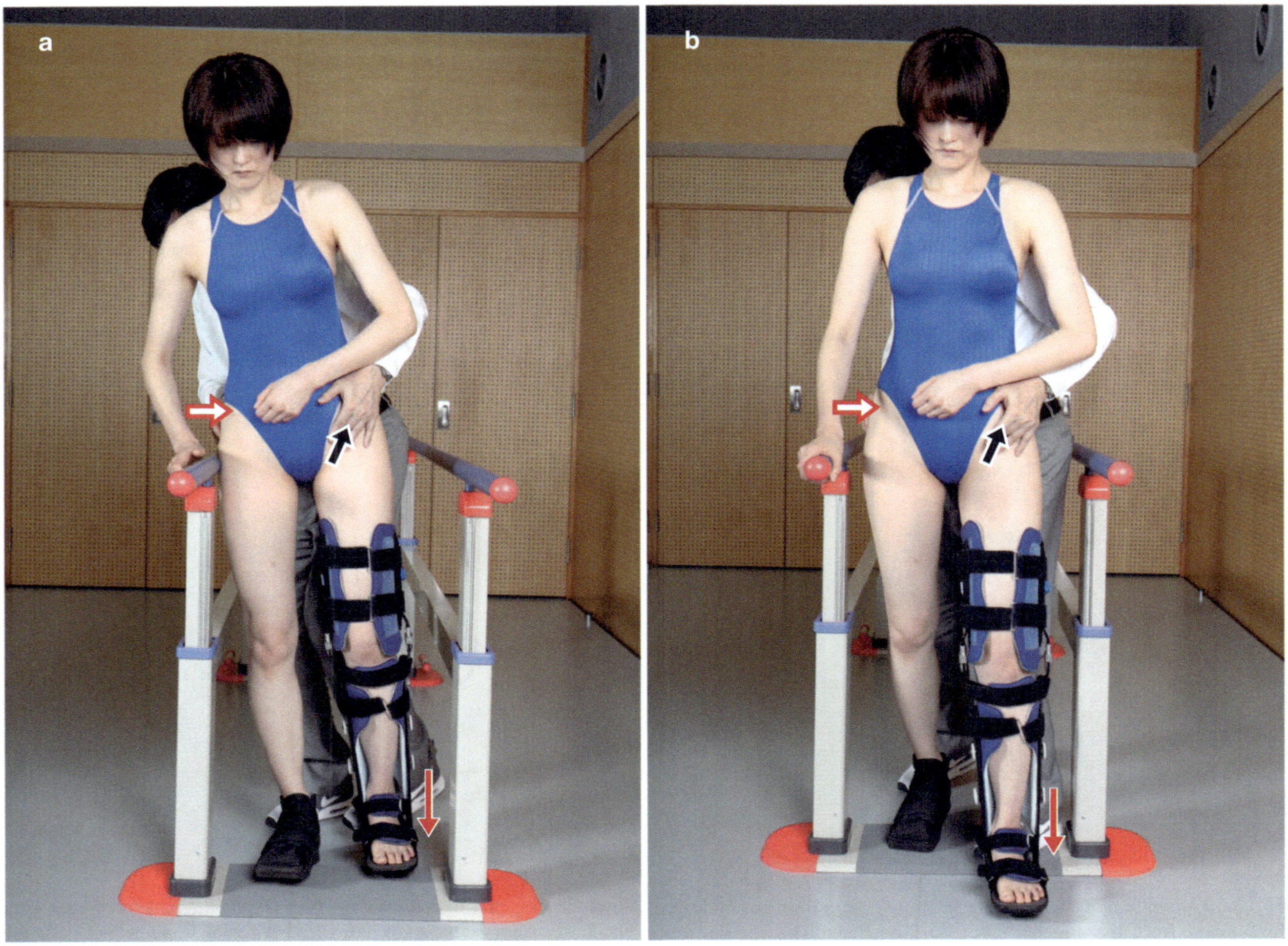

Fig. 6.10 Let's try: Exercise for the walking pattern with a long leg brace and parallel bars (front view) (Video 6.7). (**a**) Positioning for the therapist while the patient swings the affected lower limb forward. (**a**) The therapist stands behind the patient, placing one hand on the patient's inguinal ligament on the affected side, placing the other hand on the gluteus medius, turning the foot inwards, and touching the side of the little toe to the patient's heel on the affected side. The therapist should instruct the patient to "Stand on your unaffected lower limb, leave the affected foot off the floor, and swing your arm and affected lower limb together." The therapist then taps the gluteus medius on the unaffected side and rubs upwards the inguinal ligament on the affected side. Stepping out onto the affected lower limb and swinging the unaffected arm together facilitates two-point gait pattern and standing firmly on the unaffected lower limb. (**b**) Assistance while stepping out onto the affected lower limb and the arm on the unaffected side swinging in synchronization. (**b**) The therapist instructs the patient to "First, swing your arm and the affected lower limb at the same time." If the patient's step is insufficient, the therapist pushes the patient's foot with the outer side of the foot. The therapist should continue rubbing the inguinal ligament upwards so that the patient can flex the hip joint.

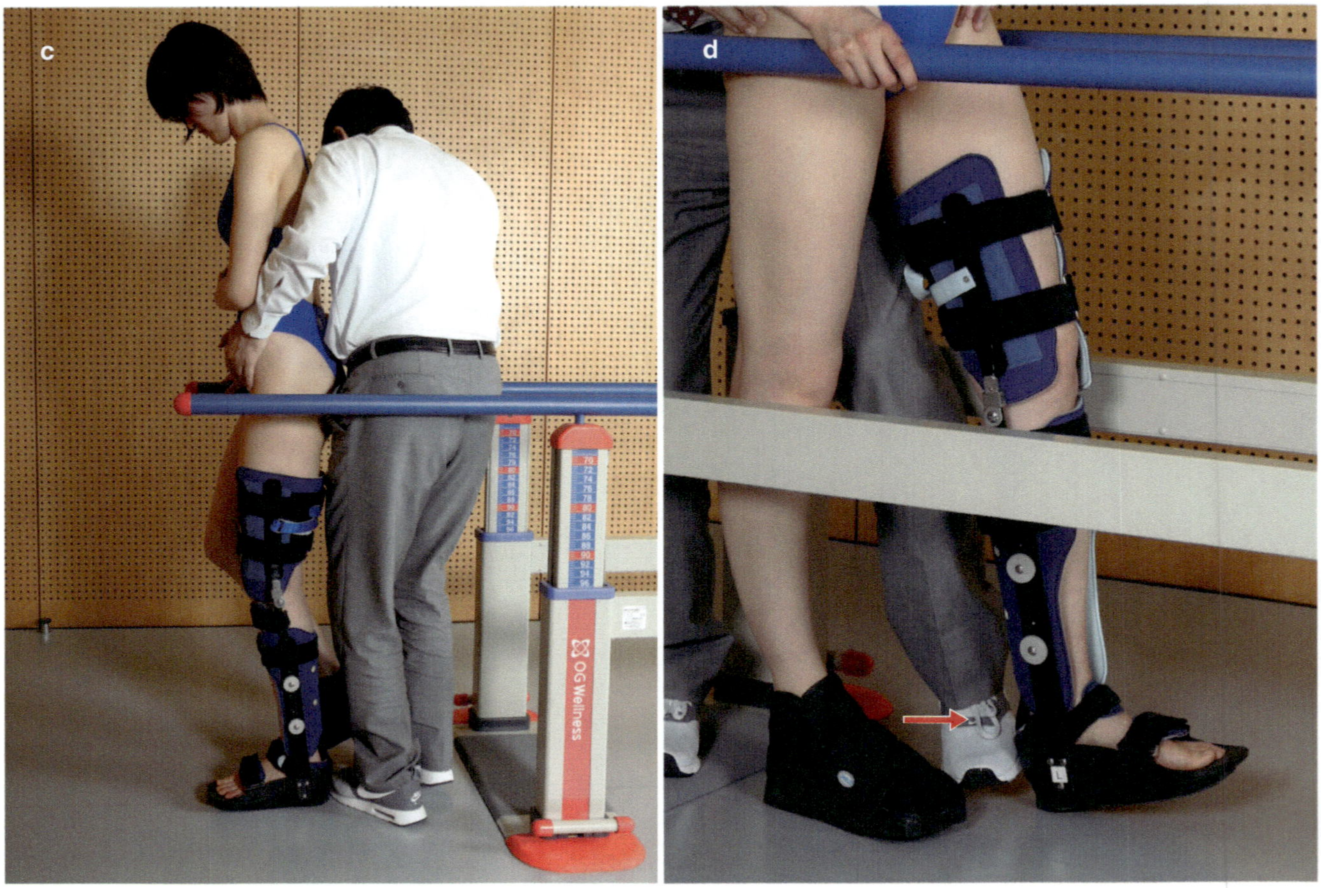

Fig. 6.10 (continued) (**c**) Assisting stepping out on the affected lower limb (view from the affected side). (**c**) The therapist turns his foot inwards to press the outer side (little toe) towards the patient's heel on the affected side. The therapist places his head on the patient's unaffected side for support; the patient's affected lower limb is not in sight. As the therapist uses his toe to support the movement, the affected lower limb is pressed outwards, which allows the patient to walk with the lower limb being abducted. (**d**) Assistance stepping out onto the affected lower limb (view from the unaffected side). (**d**) The therapist internally rotates the lower limb, then pushes the patient's affected heel with the outer foot (side of the little toe)

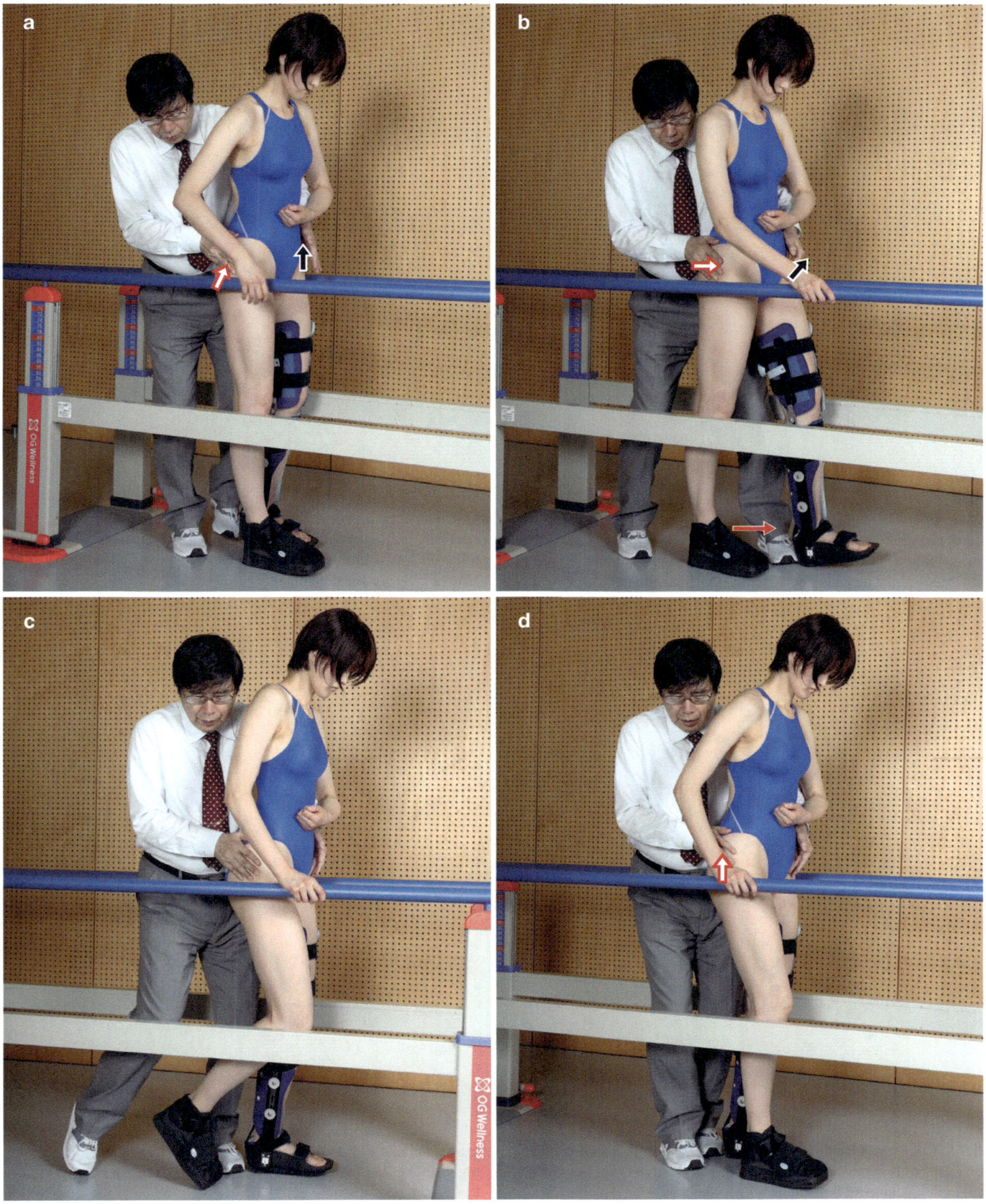

Fig. 6.11 Let's try: Exercise for the walking pattern with a long leg brace and parallel bars (three-quarter view from the unaffected side) (Video 6.7). (**a**) Assisting the patient to stand on the unaffected lower limb. (**a**) The therapist instructs the patient to "Stand on your unaffected lower limb," and taps the gluteus medius on the unaffected side. (**b**) Swinging forward with the unaffected arm and the affected lower limb together. (**b**) The therapist instructs the patient to "Swing your arm and your affected lower limb," and then rubs the patient's inguinal ligament. The therapist should assist the patient in swinging the affected lower limb forward with the outer side of the foot. (**c**) Assisting the patient to stand on the affected lower limb and pushing off from the floor with the unaffected lower limb. (**c**) The therapist instructs the patient to "Transfer your weight onto the affected lower limb lightly, pushing off from the floor with the unaffected lower limb, and return your weight onto the unaffected lower limb." (**d**) Assisting the patient to stand on the unaffected lower limb. (**d**) The therapist taps the gluteus medius on the unaffected side and instructs the patient to "Stand on the unaffected lower limb firmly"

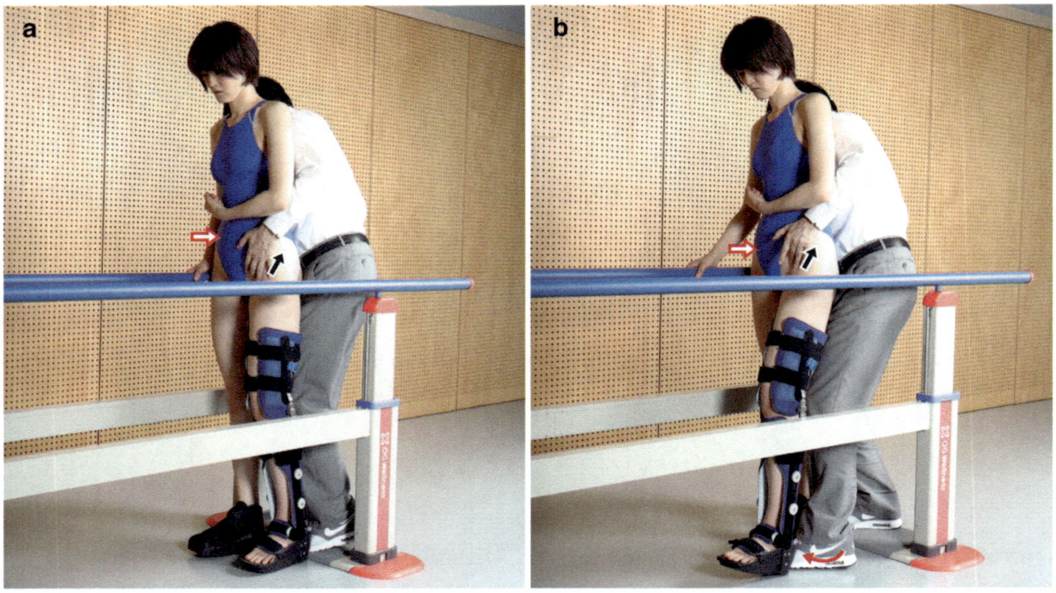

Fig. 6.12 Let's try: Exercise for the walking pattern with a long leg brace and parallel bars (three-quarter view from the affected side) (Video 6.7). (**a**) Standing on the unaffected lower limb. (**a**) The therapist instructs, "Stand on your unaffected lower limb," and taps the patient's gluteus medius on the unaffected side. (**b**) Swinging the arm and the affected lower limb. (**b**) The therapist instructs the patient to "Swing your unaffected arm and your affected lower limb together," and rubs over the patient's inguinal ligament. If the patient's step with the affected lower limb is insufficient, the therapist should use the lateral side of the foot for assistance

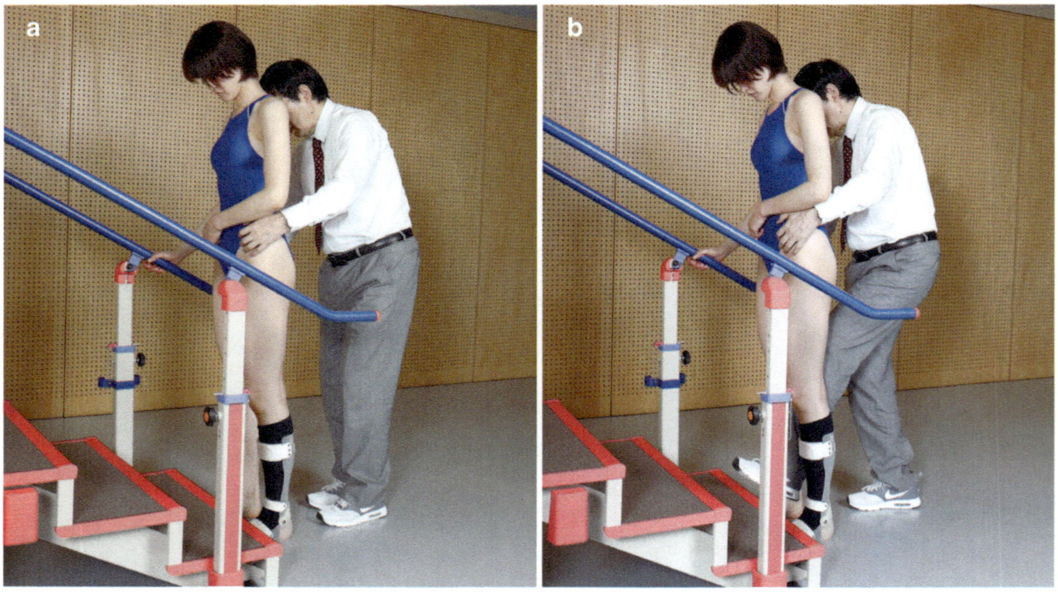

Fig. 6.13 Let's try: Assisting the unaffected lower limb stepping upstairs (Video 6.8). (**a**) Positioning for the patient and the therapist. (**a**) The therapist places one hand on the patient's lower abdomen on the affected side. The therapist then moves the foot through the patient's feet and puts it on the first stairstep, placing the thigh beneath the patient's buttocks. He grasps the bar with the arm on the patient's unaffected side. (**b**) Stepping out with the unaffected lower limb. (**b**) The therapist instructs the patient to step out onto the unaffected lower limb. The therapist should move the knee forward to keep the thigh placed under the patient's buttocks for support.

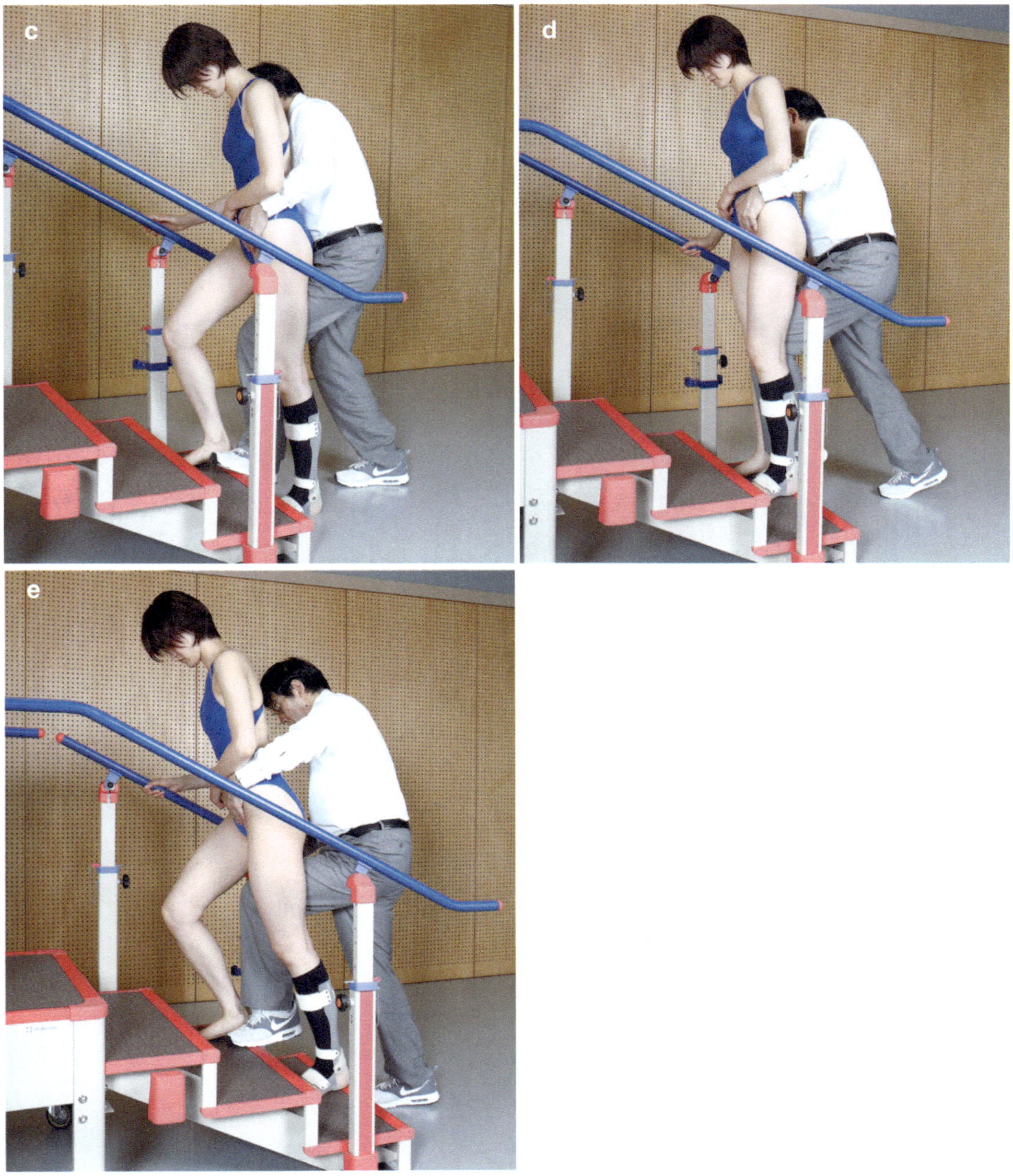

Fig. 6.13 (continued) (**c**) Movement for standing on the unaffected lower limb. (**c**) To prepare for the possibility of the patient's knee giving way, the therapist's knee movement is aligned with the patient's movement so that the thigh remains placed under the patient's buttocks. The therapist should also keep the hand on the patient's lower abdomen firmly. (**d**) Raising the affected lower limb through standing on the unaffected lower limb. (**d**) While the patient raises her affected lower limb by fully standing on the unaffected lower limb, the therapist should not remove his thigh from the patient's buttocks to prevent the patient's stance giving way. **Key point**: The therapist must be fully prepared before instructing the patient. The therapist should not step out at the same time as the patient. (**e**) Moving onto the next stair step. (**e**) For the next step, repeat the movements (**b**), (**c**), and then (**d**). The therapist should place one foot on the upper stair step as the patient stands, and keep the other foot on the step below. This position gives better stability

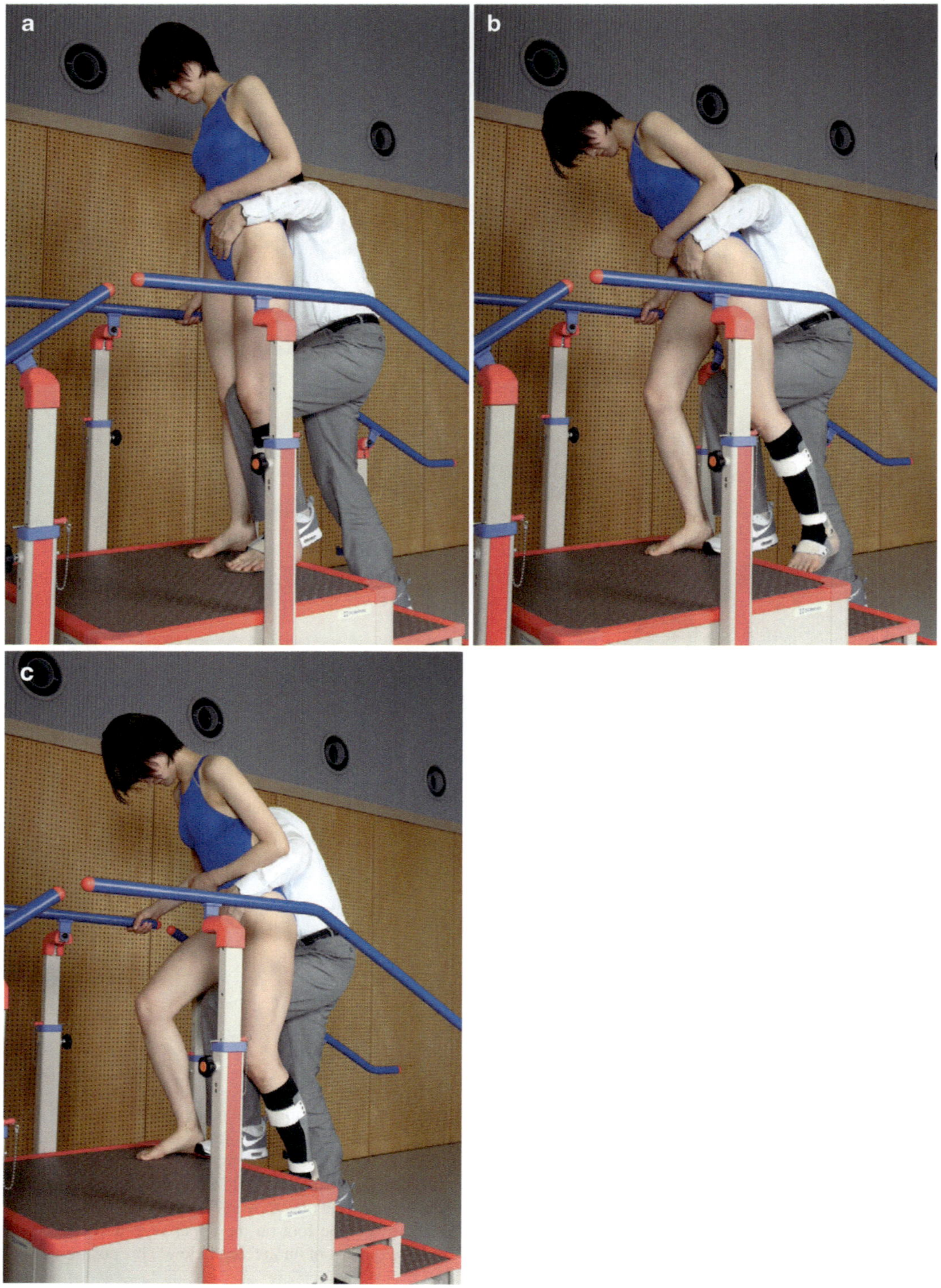

Fig. 6.14 Let's try: How to assist the affected lower limb in stepping downstairs moving backwards (Video 6.9). (**a**) Position while assisting. (**a**) The therapist places his hand on the patient's lower abdomen on the affected side. He places one foot in between the patient's feet, positions the thigh beneath the patient's buttocks, then holds the bar with the hand on the patient's unaffected side. The therapist should instruct the patient to "Keep your eyes on your affected lower limb." (**b**) Stepping down with the affected lower limb. (**b**) The therapist instructs the patient to "Raise your affected lower limb off the floor, then go downstairs backwards with your affected lower limb." The therapist should keep the thigh resting under the patient's buttocks. The patient should raise the affected lower limb off the floor and step backwards one step down. (**c**) The patient brings the affected lower limb down, adjusting the movement with the unaffected lower limb. (**c**) The therapist keeps one hand firmly on the patient's lower abdomen, and follows the movements. The therapist should instruct the patient to put her foot gently on the lower stair.

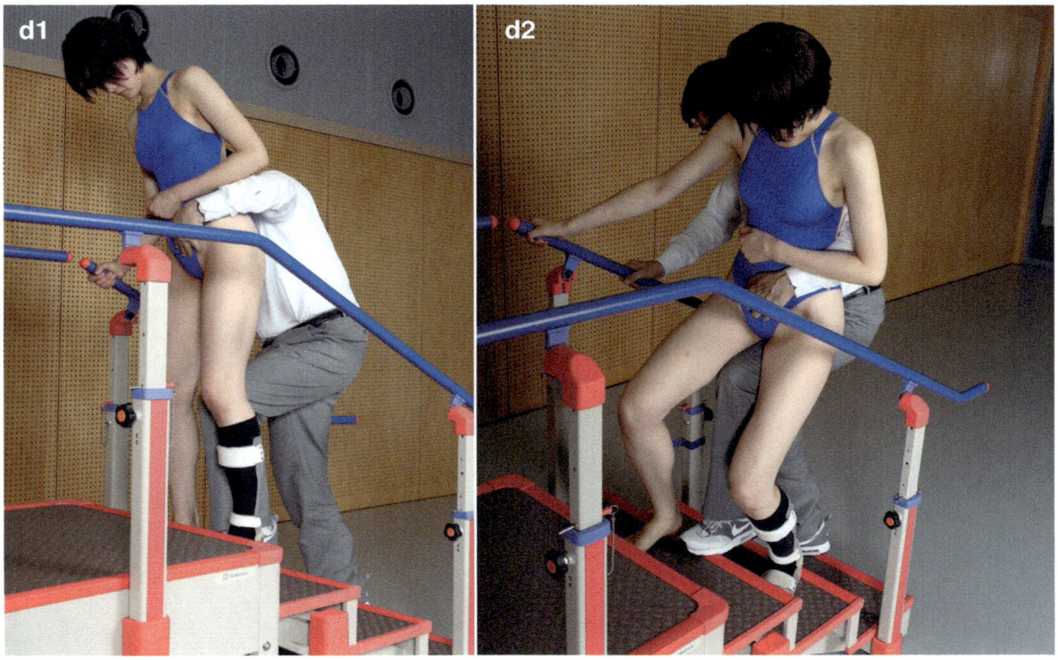

Fig. 6.14 (continued) (**d1**) Preventing the knee giving way during the stance phase. (**d1**) The patient's knee may become unstable while the affected lower limb moves downstairs, either due to insufficient muscle strength in the unaffected lower limb or lack of supporting power in the affected lower limb. The therapist keeps the thigh resting beneath the patient's buttocks while placing the affected lower limb on the lower stairstep and unaffected lower limb one step down. If the therapist holds the bar with one hand and the patient's lower abdomen with the other hand, he can prevent the patient's knee giving way and can keep her balance. (**d2**) What to do if the knee gives way. (**d2**) If the patient's knee gives way, the buttocks will land on the therapist's thigh. If the therapist follows the procedures stated above, the patient will not fall as the therapist can use both own arms and lower limb for support. The important points are: (1) The therapist must take the appropriate position and be ready to assist the patient before the movement, (2) The therapist's movements should not be aligned with the patient's movements, and (3) The therapist should rest his thigh beneath the patient's buttocks

Reference

1. Kawahira K, Tanaka N. Zuii undo no mechanism [Mechanism of voluntary movements]. Sogo Rehab. 1994;22:491–496, IGAKU-SHOIN Ltd.

Basic Natural Movements (Rolling Over, Sitting Up, Standing Up, and Sitting Down from a Standing Position)

7

Kazumi Kawahira

The movements that the patient repeats daily should be mechanically natural and not cause associated reactions. Help the patient sit up by rolling him or her over to the unaffected side, flexing and rotating the neck and trunk, and bringing the center of gravity for the body's upper half to the fulcrum of the elbow or hand. If the neck or trunk performs lateroflexion, it becomes difficult to roll over. To avoid this, you should first hold the head and lead it in the intended direction to show the course of movement. The patient should flex and rotate the head, shoulder girdle, and then lower back in a series of movements. You should not let the patient roll over to the affected side to sit up as this will hurt the shoulder and restrict the range of activities while conferring no benefits.

Key Points

1. Flexing and rotating the trunk to one side (lying on the unaffected side) (See Fig. 7.1)

   Flexing and rotating the trunk is needed in order to roll and sit up, but the hemiplegic patient's lack of strength and coordination can lead to forced movement. We need to be careful not to force the patient to roll over and sit up. Exercise is required to strengthen the trunk muscles.
2. Rolling over from lying in a supine position (Fig. 7.2)

   When rolling over, the patient flexes and rotates the neck and trunk with the unaffected upper limb controlling the movement of the affected upper limb. This prevents shoulder pain and allows the patient to sit up naturally. If the patient kicks the floor with the unaffected lower limb to lean back on the trunk, it may lead to shoulder damage and spasms in the extensor muscles.
3. Assisting in sitting up from a supine position (Fig. 7.3)

   Continuing to flex and rotate the trunk is needed to generate movement from lying supine to sitting up. The therapist should lead and assist so that the patient can rest the upper body on the elbow. You should not raise the body but lead the rotation of the trunk as you support this movement.
4. Assisting in standing up from a seated position (Figs. 7.4, 7.5, 7.6, and 7.7)

   Stability and balance while standing up relies on the strength of the trunk and unaffected lower limb. Daily movements for sitting and standing are associated with the quadriceps, gluteus maximus, back muscle, and other major muscle groups. Training aimed at standing up smoothly helps to strengthen the lower limbs. The patient should stand up while keeping his or her eyes on the affected lower limb. This will enable the patient to move the center of gravity forward and control spasms in the extensor muscles.

   Since both weighing down and putting the center of gravity in the middle of both lower limbs is ineffectual, the standing position should be balanced on the unaffected lower limb.

Supplementary Information The online version contains supplementary material available at [https://doi.org/10.1007/978-981-19-0789-0_7].

K. Kawahira (✉)
Laboratory of Repetitive Facilitative Exercise, Kirameki Terrace Health Care Hospital, Kagoshima city, Kagoshima, Japan

K. Kawahira et al. (eds.), *Exercise Therapy for Recovery from Hemiplegia*, https://doi.org/10.1007/978-981-19-0789-0_7

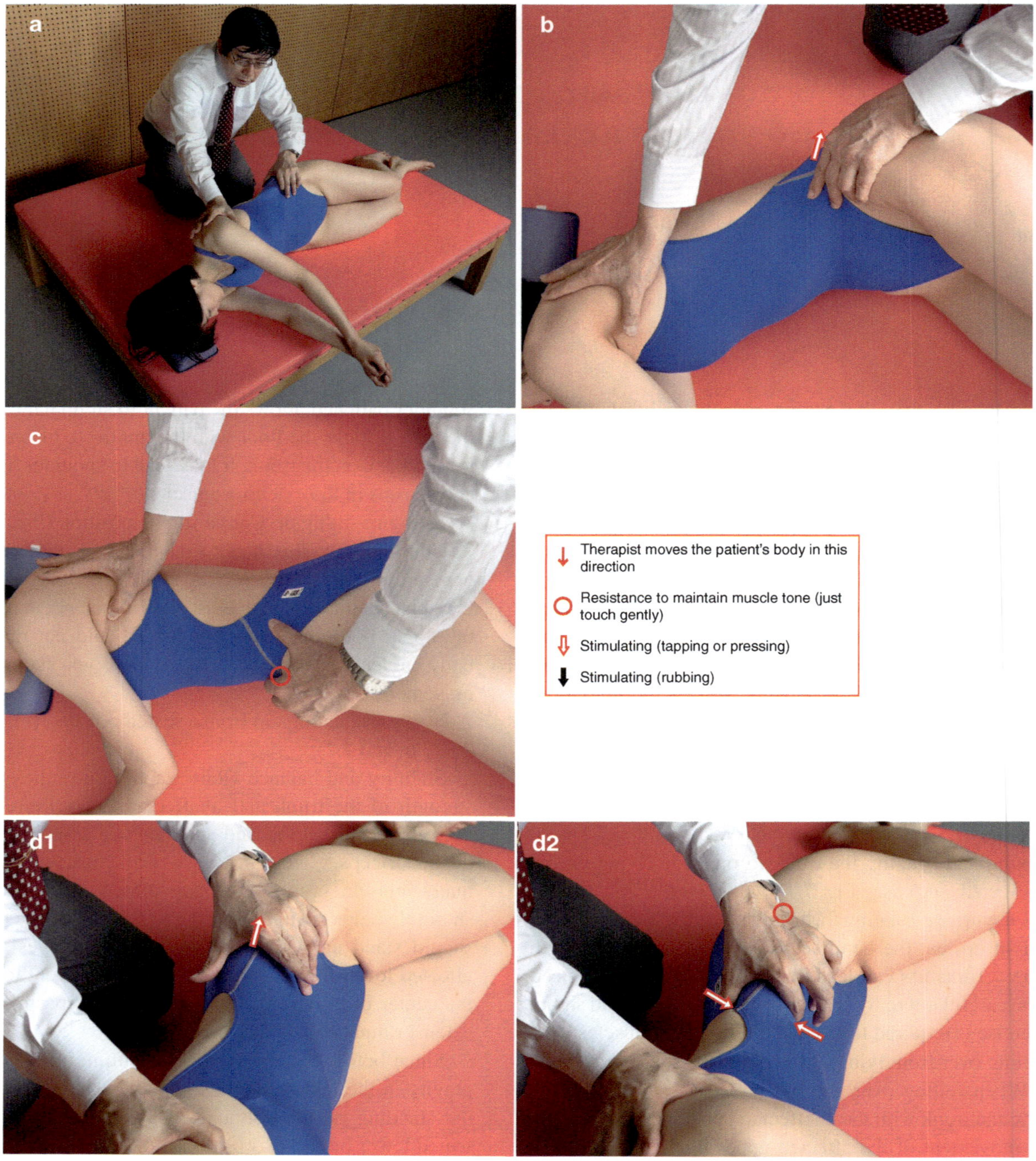

Fig. 7.1 Let's Try: Rotation and bending the trunk (lying on the unaffected side, keeping the affected side facing up) (Videos 7.1 and 7.2). (**a**) Supporting the shoulder girdle muscles and lower back. (**a**) Raising the affected pelvis side (tension reflex of hip) and rotating the trunk are important movements in preparing to walk. The therapist puts one hand on the affected shoulder girdle muscles and chest rib, and the other on the crest of ilium. When placing one hand on the shoulder girdle muscles, the thenar eminence and hypothenar eminence should be placed on the ribs to fix the chest in position. (**b**) Flexing and rotating the trunk. (**b**) The therapist says, "Turn your hip on the affected side toward the front," and quickly pulls the iliac crest of the affected side downward and back. Then the therapist taps the affected side with his index finger, as if prodding the external oblique muscle, and says, "Now move your hip in a wide rotation of." Exerting excessive force may cause lower back pain in the patient, so the technique must be performed lightly and quickly. (**c**) Continued flexion and rotation of the trunk. (**c**) To enlarge the patient's hip rotation, the therapist applies gentle resistance against the patient's movement. The resistance should not be so great as to prevent the muscle from slackening. (**d1**) Controlling the lateroflexion of the trunk. (**d1**) The therapist says, "Raise your hip toward your shoulder," pulls the crest of ilium downward quickly to move it away from the libs and says, "Now raise your hip." Exerting too much force may cause the patient to experience lower back pain, so the technique must be performed lightly and quickly. (**d2**) Continuing lateroflexion of the trunk. (**d2**) The therapist taps the trunk as if pinching with the thumb and index finger. The therapist exerts light resistance with the palm of the hand to increase lateroflexion

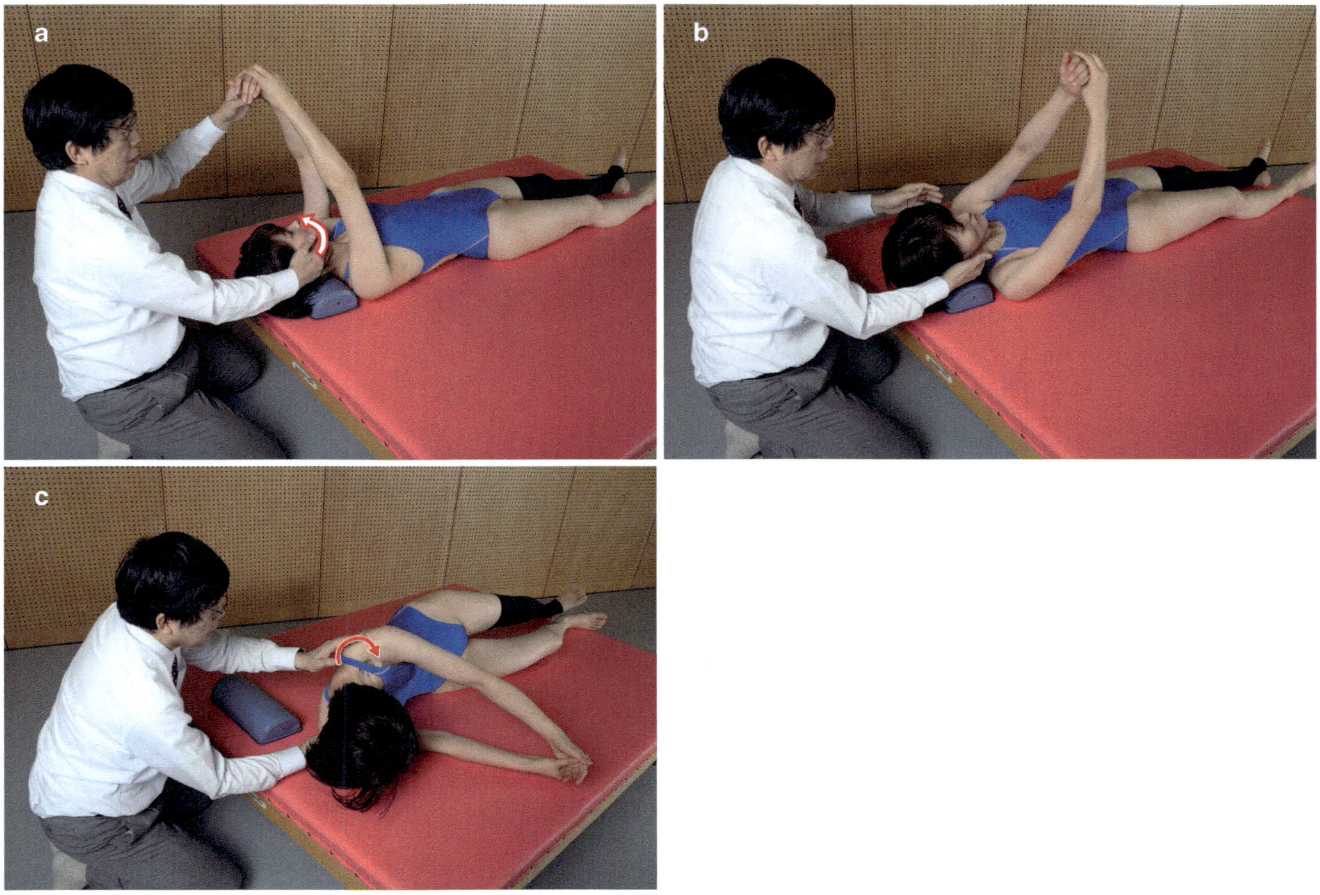

Fig. 7.2 Let's Try: Turning over (supine position) (Video 7.3). (**a**) Keeping the affected upper limb extended. (**a**) Lying in a supine position, the patient straightens his/her neck, trunk, and lower limbs, then joins the hands together and extends the arms above the face. The patient uses the unaffected hand to hold the palm of the affected side to turn the palm towards to the face (the shoulder external rotation and forearm supination). By contrast, joining every finger is troublesome and if the patient holds the wrist instead, he or she may pull and damage the shoulder. **Key point**: If the patient cannot roll over as expected, lateroflexion of the neck and trunk is likely to occur. The therapist should tell the patient to raise his or her joined hands upward to the affected side and say, "Look diagonally upward" to bring about a move in the opposite direction. Tell the patient to extend the neck and turn the face upward to the affected side before moving on to step (**b**). The patient should be instructed to look at his or her hands throughout. (**b**) Supporting head movement to induce flexion and rotation of the trunk. (**b**) The therapist sits at the patient's head and places five fingers behind the earlobe for support. Then the therapist has the patient swing both arms downward, saying, "Curl your body and move your forehead to the unaffected side of your hip (anterior superior iliac spine)." The patient should flex and rotate the neck and trunk downward toward the unaffected side. (**c**) Flexing and rotating the patient's trunk to a lateral position. (**c**) The therapist supports the patient's head with both hands and guides the neck and trunk in a flexing and rotating direction. If this is insufficient, the therapist should press on the patient's shoulder to guide it to a lateral position

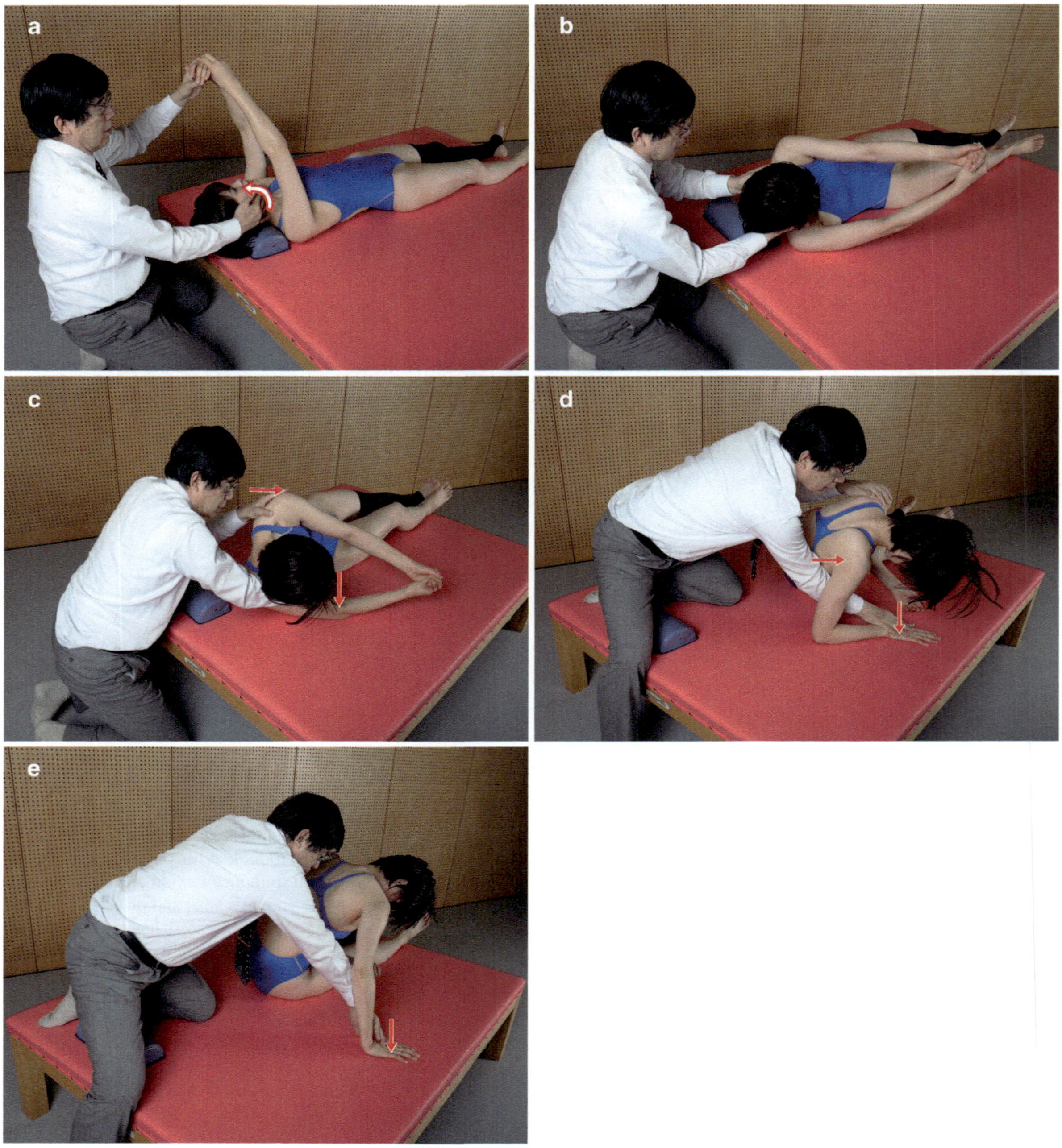

Fig. 7.3 Let's Try: Helping the patient rise from a supine position (Video 7.4). (**a**) Keep the affected upper limb extended. (**a**) The patient lies on his or her back and straightens the neck, trunk, and lower limbs, then joins the hands together and extends the upper limbs above the face. The therapist sits beside the patient's head and supports it with both hands. **Key point**: If the patient cannot turn over as expected, the therapist should say, "Look diagonally upward," allowing the patient to extend and rotate the neck upward to the affected side before moving on to step (**b**). (**b**) Flexing and rotating the head and trunk. (**b**) The therapist should say "Now bend your body to bring the forehead closer to the unaffected hip (anterior superior iliac spine)," instructing the patient to flex and rotate the neck and trunk downward to the unaffected side. (**c**) Flexing and rotating the trunk to place the upper body on the elbow. (**c**) With the shoulder flex at 60 degrees and the patient turning to the side, the therapist places a hand under the patient's head and supports the elbow. The other hand is used to push the shoulder so that the patient can place his or her upper body on one elbow. The therapist should say "Look at your elbow as you move your upper body on it." **Key point**: If the patient faces the direction of rotation (toward the elbow), the center of gravity moves toward the elbow. This makes it easier for the patient to get up. The therapist should not say, "Push hard on your elbow to lift your body." This instruction will make the patient face upward, grab the edge of the mat, and force the patient to sit up. (**d**) Flexing and rotating the trunk to move the upper body on the hand. (**d**) The therapist releases a hand from the patient's elbow, extends an arm under the patient's armpit, and holds the patient's hand. The therapist says, "Look at your hand," and pushes the patient's shoulder with the other hand to assist with the rotation of the trunk while restraining any upward movement of the body. (**e**) Extending the arms to sit up with the lower limbs extended. (**e**) With the therapist's forearm extended under the patient's arm, the therapist pushes the trunk and assists with its rotation, enabling the patient to sit up with lower limbs extended

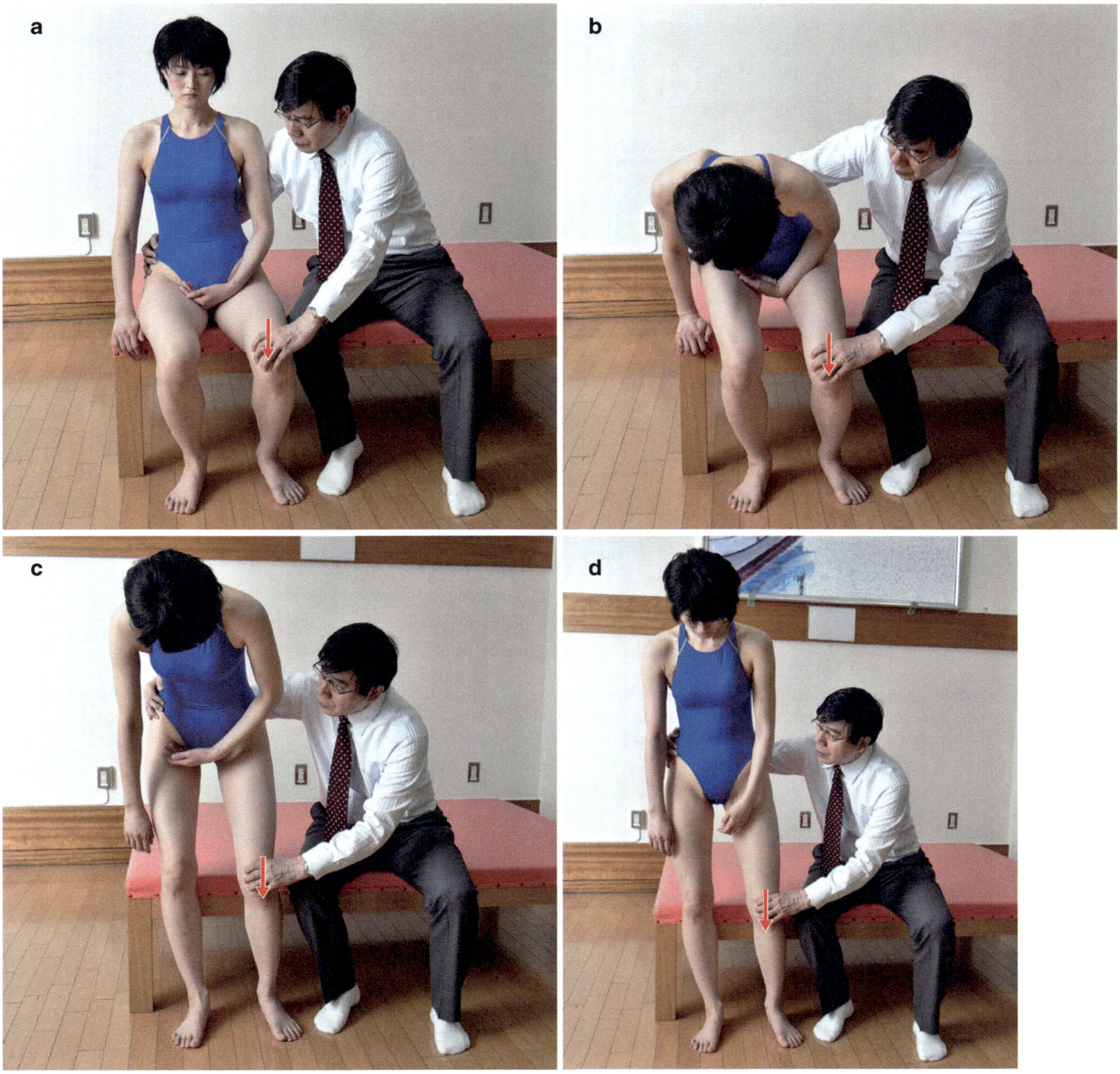

Fig. 7.4 Let's Try: Helping the patient stand from a sitting position. (**a**) Supporting the knee and hip. The therapist sits on the affected side and grasps the patient's knee with one hand while supporting the hip with the other (if the patient is wearing pants, the therapist can pass his fingers through the belt). (**a**) The patient pulls both lower limbs inward, raises her face upwards, and straightens her back. The patient places the unaffected hand on the front edge of the bed. (**b**) "Bowing" to move the center of gravity forward. (**b**) The patient flexes her neck and trunk smoothly as if bowing, continuing the movement to raise the buttocks with inertial force. If the patient stops moving halfway, it becomes harder to stand up. (**c**) Raising the buttocks by bending forward from a seated position. (**c**) When the patient's head moves ahead of his or her knees, the patient should push with the unaffected hand on the bed to raise the buttocks, then release the hand from the bed. The patient makes a continuous series of movements from moving the upper body forward to raising the buttocks. When the affected side extends, plantar flexion of the ankle joint may cause back knee, which would lift the heel off the floor. To prevent this, the therapist should press the knee downward with his or her hand. (**d**) Standing up while looking at the affected lower limb. (**d**) If the patient can keep balance with the unaffected lower limb, he or she should extend the trunk and lower limbs. If the back knee cannot be controlled, the patient should bend the hip joint slightly to suppress spasticity in the extensor muscles of the lower limb.

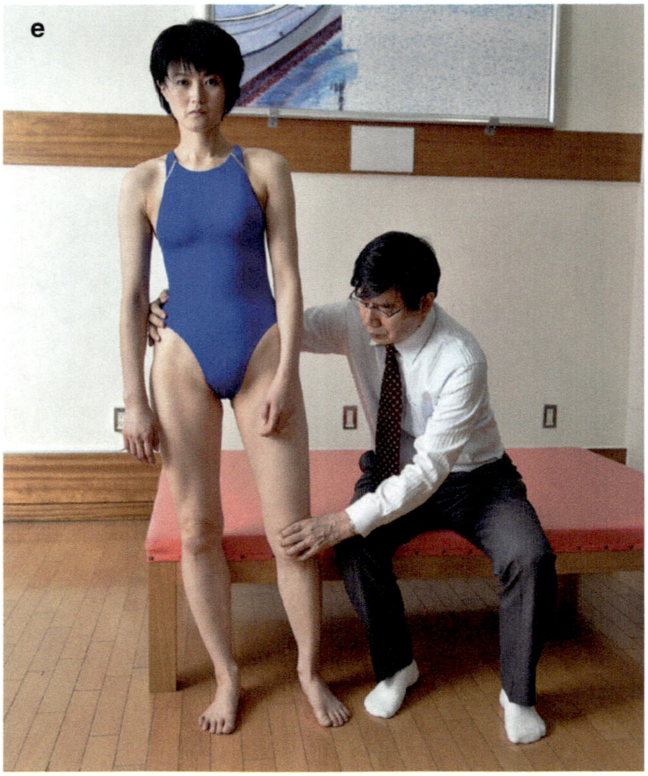

Fig. 7.4 (continued) (**e**) Standing with the knee joint slightly bent. (**e**) The patient keeps the knee and ankle joints of the affected side slightly bent. When this position becomes stable, the patient should take his or her eyes from the affected lower limb and raise the face while looking straight ahead

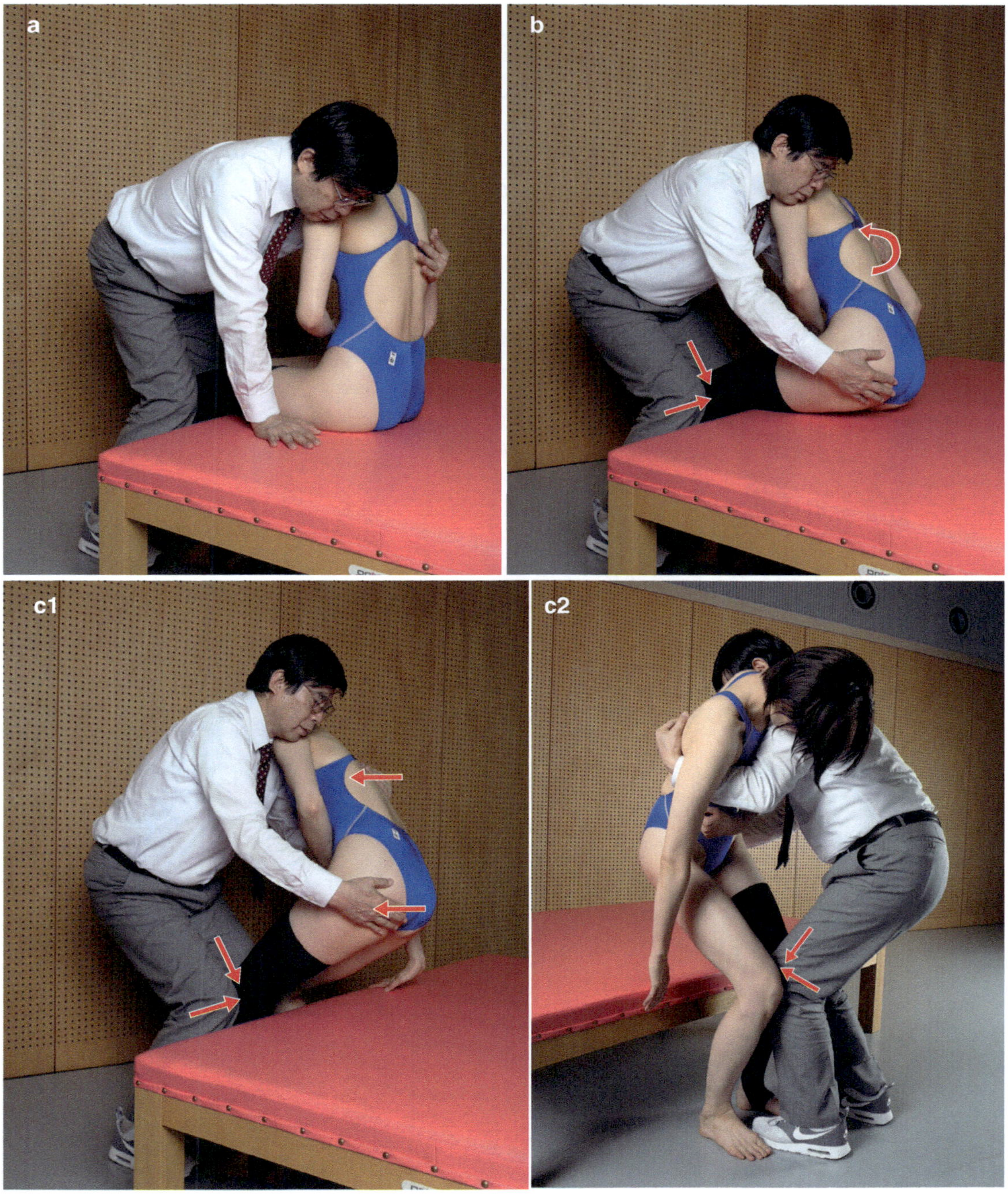

Fig. 7.5 Let's Try: Helping the patient stand up and sit down. (Severe case: unbalanced sitting position with weak knees (Video 7.5). (**a**) Supporting the trunk and affected lower limb. (**a**) The therapist holds the patient's affected knee with both knees and extends an arm under the patient's unaffected arm to touch the shoulder girdle muscle. The therapist puts one hand on the other side on the bed and moves the patient's trunk slightly backward from the neutral position to make it easy to allow the body to move forward by its own inertia. (**b**) Flexing completely forward and using inertial force to raise the affected hip side. (**b**) The therapist says, "Bend your body forward and raise your hip," then pulls the patient's upper body completely forward with one hand placed on the shoulder girdle muscle of the unaffected side. To allow free movement while flexing, the therapist's shoulder should not interfere with the patient's body anteversion. The patient should try to keep the movement flowing so that he or she can use inertial force to raise the hip. (**c1**) Supporting the standing movement. (**c1**) The therapist says, "Keep the movement flowing and stand up moving smoothly onto the unaffected lower limb." The therapist fixes the patient's affected lower limb with both lower limbs and moves the patient's upper body toward him or her with one hand supporting the buttock. This support helps the patient to achieve a smooth and stable movement in standing up. (**c2**) Movement while standing up: View from the unaffected side. (**c2**) The therapist pulls the patient's trunk from the unaffected side so that the patient leans toward that side, making it easier for the patient to stand up using the unaffected lower limb.

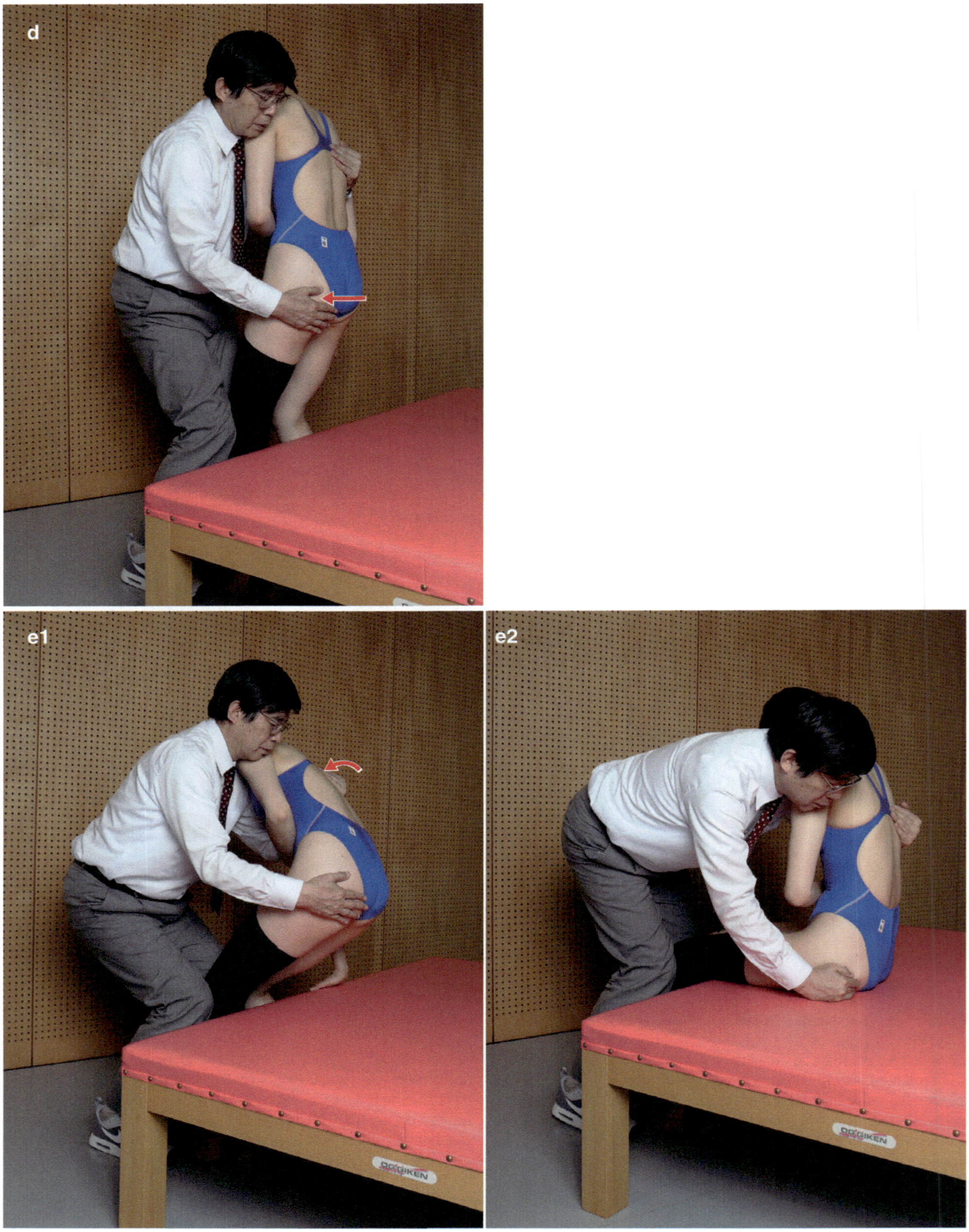

Fig. 7.5 (continued) (**d**) Completion of standing up. (**d**) If the therapist secures the patient's knees and pulls the buttocks toward him, even a severely affected patient can stand up. If the patient cannot fully extend the affected lower limb, the therapist should tap the buttock. (**e1**) Sitting down from a standing position. (**e1**) The therapist pulls the patient's unaffected shoulder girdle with one hand, facilitating forward movement of the upper body and flexion of the hip joint. This leads to a smooth sitting-down movement with the unaffected lower limb and trunk. (**e2**) Completing the sitting-down movement. (**e2**) The therapist should assist the patient in a smooth forward-and-back movement of the trunk's gravity center so that the patient does not drop heavily onto the bed

Fig. 7.6 Let's Try: Assisting in standing up from the floor (Video 7.6). (**a**) Flex the unaffected knee while sitting on the floor with lower limbs extended. (**a**) The therapist should be positioned by the patient's affected side and say, "Bend your knee with your foot under your buttock," and "Place your unaffected hand beside the unaffected buttock." It is important to place the unaffected foot right beneath the buttock and place the hand close to the buttock. (**b**) Turning the body and standing on one knee. (**b**) The therapist supports the patient's affected knee with one hand and holds the hip firmly with the other, saying, "Push on the floor with your hand to raise your buttocks." When the patient raises his or her buttocks, the therapist says "Turn your head toward the affected side" so that the patient can move the unaffected lower limb. Then the patient should flex the knee while avoiding any twisting movement. (**c**) Stabilizing the kneeling position (one knee). (**c**) The therapist says, "Place your hands to the front, a little ahead of the foot on the affected side," and "Stand using the toes on the unaffected side." The patient moves his or her head toward the affected lower limb and aligns the unaffected lower limb with the direction of the body. **Key point**: It is important that the therapist firmly fixes the affected knee. (**d**) Raising the buttocks and standing up. (**d**) The therapist says, "Lower your head, hold the floor with your hand, raise your buttocks, move your body forward, and take one step with the unaffected lower limb." The patient should raise the buttocks with the unaffected upper limb and unaffected lower limb while moving the body forward. **Key point**: It is important for the therapist to hold the affected knee firmly and keep the trunk stable with one hand holding the hip. The patient's unaffected hand should be placed on the floor using the whole palm. The patient should refrain from raising his or her head to facilitate raising the buttocks.

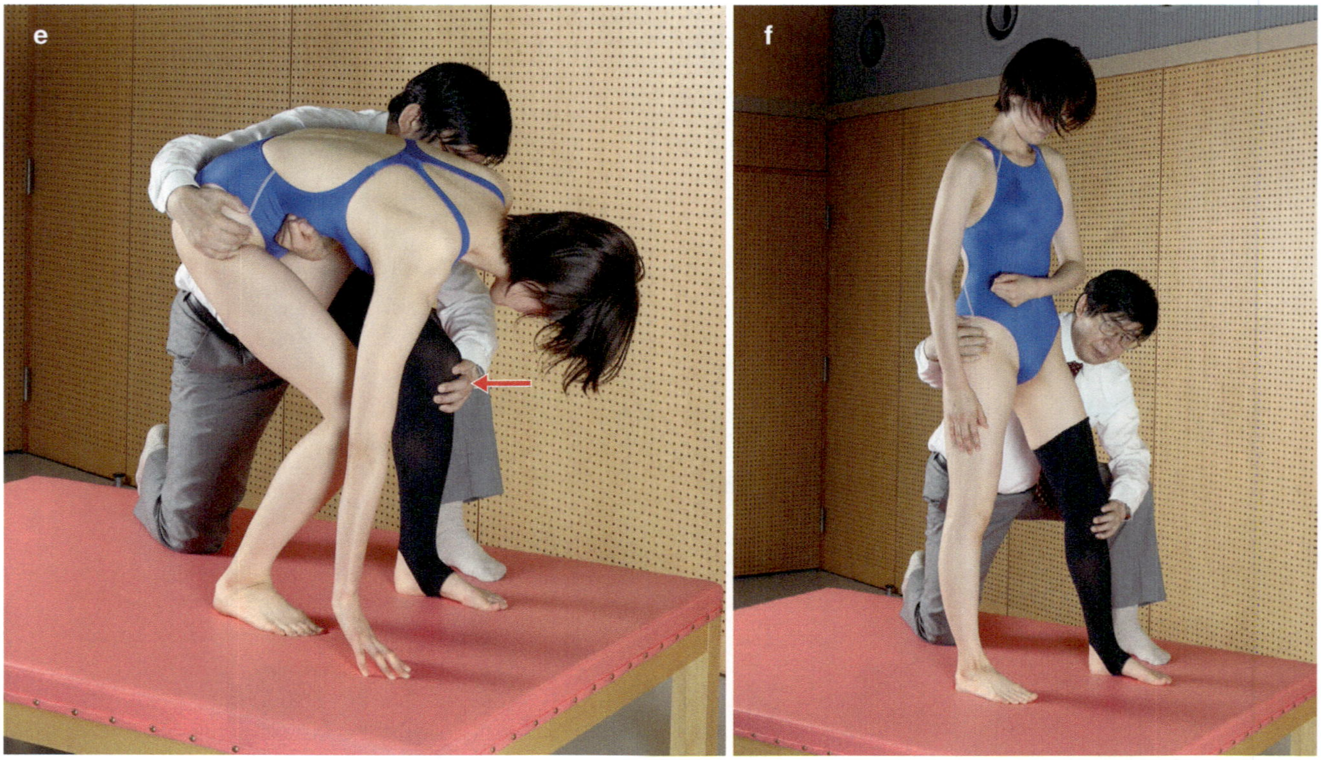

Fig. 7.6 (continued) (**e**) Making the patient look up and forward, away from the affected lower limb. (**e**) The therapist says, "Stand up while looking at your affected lower limb," and puts pressure on the affected knee with one hand. This prevents the patient's heel leaving the floor due to back knee (hyperextension) and pes equinus caused by extensor muscle spasticity. Then the therapist says, "When your posture has stabilized, raise your head." (**f**) Completing the standing movement. (**f**) The therapist supports the patient's hip and knee. When releasing the patient's hand from the floor, the therapist tells the patient to stand on the unaffected lower limb, making it easier to suppress spasticity in the affected lower limb

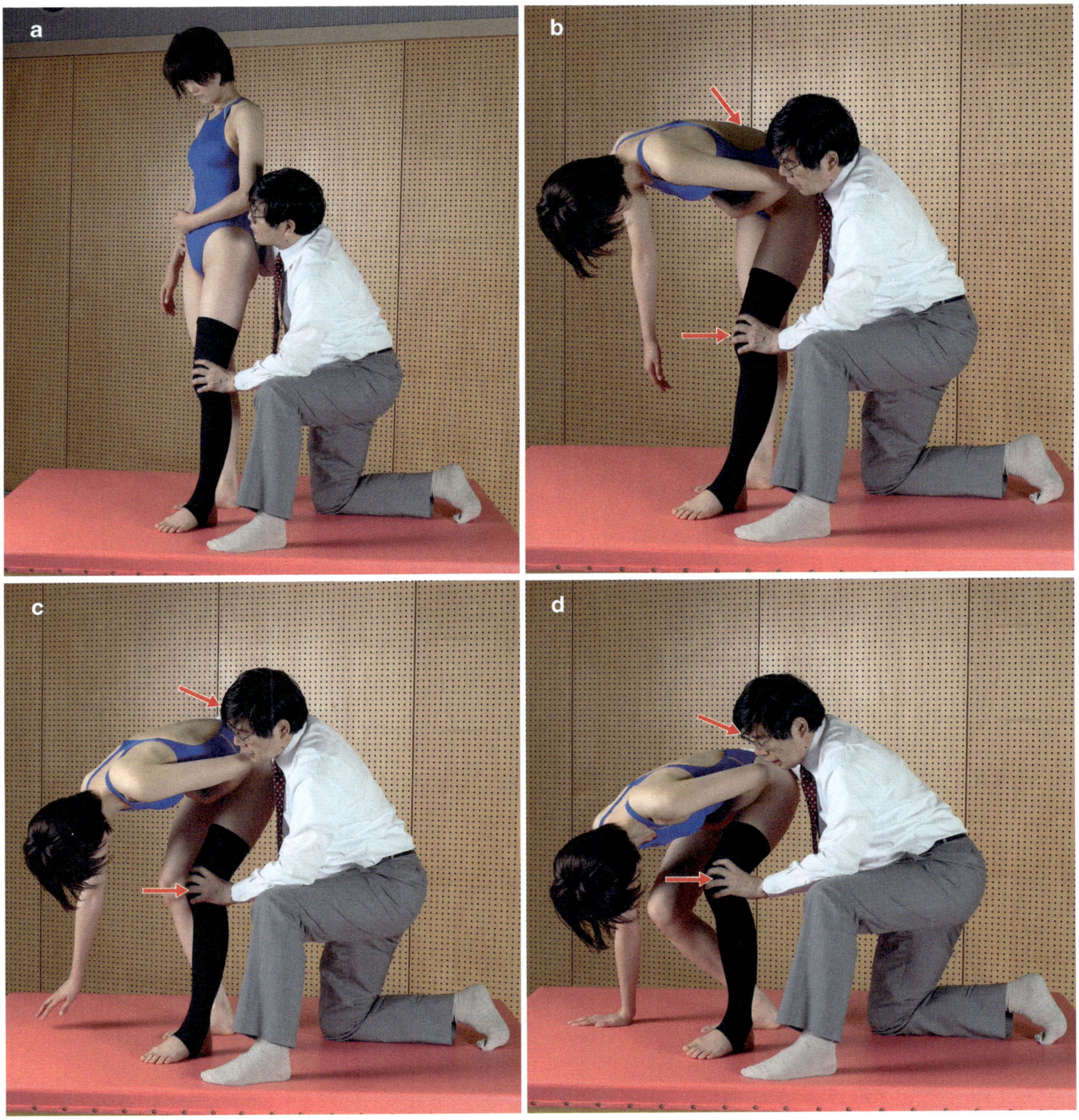

Fig. 7.7 Let's Try: Assisting in sitting down on the floor (Video 7.6). (**a**) Moving the affected lower limb forward from a standing posture. (**a**) The therapist should kneel beside the patient's affected side, placing one hand on the affected knee and the other on the hip. The therapist says, "Keep your knees straight, lower your head as far as you can, and touch the floor with the unaffected hand." (**b**) Lowering the head to place a hand on the floor. (**b**) The therapist should hold the affected knee firmly with one hand and hold the patient's hip with the other. This makes it easier for the patient to flex forward. If the patient is afraid to flex and tries to touch the floor by raising the head and bending the unaffected knee, the unaffected lower limb cannot support the body. This will result in the knee giving way and the patient losing balance. (**c**) Flexing forward to the maximum extent. (**c**) The therapist holds the patient's affected knee with one hand and the hip with his other arm using the upper arm. The patient should bend forward to the maximum extent while keeping the unaffected knee fully extended (straight), then flex the knee and touch the unaffected palm on the floor with a series of continuous movements. (**d**) Touching the floor with one hand. (**d**) The therapist holds the patient's affected knee and hip and says, "Try to move your unaffected hand close to the floor. When you reach your limit, bend your knees and touch the floor with your hand." When the patient's hand reaches the floor, the therapist should say, "When your hand is firmly on the floor, rotate your hip." Then the therapist should lightly assist the rotation with one hand supporting the hip.

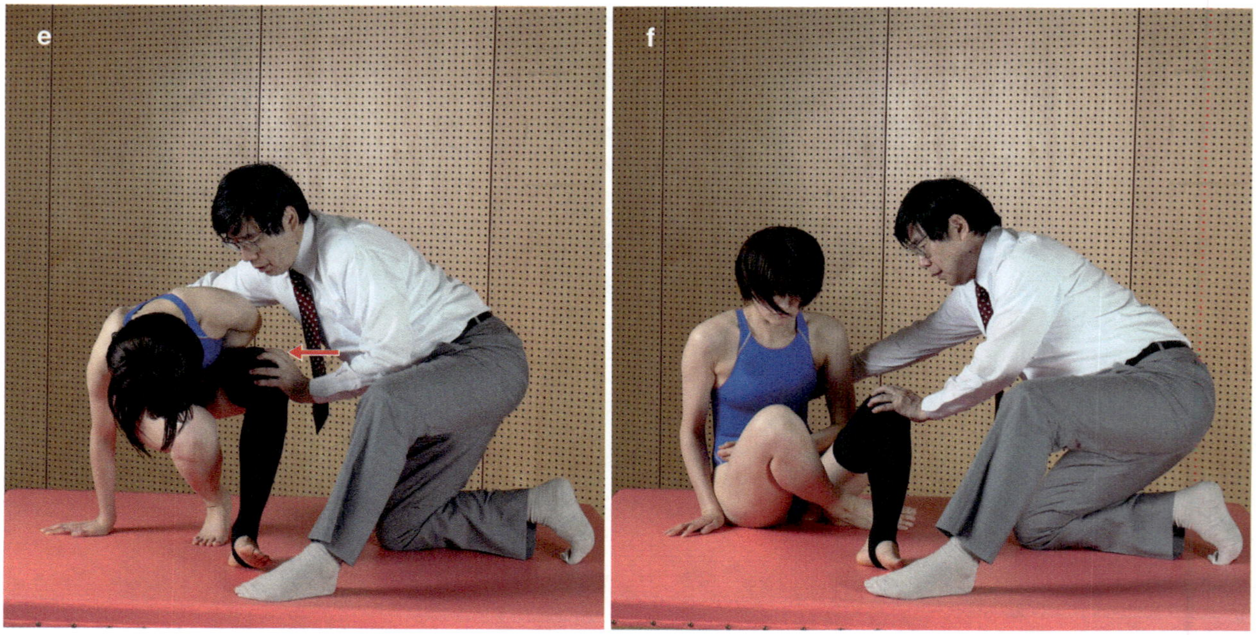

Fig. 7.7 (continued) (**e**) Rotating the hip while slowly lowering the buttocks. (**e**) The therapist holds the patient's body, rotating it while the patient's hand is on the floor as the pivot point until the patient sits down. The therapist should secure the affected knee with one hand and support the upper body so that the hip rotates within the therapist's supporting arm. (**f**) Gently lowering to a sitting position. (**f**) If the therapist supports the patient's knee and hip to stabilize the body's balance, the patient will be able to sit down gently by rotating the hip around the axis of the arm placed on the floor

Afterword

A. Challenges for the Neuromuscular Facilitation Technique and RFE

The neuromuscular facilitation technique for hemiplegia and motor disorder has never been proven to be effective, despite many studies conducted worldwide. The theory of the neuromuscular facilitation technique was considered scientific and logical, but the expected outcome has never been obtained. Both the treatment theory and technique require amendment.

RFE has cleared the first stage, the scientific proof of its efficacy, by emphasizing the development of a new facilitation therapy that achieves the intended movements and the strengthening of nerve paths (Happ's theory/learning). A more effective rehabilitation treatment is required for an aged society. Treatment which shows a significant improvement in "paralysis and walking" and "activities in daily living" is necessary. In addition, in neural regenerative medicine, which will gain momentum in the next few years, a method of restructuring the nerve paths without error is required. We need to develop and disperse the method to specify the target nerve pathway from outside and repeat the transmission of neural excitement for facilitation without errors (Fig. A.1).

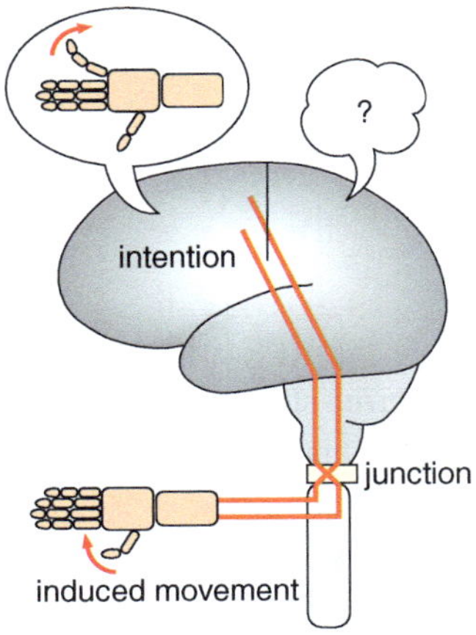

Fig. A.1 Misconnection of nerve paths in regenerative medicine

After the patients had stem cells implanted, they tried bending their little fingers, but it was actually their thumbs that bent. To make the stem cell implant more efficient, new nerve pathways need to be constructed to avoid such misconnections. Also, in cases of misconnections, the treatment technique to correct them is necessary (Fig. A.1).

We should consider what direction RFE should proceed towards, based on current neuroscience, rehabilitation medicine, and kinesiology.

B. What RFE Should Aim for

To minimize the patient's trial and error efforts, the level of neural excitement should be raised by the therapist's operation, electric/vibration stimulation and other external stimulation. The significant need for a technique that achieves and repeats intended movements will increase in the future. RFE development, where the therapists use their facilitation operations to specify the nerve pathways that achieve the movements the patients intended, will contribute to the development of robots and other training devices.

1

In order to facilitate the recovery from paralysis, emphasis should not solely be placed on normal movements and normal sensory input. We should also focus on achieving and repeating movements intended by the patient, strong sensory stimulation that reaches the primary motor area, and sensory input that assists the achievement of such movements. We need to think about transmitting the excitement to the nerve pathways that need to be strengthened and repeat it. By doing this, it will become possible to enhance and achieves nerve pathways strengthening (learning) without trial and error. Patients will be able to master the movement pattern which

K. Kawahira et al. (eds.), *Exercise Therapy for Recovery from Hemiplegia*, https://doi.org/10.1007/978-981-19-0789-0

is out of muscle synergy pattern and obtain the movement that directly connects with the ADL (activities in daily living) such as walking.

2
In order to improve walking, ADL, and other systematic movements, we need to strengthen the unaffected limb with repeated training in standing up to avoid a disuse syndrome. We also need to repeat the required walking patterns using a cane or brace to increase the volume of movements. Many studies have clarified that training without orthosis is ineffective. For walking, in particular, we should provide the treatment on the premises: Hemiplegic patients cannot walk normally. Unaffected limbs and trunk balance benefit walking and ADL. Affected limbs cannot perform small adjustments to control posture.

3
In order to enhance the RFE effects, therapy development combined with electric stimulation, vibration stimulation, magnetic stimulation, and botulinum injection therapy are necessary.

C. Closing

The long dominant concept of "Paralysis due to brain damage is irreversible" has been completely demolished. Recovery from paralysis with RFE and other new treatments has been proven scientifically. Neuroscience f-MRI and other non-invasive imaging has endorsed them in recent years. On the other hand, rehabilitation medicine and the healthcare system focus more on shortened hospital stays and the treatment of ability disorders than making efforts to apply the outcome of the actual rehabilitation practices. In spite of the evidence from newly developed techniques that have been proven by cutting-edge science, the development of treatments that optimize the plasticity and maximize the recovery from paralysis are insufficient.

In order to facilitate recovery from paralysis as well as to master walking and ADL, kinetic therapy must excite and strengthen the nerve pathways selectively which work to achieve movements intended by the patient. From this perspective, RFE is an epoch-making treatment, and we must try our best to establish combination therapies that boost the effect of RFE and prove its efficacy.

GPSR Compliance

The European Union's (EU) General Product Safety Regulation (GPSR) is a set of rules that requires consumer products to be safe and our obligations to ensure this.

If you have any concerns about our products, you can contact us on ProductSafety@springernature.com

In case Publisher is established outside the EU, the EU authorized representative is:

Springer Nature Customer Service Center GmbH
Europaplatz 3
69115 Heidelberg, Germany

Batch number: 10731528

Printed by Printforce, the Netherlands